J. Frenkel

Victor Ya. Frenkel

Yakov Ilich Frenkel

His work, life and letters

Translated from the Russian
by Alexander S. Silbergleit

Springer Basel AG

Author' address:

Professor Dr. Victor Ya. Frenkel
A.E. Ioffe Physico-Technical Institute
Ul. Politechnicheskaya 26
194021 St. Petersburg K-21
(Russia)

Oiginally published as: Yakov Il'ich Frenkel'
© 1966 Izdatel'stvo «NAUKA», Moskva - Leningrad

cover illustration:
© Yu. L. Katua, Moscow, 1949

A CIP catalogue record for this book is available from the Library of Congress, Washington D.C., USA

Deutsche Bibliothek Cataloging-in-Publication Data

Yakov Ilich Frenkel: his work, life and letters / Viktor Frenkel. Transl. from the Russ. by Alexander S. Silbergleit. - Basel ; Boston ; Berlin : Birkhäuser, 1996
Einheitssacht.: Jakov Ilich Frenkel <engl.>
ISBN 978-3-0348-9645-0 ISBN 978-3-0348-8490-7 (eBook)
DOI 10.1007/978-3-0348-8490-7
NE: Frenkel', Viktor Ja.; Frenkel', Jakov I.; EST

Ursprünglich erschienen bei Birkhäuser Verlag, Basel, Switzerland in 1996

Printed on acid-free paper produced from chlorine-free pulp. TCF ∞
Cover design: Markus Etterich, Basel

ISBN 978-3-0348-9645-0

Contents

Preface

I regard the possibility, provided by Birkhäuser Verlag, to publish in English the book about my father, Yakov Il'ich Frenkel (1894–1952),* as an honour and recognition of his contribution to the development of physics in the 20th century. Publication of this book shortly after Frenkel's centennial is of special value. I would like to express here my deep gratitude to the publishers.

More than 30 years ago, by the initiative of I.E. Tamm, a special issue of the Russian physical journal «Uspekhi fizicheskikh nauk» (UFN) /1962, v. 76, # 3–4, dedicated to Frenkel, was published. Professor Tamm, who had undertaken the composition of the introductory (biographical) paper for this issue, asked me to provide him with some materials for my father's biography. It was easy for me to accomplish this because, fortunately, a vast family correspondence was in existence, despite all the vicissitudes of life, like several wars and the siege of Leningrad. I assembled some fragments of my father's letters of 1906–1950 for Tamm. Having looked through them, Tamm told me: «But a whole book may come out of it! You must write it!» Stimulated by I.E. Tamm and by B.G. Kuznetsov, who strongly supported the idea, I began to work on the book which, in the spirit of one of M. Faraday's biographies, could be titled «The Life and Letters of Jacov Frenkel». The book was being written at the close of the Khruschev «thaw», that is, freely enough but still with the over-the-shoulder glance at the censor. Nevertheless, at the beginning of 1966, about 30 pages were thrown out of its already typeset text.

The book was addressed naturally to the readers in the former Soviet Union, and this determined the choice of material. Aiming now at the Western reader, I have made changes, withdrawn from the Russian version numerous pages which required extended comments to be understood by people unaware of many details of Russian history and the history of physics in Russia. Besides, I have been rather strictly controlled by the publishers with regard to the book's volume. A large amount of essentially new material could be included in the book because it was prepared while democracy was beginning to take hold in Russia and censorship relaxed. Moreover, in the quarter of a century since the publication of the Russian edition, I, continuing to study my father's biography (and to publish his books),

---

*) It is necessary, from my point of view, to give here a small explanation about various ways of spelling of my father's first name, — in connection with specific features of the Russian language.
His official name — as it appears in Russian — must be written as «Yakov». If one likes to add to this name his patronomic — «Ilich» (or Il'ich), it becomes «Yakov Ilich Frenkel». I decided to use this form in the title of the book. In the text one will find the other variants: Jacov, Jacob, or — most frequently — «J.I.» (in the 1920s Frenkel himself used even one more version: «James»).
When Frenkel was a little boy and teenager, he was «Yasha»; this was also how his relatives or close friends called him. This affectionate version of my father's name can be found in the book as well (remembering how my mother, grandparents, closest friends and relatives had addressed him, I counted more than 10 versions, but they are not used in the text).

learned many new things both about him and the destiny of his works. The process resulted in the reworking of about half of the text. What has remained invariant for both editions is the principle of exposition. I tried, wherever possible, to give the floor to Frenkel himself, so the evaluation of his statements, his image, would be formed by the reader without my mediation, so to say, on the grounds of personal impressions. In this respect I gained much from work in state and privately owned archives, in Russia and abroad. I also strived to avoid, when possible, making judgement on the importance and value of Frenkel's papers, quoting instead his outstanding colleagues.

In the process of my work I had dozens of interviews with the students, co-workers and friends of my father. I greatly appreciate their collaboration. I do not give the list of their names here because almost all of them are mentioned in the following pages. However, I would like to make three exceptions for I.E. Tamm, B.G. Kuznetsov and Ya.A. Smorodinsky (the editor of the Russian edition). Alas, none of them is among the living now, and I state here the debt of gratitude to their memory.

My cordial thanks to Dr. E. Jurkowitz for his kind help in checking the galley-proofs of the book.

V.Ya. Frenkel
St. Petersburg, 31 December 1992

Chapter 1
The Beginning: Family, Gymnasium, Universities

Jacov Il'ich Frenkel was born on 10 February 1894, in the city of Rostov-on-Don. His father Il'ya Abramovich Frenkel played a major role in the formation of his character, so it is worthwhile to tell more about him.

I.A. Frenkel started working as early as age 12, rewriting documents in a notary's office. He had had only 2 years of the Azov grammar school and tried to educate himself while working. A student who lived in Azov took part in his destiny; from him Il'ya Abramovich learned the ideas of the participants of a revolutionary organization «Narodnaya Volya» («Peoples' Volition»). In 1879, the 16-year-old youth was arrested for the possession of illegal literature and was released only after several months spent in a solitary confinement cell. His case was closed on the grounds of «the criminal's youth». He was unemployed after leaving prison and participated actively in the revolutionary circles of Azov and Taganrog, travelling to St. Petersburg for revolutionary literature which was delivered through V.G. Bogoraz-Tan[1].

In 1882, I.A. Frenkel came to Moscow. In his unpublished memoirs «The Pass to the Jail and Siberia» (in Russian) he wrote that he had shared, rather carelessly, his views on the necessity of mass terror in the struggle against the autocracy with several young people from, in particular, the Petrovsko-Razumovskaya Agricultural Academy. This resulted in his second arrest in January 1884. He was kept in a solitary confinement cell at the notorious Butyrskaya Prison in Moscow for about a year and then, with a group of other convicts, deported to Siberia, to the town of Beryozov, where he spent about 6 years. In Beryozov, a typical rat's corner then, there were several intelligent people, convicted and condemned to live there, and it was in their company that Il'ya Abramovich spent all of his time. These people, who later did a lot for the cause of the revolution, had a great influence on I.A. Frenkel.[2]

The living conditions of a deportee with practically no help from home were extremely hard. Il'ya Abramovich learned how to make shoes and made his living as a shoemaker. Also in Beryozov he was recruited to the army where he served the required term under the local military command.

The year 1890 brought the desired freedom to Il'ya Abramovich. He returned to Azov and soon married Rosaliya (Rachel) Abramovna Batkina. J.I. was their eldest son.

---

1) Vladimir Grigor'evich Bogoraz-Tan (1865–1936) was a member of «Narodnaya Volya», later a specialist in ethnography, a linguist, a writer and a professor at Leningrad State University.

2) The name of I.A. Frenkel figures in the dictionary of the «Narodnaya Volya» members; see: Activists of the Revolutionary Movement in Russia. Bibliographical Dictionary, v. 2, # 4, p. 1853.

Ya. Frenkel's parents: Ilya and Rosaliya (Rachel) Frenkel

Il'ya Abramovich continued his communications with the friends he had acquired in exile. He became acquainted with some other revolutionaries, in particular with A.M. Lezhava[3]. Later, while living in St. Petersburg, by the revolutionaries' request he many times hid people connected with the revolutionary movement. However, I.A. Frenkel removed himself from direct revolutionary activity.

Some important information about I.A. Frenkel and his family was revealed not so long ago in the State Archives of the October Revolution in Moscow. It turned out that, after being released from prison, he remained under the surveillance of the secret police. Let me cite a sample of the corresponding documentation kept in the Special Department of the Archives.

«On the reserve rank-and-file Il'ya Abramovich Frenkel.

To the temporary in charge of the June 29, 1901

Yekaterinoslav Gendarme Division, # 2909 Secret

I have the honour to inform the Department of Police that the reserve rank-and-file Il'ya Abramovich Frenkel, staying under secret police supervision, in the

3) The friendship between I.A. Frenkel and Andrey Matveevich Lezhava (1870–1937) lasted many years. In the Soviet years A.M. Lezhava was the Deputy People's Commissar for Foreign Trade and Crop-Growing.

middle of June of this year went abroad with the passport issued to him by the Yekaterinoslav Governor on the 13th of this June under # 865 for a half-year term.

Undercolonel (signature illegible).»[4]

While not wishing to list the many investigation files wherein I.A. Frenkel figures, we would like to make one exception. The file was opened in 1914 on the grounds that in the beginning of that year O.P. Ryss, a relative of the journalist P.Ya. Ryss[5] who was being kept under secret police surveillance should have come to him to St. Petersburg from Germany. In her letter to P.Ya. Ryss of 10 February, 1914, opened and examined by the police, the name of I.A. Frenkel is mentioned, because O.P. Ryss had asked him to send I.A. Frenkel's correct address to her. The address was immediately relayed not to O.P. Ryss, of course, but to a gendarme officer who asked for it. We quote: «The 2nd guild merchant I.A. Frenkel, age 50, lives in St. Petersburg on Kolomenskaya Str., # 32, apartment 23.» No special efforts were needed to obtain this information, since all the information on I.A.Frenkel, including his apartment telephone number, could be found in the directory «Ves' St. Petersburg», 1913.

It remains unclear of what importance O.P. Ryss was to I.A. Frenkel. But she approached him because back in Azov, J.I.'s father had been in the position of the manager of the 1st guild merchant B.I. Ryss, to whom O.P. Ryss was related.

Materials on the O.P. Ryss visit to St. Petersburg — about 100 pages! — read like a detective novel. One detail there is most curious for the Frenkel biography. When O.P. Ryss arrived in St. Petersburg, surveillance of her was immediately established «at the living places of P.Ya. Ryss and I.A. Frenkel», as stated in one of the reports to the Gendarme Department. After her first visit to the I.A. Frenkel flat (17 February 1914), one of two detectives spying on O.P. Ryss reported that «she left the apartment accompanied by I.A. Frenkel and a young man.» The latter probably was J.I. Frenkel. Anyway, on the very next day information on all the family members of I.A. Frenkel, namely, on the elder sons Jacov and Vladimir and on their mother, Rachel Frenkel, was directed to the Gendarme Department. Strangely, she had been marked by the police, and a file had been opened back in 1912 because her name had been found in the notebook of some person involved in the investigation of the socialists-revolutionaries party.

Let us return, however, to the events at the onset of the 20th century. By this time I.A. Frenkel had worked first in Azov (for B.I. Ryss) and then in Lougansk.

In the beginning of the 1900s, Il'ya Abramovich and his family, including by then four children, J.I., two of his brothers and his sister Vera, went to live in Kazan.

---

4) The Central State Archives of the October Revolution. Special Department. Year 1901. File # 460.

5) The interest of the police in P.Ya. Ryss, judging by the materials from his file, was due to the fact that his two brothers were active members of the «Narodnaya Volya». One of them was executed in the 1900s, the other sent to penal servitude.

The meetings of the Kazan RSDRP (Russkaya Social-Demokraticheskaya Rabochaya Partiya, i.e. Russian Social-Democratic Party of Workers) Committee often took place at the Frenkel apartment, and during the 1905 revolution the colonial consumer goods store managed by I.A. Frenkel housed their illegal printing office.

A squalid wave of Jewish pogroms reached Kazan soon. That is why Il'ya Abramovich sent his family to Switzerland. By this time Yasha (Jacov) Frenkel had finished the first grade of the gymnasium, having had moderate success in his studies. In particular, he earned only a C in arithmetic.

J.I.'s first letters are from exactly this period of his life. Here is an excerpt of a letter to his father (from Switzerland, 1 January 1906): «Best wishes for the New Year, which will be more happy, I think, than 1905 was. I wish that peace and prosperity may come to Russia, and that the government be suppressed and demolished by the revolutionary forces. May God provide that all of ours be happy and healthy!»

If we continue to read this letter, we see that the 11-year-old boy moves rather soon from general political problems to questions of an absolutely different nature. He is extremely worried by the state of health of his and his brothers' beloved pet dog staying in Kazan, and the major part of the letter is dedicated to this concern.

By including this quotation, I am, naturally, far from stating that Yasha Frenkel was obsessed by revolutionary ideas «from the very beginning». But altogether it is quite characteristic of the situation in which Il'ya Abramovich's children were raised.

By the fall of 1906 I.A. Frenkel's family had moved to Minsk. Il'ya Abramovich himself worked then at the Russian Society for Matches Trade. The family income was rather modest, and all free means were put into the childrens' education. I.A. Frenkel wrote to his wife during one of his long (and frequent) trips across Russia: «I am fed up with this gypsy life, but when you think about how much is needed and how little we, comparatively, own, a surge of energy instantly wells up in me, as in a horse lashed by a knout.» And he added: «Remember, the more our children know, the easier their life burden will be.»

A French woman, Mademoiselle Robaglia, lived with Frenkel's family. She was called Aunt Jean in the house and cared for the children. That is why Jacov and Vladimir had a perfect command of French and were thus able to continue studying at a Swiss school.

In his autobiography (1946)[6] J.I. wrote: «As a child, I was very talented at music and painting. This impelled my parents to arrange for me to have lessons in playing the violin (as of age 8) and drawing (as of age 12). Both of these activities have remained my favourite hobbies up to the present time.»

---

6) It was written for a directory «Who is Important in Science» and is found in the book J.I. Frenkel: Memoirs. Letters. Documents. Leningrad, Nauka, 1986 (in Russian).

A beggar from Minsk. Oil portrait by Ya. Frenkel, 1908

His teacher of painting in Minsk was Ya.M. Kruger[7]. Kruger, a rather talented person, was a very funny man, judging by Frenkel's letters. Several pictures by J.I. from this period have been retained, and some of them were exhibited later at the Leningrad scientists' exhibitions organized at the Gorky House of Scientists in Leningrad. Among them was a portrait of a beggar (critics would say that it was painted in a manner influenced by Rembrandt), painted very professionally.

In 1909 the Frenkel family moved to St. Petersburg, and the brothers entered the famous private K. May Gymnasium.

The May Gymnasium, founded in 1861, was one of the best in St. Petersburg. A strong body of teachers worked there. Within its walls A.N. Krylov in 1906 taught for the first time his famous course on approximate calculus; his audience was a small number of persons who formed a free mathematical department headed by the former gymnasium teacher N.M. Gyunter, later a corresponding member of the Soviet Academy of Sciences. Physics was taught by F.N. Indrikson, a laboratory assistant at the Physics Cabinet of St. Petersburg University, who assisted O.D. Khvol'son at his lectures on general physics. The «Handbook of Physics» by Indrikson, as well as his «Abridged Handbook of Physics», were printed several times and very popular in the 1910s.

Also among the teachers at the gymnasium during those years when J.I. studied there was E.E. Essen (1879–1931)[8], who was still involved in revolutionary activity.

Frenkel's interest in mathematics and physics manifested itself shortly before he moved to St. Petersburg. It appeared without any particular stimuli, just from an acquaintance with the standard courses.

In one of the early biographies (1925) of J.I. we read: «I first felt an interest in mathematics and physics at the age of 14. By the end of the 5th grade I had learned the whole gymnasium course in mathematics, and by graduation from the gymnasium, the best part of the university courses in mathematics, mechanics and physics. Unfortunately, all along there was no one to guide me, so I had to study on my own.»

In the summer of 1910, J.I., a 16-year-old youth, studied both volumes of the «Elements of Higher Mathematics» by H. Lorentz, along with K. Posse's calculus book. The first of these books contained, by the way, a thorough chapter on the history of mathematics[9]. J.I. highly valued this book and often thought of it.

7) Ya.M. Kruger (1868–1940) graduated from the St. Petersburg Academy of Arts under the well-known painter V.G. Makovsky. It is interesting to note that one of the few pictures by Kruger kept now in the Minsk Painting Gallery is a portrait of a violinist Yu.G. Zhukhovitsky, who taught J.I. to play the violin.

8) He was an active participant in the October Revolution in Petrograd, then did revolutionary work in the Crimea where he met, in particular, J.I. After the revolution E.E. Essen was the People's Commissar of the Goskontrol'; in the last years of his life he was the Rector of the Academy of Arts in Leningrad.

9) Frenkel read the Russian version translated by V.P. Sheremet'evsky.

Referring to it, he liked to tell about Cardano, whose name is borne by the formula for the roots of a cubic equation. Cardano was not only a mathematician but an astrologist as well. He prepared a horoscope for himself, and the stars showed he would die in 1576 at the age of 75. On the predicted day of this year Cardano committed suicide, «giving his life for astrology», as J.I. used to say.

The beginning of Jacov's own work in mathematics, dedicated to the generalization of the theory of arithmetic and geometric progressions, is also associated with the year 1910. The work was finished by the fall of 1911, when he was in the 7th grade.

During the whole academic year 1910/11, Frenkel studied a lot, and his parents saw this as a reason why his cardiac sickness started then. Be this true or not, when the Frenkel family went to rest in Zoppot in the summer of 1911, J.I. with great difficulty pleaded the right for at least some study. He wrote to his father:

«Zoppot, 12 July 1911.

... In the meantime I am doing nothing, more than ever, and despite the completely incomprehensible pleasure this brings to my mother, I am definitely going to return to the sources of wisdom and distress, Lorentz, Posse and Co. Generally, please accept the absence, on my part, of any concessions in the next year, because it is much more worthwhile for me to dispose freely of my time than being present in any Zoppots would ever be, and it is just this freedom I am completely deprived of. I recall with sadness the Terioki pigeon-loft where so little was done, as it had seemed to me then, and so much, actually, as I see it now»[10].

Frenkel showed a voluminous mathematical paper, of about 100 pages, entitled «Progression Calculus», to a young mathematician, Jacov Viktorovich Uspenskii, a student of the famous A.A. Markov. Exactly in the described year 1910, Uspenskii brilliantly defended his Master Thesis (in 1921 he was elected member of the Russian Academy of Sciences). In young Frenkel's paper Uspenskii discovered all the principal results of the calculus of finite differences, developed simultaneously and in parallel with the usual calculus in the treatises by Fermat, Leibniz, Newton and others (the calculus of finite differences then, as now, was absent from the general books on higher mathematics and thus remained unknown to the young author). Uspenskii told Frenkel all of this after attentively looking through the paper, which was thus approved.

In October of 1912, in the 8th (last) grade of the gymnasium, Frenkel completed his first work in physics, which attempted a construction of the theory of the Earth's magnetism and atmospheric electricity. Considering «The Statics of the Solar System» (as the first chapter of his paper was called), he demonstrated that the Sun possesses some positive charge, due to the emission of electrons from

---

10) The Terioki pigeon-loft is the attic of the summer house in Terioki, a country place near St. Petersburg the Frenkels rented in 1910. J.I. had established his office there.

its scorching surface. Rotation of the charged Sun gives rise to a magnetic field which, though very weak at the distance of the Earth's orbit, may provide magnetization of the terrestrial globe, if one assumes that the liquid earth core is (or was, at some distant stages of its history), a ferromagnet.

In the same paper J.I. also gives a theory of atmospheric electricity, though a rather naive one, as he later pointed out. Nevertheless, even then the young gymnasium student expressed a number of correct and far from naive ideas. For instance, he observed that the condensation centres of water vapours in the atmosphere, as negative ions, contribute essentially, when coming to earth (with the rain drops they help to form), to supporting the earth's stationary negative charge. This component of the general current balance in the atmosphere is accounted for in the modern theory as well. Reading a thick exercisebook in a black cover retained in J.I.'s papers, which contains his first work in physics, reveals that at that time he was — not at all superficially! — acquainted with physics (which he had studied back at the gymnasium by the multi-volume book by O.D. Khvol'son), celestial mechanics, astrophysics and geophysics, and, of course, mathematics.

It is interesting that in this early paper we also find for the first time his primary consideration of not only geophysical, but also purely physical questions, such as, for instance, the problem concerning the charged rotating sphere to which Frenkel would return in 1918 and at the end of the 1920s. There he also invents the term «wandering electrons», which he would use later and which would gain «the right of citizenship» by the end of the 1920s and early 1930s, after Frenkel's work on the electron theory of metals, and so on.

B.R. Abramson, a distant relative of the Frenkel family and member of the Russian Physics-Chemical Society, told A.F. Ioffe, then a professor of St. Petersburg Polytechnic Institute, about Frenkel and his work. Ioffe was ready to meet the young man. The meeting took place at the end of 1912; by this time Frenkel already knew Professor Ioffe by his works.

In his 1946 autobiography J.I. wrote the following about his first paper on physics: «I showed my memoir to Prof. Ioffe in 1912 who has kindly read it despite its great length (250 pages), and benevolently and critically commented on it; I, however, at that time decisively rejected his criticism. This acquaintance with Prof. Ioffe later led to my friendship and collaboration with him.»

In J.I.'s archives one more exercise-book is retained, with a composition written on the topic, suggested to the 8th grade students, which was very far both from physics and mathematics: «War or peace?» By the time he wrote it, Yasha, guided by his father, had already studied several books on philosophy (in particular, he was greatly impressed by Spencer's «Psychology»[11]) and political economics (mainly a book by V.Ya. Zheleznov, which was prohibited by the tsar's censorship). And his composition differed essentially from what other students

11) St. Petersburg, 1876, v. I–IV (in Russian).

wrote and from what had been expected by their teacher of Russian and Russian literature, V.N. Korablyov.

Frenkel, with all the passion of youth, came down on the pretentious servants of autocracy who formed the majority among the officers of the tsarist army, and spared no epithets when characterizing these «loafers and involuntary killers» transforming them from «heroes» into «parasites living off society and bringing it only harm». «Therefore,» he concluded, «intelligent and talented people avoid military service if they can, and the contingency of «fighters» is recruited from either ignorant or incapable people, causing thus the drop of the authority of the military class, this survivor of barbarism.»

Many years later Frenkel recalled how Korablyov, striking at his own chest, cried pathetically that only the state of his health had made him part with his dream of military service, etc. It goes without saying that the young pacifist got the lowest grade for his composition, to which a categorical remark was attached: «The paper testifies to some erudition on the author's part, but on the other hand that all he read has turned into a kind of porridge in his head.» The question of keeping Yasha Frenkel in the gymnasium was raised at the meeting of the Pedagogical Council (and, in the end, decided in the affirmative).

In the State History Archives of the Leningrad region, verbatim records are kept of the meetings of the May Gymnasium Pedagogical Council. The name of Frenkel appears there first in an unflattering context: «The weakest pupils ... [including] Frenkel ... are behind academically because of their lack of capability» (31 October 1909). «Frenkel is a student of mediocre abilities, insufficiently diligent» (19 December 1909). However, for March of 1912 one reads:«Frenkel is very capable, perhaps a future scientist.» But though by this time none of the teachers denied his capability, many of them, especially humanities teachers, regarded him in more than reserved a manner, accusing him of arrogance and stressing his unfriendly attitude towards school.

If we turn once more to the records of the May Gymnasium, we find the following opinion by V.N. Korablyov about Frenkel's classmates: «A developed, capable, intelligent form ... the general complexion remains undetermined ... children ... Frenkel is the exception.» Naturally, Frenkel, who was quite mature by then, was not content about being regarded as a featherless nestling. Therefore the gymnasium director, A.L. Lipovsky, summing up the discussion on one of Frenkel's many misdemeanors, which occurred approximately two months before the graduation exams, appealed to the Council to demonstrate «all the calmness and indulgence possible. It is now too late to improve Frenkel,» he added.

Frenkel passed the graduation exams in 1913 with distinction and was awarded the gold medal.

As of the 7th grade he started to teach gymnasium lazy-bones mathematics, physics, chemistry, even the Bible (God's Law, as it was called in Russian schools), which he studied especially for this purpose, since he had not studied this discipline at the gymnasium.

The money earned helped Frenkel to go to the USA. He went not only because of the passion for far travels he inherited from his father, but also because he wanted to find out about the possibility of getting higher education at one of the North American universities in case there would be no such chance back in tsarist Russia. (And precisely in May of 1913, the competition of grades, by which persons «of the Jewish religion» were accepted to universities, within some quota, was replaced by a casting of lots, reducing once more Frenkel's likelihood of being accepted.)

In the summer of 1913 he wrote from New York to his parents that one American acquaintance, a Russian emigrant, guaranteed him a job as a violinist in a large orchestra playing in one of the fashionable New York restaurants, with a salary of $100 per month which would allow him to pay for his education and to feed himself (though would not leave much time for his studies).

But, fortunately, J.I. was accepted by the Physics and Mechanics Department of St. Petersburg University, and, upon returning at the end of August 1913 from a one and a half month visit across the ocean, young Frenkel took to his studies.

O.D. Khvol'son (1852–1934), I.I. Borgman (1849–1914) and N.A. Bulgakov (1867–1931) were then professors of physics at St. Petersburg University. Khvol'son and Borgman contributed a lot to the development of physics in Russia. The former was the author of a six-volume book on physics recognized not only in this country but also abroad, where it was translated into French, German and Spanish. In Khvol'son's lifetime, after the revolution, the 7th edition of the 1st volume of his book was published. Borgman was the organizer of the physics seminar at St. Petersburg University; the Physics Institute of the University was built on his initiative. But one should stipulate that neither of them were creative scientists. Khvol'son, for instance, in every way greeted the establishment at the University in those years as a (in his own words) «splendid tradition to reproduce the best of the foreign scientific works», and in reply to Ioffe's question as to whether it would not have been better to formulate and solve new problems, Khvol'son argued that one has to be a J.J. Thomson in physics to invent something really new in it. Academician I.E. Tamm recalled one more episode, very characteristic of that period, associated this time with Borgman, which Ioffe had told him. When Ioffe explained to Borgman the plans for the work he intended to do (related to the study of the electrical properties of crystals), Borgman, convinced that no such work had been or was being done anywhere, exclaimed: «What a conceited person you are, young man! We have nothing to do with the search for anything new; it would be just very good if we were able to reproduce here measurements carried out abroad!»[12]

It is true that a brilliant mathematical school headed by A.A. Markov and V.A. Steklov existed at St. Petersburg University. The first steps aimed at closing the gap between physics and mathematics are associated with the names of Steklov

12) I.E. Tamm. UFN, v. 76, # 3, p. 401, 1962 (in Russian).

Ya. Frenkel, while a student of the Physico-Mathematical Faculty of the Petrograd University and his student V.I. Smirnov, the author of the brilliant book on higher mathematics which served as a textbook for all the physics students in our country from the 1940s to the 1970s.

As for theoretical physics, it was represented then by N.A. Bulgakov who had replaced Borgman, the Physics Institute Director. Bulgakov, however, was more of a mathematician than a physicist. Frenkel recalled with a smile his lectures on electromagnetism. The lecturer usually started to speak outside of the auditorium and continued at the blackboard, not a bit embarrassed by the fact that he was actually heard by only those few people whose «turn» it was to be present, to keep the lecturer going.

Bulgakov had known Frenkel when the latter was a gymnasium student; like Ioffe, he had read Frenkel's paper on atmospheric electricity, and provided J.I. with literature on geophysics. Through all the years of J.I.'s stay at the university, Bulgakov paid his due to Frenkel's capability and supported him in all the ways he could.

Starting in the summer of 1916, I.A. Frenkel mostly stayed in Moscow, on business matters. This circumstance enables us to obtain more detailed information about J.I. from his letters to his father.

We learn from them that the fall of 1916 was saturated with work. Having completed the whole university programme in 3 years, J.I. passed the state examinations in September and October and simultaneously wrote his thesis on the structure of atoms in light of the work of E. Rutherford and theoretical papers of his student, the Danish physicist N. Bohr. The presentation of a thesis was necessary, according to the regulations in force, to get the 1st Grade Diploma which allowed the possibility of staying at the university.

Here are two of J.I.'s letters to his father.

«Petrograd, 14 September 1916.
... Baby sat for me today for the whole hour and a half[13]; I told him various tales about natural history and even sang songs. Tomorrow I have the second exam. I have just started to prepare for it: it is not difficult to leaf through 200 pages. I met Ioffe (at the meeting of the Physics Society) and even drew a sketch of him, using this convenient occasion. I entertained myself by drawing because the meeting was extremely boring.»

«Petrograd, 17 September 1916.
... No special news on my part, except maybe for the fact that yesterday I passed the second exam. Thus 4 more of them are left, of which one is on the 21st, the other on the 29th. ... The day before yesterday I had a conversation with Ioffe and was honoured to be invited to participate in a circle he organizes at the Polytechnic Institute, of 6 to 7 persons, some of them students but mostly not. So

13) Baby is Victor, the fourth and youngest son of the Frenkels, born in 1910. J.I.painted his portrait.

A.F. Ioffe's physics seminar. From left to right (seated): P. Kapitza, Ya. Frenkel, N. Semenov, K. Nesturkh, P. Lukirskii, I. Bobr, M. Kirpichova; (standing): Ya. Schmidt, N. Dobronravov, A. Yushchenko, Ya. Dorfman; (at the blackboard): A. Ioffe. Petrograd, 1916

I will have to go to the Institute two times a week more, which I shall certainly do with pleasure.»

Frenkel, just briefly touching in this letter on the seminar at the Polytechnic Institute, probably could not foresee the outstanding place which this seminar would occupy. One can state, on absolutely sound grounds, that Soviet (post-revolution) physics was started exactly in Petrograd and precisely at this seminar, by the State Rentgenovsky and Radiologichesky Institute, organized two years later (in 1918), the core of which was formed by the participants of this seminar. Exactly 30 years later, in the already mentioned autobiography (1946) Frenkel wrote:

«In the fall of 1916 Ioffe organized a physics seminar at the Polytechnic Institute and invited several young people interested in the newest physics development and desiring to work in this field to participate in it. Among them were P. Kapitza, N. Semenov, P. Lukirskii, Ya. Dorfman, the late M. Kirpichova, the late Ya. Schmidt and myself. Ioffe's interests determined the choice of material to be discussed, and his stimulating comments made a deep impression on all the young participants of the seminar, including me.»[14]

In the middle of October 1916, J.I. obtained a certificate stating that he, as «having completed the course in the mathematical division of the Physics and

14) See Frenkel. Memoirs, p. 468 (in Russian).

Ya. Frenkel. Self-portrait, Crimea, 1919

Mathematics Department of the Petrograd University, according to the test performed by the commission at this University, was recognized as having a legal right to receive the 1st Grade Diploma.»[15] Soon after that at the meeting of the Department, a resolution was taken, after a presentation by professors O.D. Khvol'son, N.A. Bulgakov and D.S. Rozhdestvensky, to keep J.I. Frenkel at the university «for preparation in a professorial and teaching capacity» (this corresponds to the present doctorate).

Only two weeks after Ioffe's invitation, J.I. presented a report on his thesis at the seminar. Ioffe, who was a member of the Editorial Board of the Journal of

15) State Historical Archive of the Leningrad Region, Record group 14, inventory 1, file 11270, p. 30 (in Russian).

the Russian Physics-Chemical Society (ZhRFKhO), proposed that Frenkel enlarge this work and to submit it as a series of papers to the journal.

And in the middle of November, J.I. started to write a paper on the double electrical layer, the main idea of which had arisen from one of the Ioffe seminars (and by the end of November the paper went to the journal).

The idea in Frenkel's paper was extremely simple and reduced to the application of the Rutherford planetary atom model to solids and liquids. J.I. wrote about the outer surface of such bodies which goes through the nuclei of the «most outward» atoms. Electrons belonging to these nuclei, rotating along circular (or elliptic) orbits, form a layer of negative charge on the surface. This layer, together with the positive nuclei, constitutes the double layer whose electrical field is directed in such way that it keeps free electrons inside a metallic body. The energy transferred to the electrons in the process of heating goes just for the work against the forces of this double layer. As a result, electrons come out of this surface («emigrate», as J.I. wrote in his early work in 1912), and electron emission occurs. Frenkel clothed this idea in a simple mathematical formulation and calculated interior potentials of metallic bodies which agreed very closely with the experimental data for sodium and lithium.

Knowing the interior potentials, Frenkel arranged metals in a series, each term of which has to be positively electrified by contact with any of the following ones. This series proved to practically coincide with the «Volta series» established experimentally for the first time by Volta himself. According to J.I.'s theory, the mechanism of such electrization is evident. «It always seemed to me», J.I. wrote in his paper, «that Volts, the units invented to measure contact potential differences (i.e., the difference between the interior potentials), were also suitable for measuring ionizing potentials, which are equal to several Volts in the case of the metal vapour. This random, according to the first view, coincidence gains a simple and rather natural explanation on the grounds of the above theory.»[16]

On 8 November 1916, J.I. gave a report at the meeting of the Russian Physics-Chemical Society entitled «On the origin of the contact potential difference». His presentation gave rise to a hot discussion in which A.F. Ioffe, D.S. Rozhdestvensky and G.G. Veikhardt took part, among others. By their commendation, J.I. Frenkel was on the same day elected a member of the Society.

The paper by Frenkel was published in the ZhRFKhO and in the English «Philosophical Magazine». Its worth may be formally judged by the fact that it was included in the collected papers of Frenkel published after his death by the Academy of Sciences of the USSR. Informally, its results retain their value even now. A simple physical model of the double layer is found in text-books on physics and remains in the explanation of some surface phenomena in physical electronics.

Along with this work J.I. prepared for publication a whole series of papers under the general title «The structure of atoms from the point of radioactive ir-

16) J.I. Frenkel, ZhRFKhO, part phys., v. 49, # 2, pp. 110–111, 1917 (in Russian).

radiation», where he gave, in an original manner, a fairly complete survey of the achievements of atomic physics.

It is interesting to note that Frenkel's first paper was published in the journal «Automobile» back in 1916 and was dedicated to the car differential. J.I. explained the kinematics of this unit, and derived the equation of motion for the car wheels connected with the driving shaft through the differential. The equations happen to be identical to those determining electrical currents in two inductively connected circuits. The angular velocities of the wheels correspond to the currents, moments of their inertia — to the self-induction coefficients and so on. J.I. carried out all the analysis of the differential operation with the help of this similarity.

Living conditions in Russia in the third year of an exhausting war grew worse by the day. In Petrograd the shortage of food and fuel became greater and greater. These hardships, however, were not reflected by Frenkel's work, which was then making rapid progress.

But in the beginning of January 1917 deep sorrow descended on the Frenkel family: having been sick for only one day, the youngest of the brothers, Victor, died.

J.I. wrote to his father: «I have to pull myself together and finish my philosophical paper[17], though presently I am far from being in a philosophical spirit. Have not yet started my studies. Am entertaining myself with newspapers and pure mathematics.»

J.I. made himself return to his work. The proofs of his papers on the atomic structure began to come in, he resumed attending the meetings of the Physics-Chemical Society at the University and the seminar at the Polytechnic.

Frenkel greeted with admiration the news of the fall of tsarism. His letters to his father testify that he oriented himself correctly in the rapidly changing political situation of those days and saw ahead of the liberal Russian intelligentsia which welcomed the Provisional government and regarded suspiciously the activity of the Soviets of workers' and soldiers' deputies. He wrote:

«Petrograd, 6 March 1917.
... Long live freedom and those to whom we are indebted for it! This is the first freedom in Russia achieved by the third Russian revolution. The first one (1825) was the military without the people, the second (1905), the people without the military and, finally, the third was the revolution of the people and the military, united tightly. Therefore the third revolution has come with the speed of lightning, in one powerful and joint effort which scattered the terrible ghost of the old power. And people who suffered for so long under the yoke of this power, with indescribable surprise saw that they had been scared by a ghost whose whole force lay in the fact that the people were separated; having united, the people removed

17) «On the general nature of the life processes»; for more about this paper by Frenkel see further on.

from power, with a powerful hand, all those contemptuous and vile persons who, scaring the people with this old ghost, had tried to subdue the people to a new one. And the ghosts are disappearing in the mist they were fussing in, leaving people to strike out for their own freedom.»

«Petrograd, 14 April 1917.
... There is no news here. Nor do I have any myself, except for three brochures (on imperialism, militarism and education in a democratic state) that I have written these days. Hope they will be published.»

«Petrograd, 18 April [1 May, new style], 1917.
... Volodya wrote you with the details of what has usually been called «the May disorders» but is now turned into a celebration of order, freedom and international solidarity. I'd like to add only a few words about my own feelings and adventures.

At 10 in the morning I escaped from home and went to Alexandrovskaya Square, the gathering point of the Rozhdestvensky district. Listened to the orators for a long while; some were good, some bad. The public's attitude to them was very characteristic. Particular protests were immediately suppressed, and it was proposed that an unsatisfied person express all the arguments from the podium, instead of interrupting the speaker. The password «freedom of speech» worked magic. I could not stand one of the speaker's nonsense (an intellectual, of course) and for the first time in my life acted as an orator-agitator. I tried to convince the public that one should not mix the people with the ruling classes; that the German people are as unwilling to have war as the Russians; that they were deceived that «they had been attacked»; that to make peace as soon as possible it is necessary to convince the Germans that not only the Russian people but the Russian government as well have no intention of conquering anybody; that to do this, our government must appeal to the alien and friendly states with an official programme of peace, etc. At first some people cried: «Down with him!», but then they started to cry: «Right!»»

At the end of the spring of 1917, J.I.'s mother, because of the state of her health, which turned sharply for the worse after the death of her youngest son, went to Yalta (in the Crimea) with her son Sergei, who was in the last grade at the gymnasium. In the summer the elder brothers, the medical student Vladimir and Jacov, together with their father also went there.

In the Crimea Frenkel learned that back in 1916 a question had been asked before the State Council about the organization of a university for people weak health (mainly suffering from tuberculosis) on the South Bank of the Crimea. They had not managed to find an appropriate place for the university. The February Revolution and the fall of the tsar's power opened new possibilities.

In his 1936 autobiography J.I. wrote: «In the summer of 1917 a group of professors staying at this time in the Crimea (N.M. Krylov, N.I. Kuznetsov, S.I. Metal'nikov) decided to organize a new university with a boarding house on the South Bank of the Crimea.»

J.I. returned to Petrograd on 18 October 1917 and on the next day he wrote to his mother:

«Petrograd, 19 October 1917.
... This morning I went to the university to talk with Khvol'son, Bulgakov and Rozhdestvensky with the result that on Thursday, i.e., in a week, I shall have an examination [Masters — V.F.] in physics. I arrived just in time. Tomorrow or the day after it would have been too late to be put on the list.

Khvol'son stated that he not only read, but studied my (English) paper[18], and that he found it to be perfect, written in the spirit of the old classical physics, though in the most modern form. In a word, he paid me a lot of compliments and said that he regards it as inappropriate to examine my knowledge and suggests that I explain to him whatever I want.»

J.I. recalled his first examination time and again. In the early morning of 26 October he went out of the house at the corner of Mytninskaya and 2nd Rozhdestvenskaya streets, where he lived together with his brother, and headed for the university, but did not get farther than the Dvortsovy Bridge; he was stopped by the patrol of the revolutionary sailors there who did not let people get to Vasil'evsky Island. Several days later he met the Chairman of the Examination Commission, Professor Khvol'son, who reproached J.I. angrily and chastised him because three professors had specially gathered and waited an hour for him. «But the revolution happened», said J.I. shyly. In reply he heard: «Remember, young man: scientists recognize no revolutions.»

Probably, Frenkel did not at once feel the scales of the revolution. He was completely absorbed in his work. And it is worthwhile mentioning that revolution did not disrupt the course of education and scientific life in Petrograd. As before, lectures were given and exams accepted. Here is the letter of J.I. to his mother of 14 November 1917.

«I am writing this at 5 in the morning on the main staircase of the house # 4 in Mytninskaya street, which I have to guard in person between 3 and 6 a.m.; I do not know from whom. They are going to cancel these watches on the main staircases the day after tomorrow. However, I was glad about this occasion to stay awake since I used it to write a great number of letters, having little time for that during the day. So, yesterday I passed my first master exam. It took place in the apartment of Khvol'son who was not well. Rozhdestvensky and Bulgakov came. The four of us sat around a small card table with a stack of paper, as if we were going to play cards, but I had time only to open my cards a little, since they, afraid of losing, hurried to let me free after only $1\frac{1}{2}$ hours. Of course, I was able to tell each of them only trifles in half an hour. Generally, this exam is just a stupid formality, or, if you like, an excuse for studies, like all examinations are. I

18) The paper on the double layer in «Philosophical Magazine», v. 33, p. 297, 1917.

could easily have passed it a week after coming back [from the Crimea — V.F.]. Though in terms of acquired or solidified knowledge these two weeks were rather fruitful for me; I still begrudge the time spared. There is only a month left until Christmas, and I yet have to pass 3 more exams.»

As is clear from the verbatim record of the meeting of the Physics and Mathematics Department,[19] the following topics were suggested to Frenkel:

1. Equations of electromagnetic field by Minkowski and Einstein.
2. The reasons for the crystal birefrigency.
3. Foundations of statistical mechanics and the theory of fluctuations.
4. Linear differential equations.
5. The problem of the cooling of a solid homogeneous rod.
6. The motion of a gyroscope over a smooth surface.

J.I. answered the first three questions at the card table, as mentioned in his letter. For the last three questions he was examined by Academician V.A. Steklov and Professor G.V. Kolosov. Two weeks later J.I. passed the exam on meteorology given by Professor P.I. Brounov.

Frenkel had no grounds to be worried about the result of the examination in mathematics, a subject he commanded well. This exam was given by Professors Yu.V. Sokhotsky, A.A. Adamov and Ya.V. Uspenskii; the latter probably remembered J.I. as «the pioneer» of the calculus of finite differences.

The relevant record reads:[20]

«On 11 December 1917 the test was taken by Jacov Il'ich Frenkel for the degree of Master in Physics in the theory of definite integrals, higher algebra, the theory of elliptic functions and the calculus of finite differences. The following topics were suggested:

1. The proof of some basic properties of definite integrals.
2. Root determination by the Fourier method.
3. The periods of elliptical integrals.
4. The convergence of the Gauss formula.

In probability theory:

1. The Poisson theorem (the law of large numbers).
2. The problem of the hypotheses probability determination [the Bayes formula — V.F.].»

On 12 December, Frenkel wrote to his parents:

---

19) State Historical Archives of the Leningrad Region, record group 14, inventory 1, file 11270, p. 30.

20) See the previous footnote.

«... I have passed my exams. Yesterday I went directly from the hospital [where a close relative of Frenkel died — V.F.] to the university. I was in the most unpleasant mood. My salto-mortale turned out to be quite successful; however, I enjoyed no gladness from it. It seems that I broke all previous records by passing my three Master exams in a month.»

J.I., as one can see from his letters, became closer and closer to Ioffe. At the end of October Ioffe presented a report on the work he performed with Frenkel[21] at the meeting of the P.N. Lebedev Moscow Physics Society.

On the evening of 30 November 1917, J.I., at Ioffe's invitation, was at the Polytechnic Institute, where «the draft of the law on the establishment of the Physics and Mechanics Department within the Polytechnic» was discussed. A year later, on 27 November 1918, a group of professors from the Petrograd Polytechnic Institute led by A.F. Ioffe made a corresponding presentation to the Council, and 3 months later, on 1 March 1919, the decision on the organization of the Physics and Mechanics Department was made.

One of the last letters written by Frenkel during this period, a very remarkable one, is given below.

«Petrograd, 27 December 1917.
My dears! We continue to write you, though we have only a weak hope that you indeed receive our letters. If the situation does not change in one more week, then I'll gather my scarce belongings and try to make my way through to Yalta to inform you about what is happening here and thus to moderate your unnecessary but inevitable worry ...

About the near future ... the following opinions were heard. Peace negotiations are to be moved to Stockholm, to be held in the view of the whole world and thus to influence more European democracies to become involved which in these negotiations is the Bolsheviks' principal objective. They are, however, taking account of the possible failure of the negotiations; so, while trying to drag them out, they are simultaneously preparing a new army whose core is formed by the Red Guards to replace the old army, which is absolutely fallen apart and unable to fight.

Do not conclude from the above that I have turned Bolshevik. I still consider their attempt to be utopian and their power to be ephemeral. But I cannot deny that it has a certain stature, and that they themselves have an unusual courage, logic and consequence; I also cannot help wishing them success, not believing, however, that it is possible.

I think that it is already too late to return back on the long roundabout way, and therefore I cannot help blaming the Russian intelligentsia for attempting to sabotage Bolshevics instead of helping them along the road from which it is now impossible to deviate. It is late now to struggle with the Bolshevics; we have

21) It seems that either the paper «On atomic models» or «Magnetic origin of mass» is meant.

to help them to diminish the negative results of their policy and to enhance the positive ones.

On the other hand, my dears, I am too far from active politics and practical life in general and am not at all inclined to exchange my sciences for it.

And the matter of sciences is progressing very slowly ... This evening I am going to visit the Fauseks, and tomorrow the Polytechnic Institute, for several days, to turn to my studies just a little ....»

This letter is typical enough for certain circles of the Russian intelligentsia of that time which did not take an active part in the revolutionary events and tried to understand them by reading extremely controversial and prejudiced newspaper information. For instance, Ioffe wrote: «I did not at once understand the meaning of the October Revolution. I at first regarded the coming of Bolsheviks to power as one of the episodes of the revolution, caused by the desire to finish the war.»[22]

Such was the spirit of J.I. at the beginning of 1918 when he, together with his brother Vladimir and the father of his gymnasium comrade M.D. Gurari, whose family had also got stuck in the Crimea, headed for Yalta.

It took more than a week for the Frenkel brothers and their travel companion to get to Simferopol' where they arrived on 15 January 1918. A revolutionary committee was in power for the second day in the city. And in Yalta power had changed hands twice in 10 days.

Greatly worried, the Frenkel brothers decided immediately to go to Yalta. V.I. Frenkel remembered in detail all the circumstances of this trip. They hired a cart with great difficulty. The gloomy master of the cart, who assured them that they would not manage to get far from Simferopol' anyway, commenced the journey.

However, despite these predictions the travellers reached the Pereval (Mountain Pass) and started to proceed slowly in the direction of Yalta along the absolutely deserted road. It was already dark when they first heard and then saw a mounted detachment not far from Alushta. Jumping off the cart, J.I. asked one of the military men: «Tell me please, comrade, are we far from Alushta?», using, without thinking, the form he had got accustomed to back in Petrograd. The polite question of J.I. caused an absolutely unexpected response: curses started falling onto his head. In a single instant he was surrounded by several unmounted soldiers. Having seen this, M.D. Gurari, who was with the Frenkels, came to help and forced his way to J.I. «Quiet down, please, sirs, it is just a misunderstanding. You are forgetting that the word «comrade» was not invented by the Bolsheviks. Or maybe you think that each comrade of a Minister in the good old days was also a Bolshevik. You see, «comrade» means «friend», good fellow, and Professor Frenkel has just appealed to you with this primary meaning of the word. We are very worried about our relatives in Yalta and are making our way there ....»

22) A.F. Ioffe. «My Life and Work». Moscow-Leningrad, GTTI, 1933, p. 20 (in Russian).

They had to explain, for quite a period, to the detachment of White officers where they were coming from and going to, and also had to present documents. Anyway, they managed to arrange things satisfactorily, but Vladimir Il'ich thought that J.I.'s life had been in mortal danger.

The political situation in the Crimea was extremely complicated in those years. It suffices to say that in 4 years (1917–1920) a change of power occurred 7 times.

J.I. came to Yalta on 18 January, when the Soviet power had been already established there. Its organs supported the idea of organizing a university-sanatorium and put several rooms in the Livadia Palace and in Oreanda at its disposal. In the printing house of the Yalta Soviet of Workers', Soldiers' and Peasants' deputies a booklet was printed, composed by the professors of Yalta College, which announced that «Tavrichesky philial [department] of the St Vladimir University [Kiev] is being opened in Livadia on 11 May 1918, and will incorporate the first year of the natural and mathematical division of the Physics and Mathematics Department.» While waiting for the opening, Frenkel made his living out of private teaching as he used to long ago. Among the advertisements in the local newspaper issue of 12 March 1918, one can find an item announcing that «Associate Professor of Livadia University, J.I. Frenkel, will teach lessons in physics and mathematics to single persons and groups.» Since such advertisements appeared rather frequently, the demand for the services was low.

It was assumed that the first semester lectures would continue until the middle of August. In the booklet mentioned it was also stated that physics would be taught by Professors A.F. Ioffe and L.I. Kordysh and Associate Professor J.I. Frenkel.

However, before the assigned day of the university's opening in Yalta, the political situation in the Crimea changed drastically. After the Brest Peace Treaty (March 1918) the Ukraine was occupied by the Germans. This gave rise to the danger of their penetrating the Crimea. In March 1918, the 1st Congress of Soviets in Sebastopol declared the formation of the Tavrida Republic on the territory of the Crimea, thinking that this measure would stop the German troops. But General Kosch's division took over the Perekop in April, and the whole the Crimea by the 1st of May.

In the first months of the occupation the Crimea was not yet cut off the rest of Russia, and individuals were able both to penetrate and try to leave it. In the summer of 1918 Ioffe came to Yalta to his family. It became clear at once that he would not teach at the University since he had started to prepare the opening of a new Physics Institute back in Petrograd. Ioffe invited Frenkel to join this work, the more so because the situation in the Crimea was far from promoting any scientific activities. J.I. understood all this perfectly, but family circumstances forced him to decline the proposition, since his mother's illness continued. Yielding to her requests, he decided to stay in the Crimea for a year.

At the beginning of September 1918 Frenkel went to Petrograd on a commission from Tavrichesky University to buy materials for the demonstrations for the course on general physics. J.I. stayed in Petrograd for about a month and during this period became again involved in the usual atmosphere of intensive work. During the day he made efforts to get the facilities and books for the university; in the evening he wrote the paper «Rotational motion of an elecrified sphere» (we will discuss this later in association with Frenkel's works on the rotating electron and classical electrodynamics which follows later).

What Ioffe had tentatively spoken about in the Crimea, was implemented on 23 September 1918. Narkompros (The People's Commissariat for Education) decided to establish the State Rentgenovsky and Radiologichesky (X-Ray and Radiology) Institute in Petrograd.

And three days earlier, identification papers had been issued to J.I. in the Narkompros office: «This is to certify that citizen Jacov Il'ich Frenkel, Associate Professor of Tavrichesky University, is really going to his place of work, the city of Simferopol' in Tavricheskaya Province, having in his possession the appropriate materials, such as: scientific books and manuscripts, microscopes, etc., as well as his personal belongings.

«The Education Commissariat of the Union of the Communities of the North Region requests thus that all local Soviets and civil authorities provide citizen J.I. Frenkel with all possible help in the matter of his free passage, with all the goods mentioned, to the city of Simferopol'.»

One can imagine J.I.'s sadness at leaving Petrograd and his colleagues. By the middle of October he had made his way to Simferopol', where Tavrichesky University had finally moved by this time. The Chair of Physics was headed there by the Kiev University Professor L.I. Kordysh (1878–1932), who even before the revolution was, in the words of Khvol'son, «one of our most talented specialists in theoretical physics». However, J.I. had practically no communication with Kordysh; they had some kind of a quarrel. In J.I.'s letter of 4 February 1919 he makes the following remark in passing: «My relations with Kordysh are rather cool. I am going to speak to him and either settle everything or break with him completely.» It is a very characteristic remark of Frenkel who did not like compromises. He eventually parted from Kordysh completely.

At first only J.I. and his 18-year-old brother Sergei, who had graduated from the Yalta gymnasium with a gold medal in the spring of 1917, worked at the chair of Physics, along with its Chairman. At the same time Sergei Frenkel entered the Physics and Mathematics Department of Tavrichesky University, as well as holding the position of Assistant Professor.

Sergei Il'ich Frenkel was, as testified by people who knew him, an exceptionally gifted young man with great capabilities in technology and «golden» hands. Due to his efforts the Physics Cabinet was properly organized in the university; he assisted Kordysh and his brother in their lectures, and invented and constructed several original mathematical machines (mechanically driven) for the Mathematics Cabinet at the N.M. Krylov Chair.

In 1919 the Chair of Physics was somewhat enlarged. I.E. Tamm, L.S. Vagin and some others joined it.

Frenkel became friendly with people from the Chair of Mathematics, with the Chairman N.M. Krylov in the first turn (he became an Academician in 1929). J.I. probably got to know him back in Yalta in the summer of 1917. Krylov was 15 years older than Frenkel, but the difference in age did not disturb their friendship. Even back in those years Krylov was known for his work on the interpolation theory, approximate integration of differential equations of mathematical physics and variational calculus[23].

N.M. Krylov was close to that group of the St. Petersburg mathematical school which inherited from P. Chebyshev and A. Lyapunov the task of founding modern mathematical physics.

At the initiative of Krylov and his coworkers at Tavrichesky University, the Mathematical Society was organized there. They even managed to provide the Mathematics Cabinet with paper and printing facilities and to organize publication of scientific articles and the verbatim records of the Society meetings held regularly, twice per month.

Frenkel was, among Krylov's other staff members, one of the founding members of the Society, a participant and reporter at its meetings, though he mostly presented the latest achievements in physics: the theory of quanta, the theory of relativity and the like. He also gave lectures on applied mathematics (for instance, on the applications of complex numbers to vector analysis, etc.).

Some papers were published at the Mathematics Cabinet as brochures; in appearance they were similar to modern preprints.

In particular, two papers by Frenkel were published in the form of such preprints. One of them was a brief summary of his paper on the rotational motion of a charged sphere; the other was dedicated to a new question concerning the foundation of electrodynamics without «magnetic charges». It opened a whole series of Frenkel's works completed between 1926 and 1934 by publishing his two-volume «Electrodynamics». We will speak of these investigations further on.

N.M. Krylov had his own funny ways, the kind which, as J.I. used to say, decorate and enliven a personality. Krylov had a sharp tongue, many people were afraid of him, and he did not have good relations with everybody.

Once at a session of the Ukrainian Academy of Sciences Krylov came up to one of his colleagues whose scientific services he did not value much. They had a conversation; soon a small circle formed around them. Discussing the last (or perhaps next) Academy elections, Krylov remarked skeptically that many people were elected without any sound grounds for it. «There is no need to go far for an

---

23) N.M. Krylov became world famous for his works on nonlinear mechanics, mostly done with his student, future Academician N.N. Bogolyubov. N.M. Krylov strongly encouraged the transition from the «linear prejudices» to «nonlinear psychology», to use the terms of Professor S.E. Khaykin.

example», he said, «just look at you and me: well, what kind of Academicians are we really? We are, pardon me, just shit.» Absolutely disappointed, the Academician began to praise Krylov's scientific achievements, but was in no position to speak about his own. Krylov let himself be persuaded and left, quite pleased.

In November 1918, J.I. started lecturing regularly to the students of the Medical Department of the university (the physics students were lectured by Kordysh). He lived in Simferopol' together with his brother. Their parents stayed in Yalta, where the brothers wrote regularly. Below we give several excerpts from these letters of J.I.'s.

«Simferopol', 10 November 1918.
... This morning, despite the holiday, I arranged lessons for my medical students who arrived in a rather large group. The lessons consisted of their writing down, as I dictated, the contents of my first lecture, with some additional explanations on my part. The aim of these temporary lessons is to help them understand the basics of physics and to learn how to write down lectures. I have already agreed with the Student Publishing Committee, which asked me to publish my lectures. As their printing is not bad at all, and as I would like to get my course fixed, I agreed to the publishing committee's proposition and started to compose my lectures, which take me much time.»

Frenkel's introductory lecture was actually published[24]. He presented there his line of thought concerning the position occupied by physics in the system of natural sciences. Keeping his audience in view, future medical workers, Frenkel analyzed in detail the relationship between physics and psychology. «Our attention,» he wrote, «is attracted by a change of observed phenomena, by the variety of its manifestations. We easily get used to uniformity and constancy; we stop noticing it. The usual seems to us natural, understood, while the unusual appears unnatural and not understood. Having found «the invariants» of the phenomena studied, we get accustomed to them, and it seems to us that we thus have understood them.» «In fact,» Frenkel continued, «we cannot understand, but actually can only get used to.»[25]

By means of our «inexperienced» feelings, we see nature as being infinitely diverse and always changing. Our mind uncovers uniformity and constancy behind this diversified picture: there are invariant things (particles) related by invariant mechanical laws. The objective picture of nature lacks colours and sounds, tastes and flavours; all its variety is transposed onto our mentality by physics — these are the foundations of the mechanical picture of the world. «One has to stress once again,» concluded Frenkel, «that the mechanical world view and psychophysical interpretation accompanying it are based not on the instructions of the

24) J.I. Frenkel. «The General Course of Physics. Introduction. 1. The position of physics in the system of natural sciences». Publishing Committee of Students of Tavrichesky University, pp. 1–15.

25) J.I. stressed this thought many times later. See, for instance: A.I. Kitaygorodsky. «Physics is My Profession». Moscow, Molodaya Gvardiya, 1965, p. 85 (in Russian).

philosophizing mind, but on the clear and accurate facts discovered by experiment and observation; and in the cases of noncorrespondence (very rare, fortunately) between the requirements of the mind and the facts, reason must adjust to the facts, and not vice versa.»

Our feelings, while subjective, are caused by objective reasons. Frenkel illustrated this with the help of examples. What is sound? From the subjective view point, it is a feeling in our hearing apparatus caused by air oscillations acting on it. The study of sound (acoustics) is reduced to «1) establishing the psycho-physical relationship between the auditory sensations, on one hand, and the oscillatory motion, on the other, and 2) investigating this oscillatory motion from the purely mechanical point of view, i.e., the mechanics of the vibrational motion of material bodies.» Frenkel noted that the «mechanical» (second) point of view is «accessible» to muscular feeling, «the only one of our feelings that does not know any differences except for those in quantity, intensity, magnitude.»

The questions associated with heat, the feeling of warmth and cold and their «mechanical» (objective) meaning, were treated in a similar way, as was the study of light.

J.I. divided his course into two parts. In the first one he was going to address general mechanics, ignoring the real microscopic nature of material bodies, with consideration to micromechanics (mechanics of atoms and molecules), including the molecular theory of heat.

The second part of the course was to start with macroscopic electrodynamics (the general mechanics of electricity, as J.I. called it), and to end with the micromechanics of electricity, including magnetic phenomena, the structure of atoms, the theory of the electrical discharge in gases and so on.

«Simferopol', 25 November 1918.
... The question of my reelection was resolved safely, that is, the faculty ruled to recognize the old elections, ignoring the Kiev ballot. [Professors and associate professors elected at Tavrichesky University had to be approved in their positions by the Council of Kiev University; J.I. was not elected by this, mostly anti-semitic, Council. — V.F.]

... Tomorrow the meeting of the Physics and Mathematics Department will be held, where the question of the «election of student S.I. Frenkel, Assistant at the Physics Cabinet» will be solved, by the way.

I am proceeding successfully with the lectures given in front of a large enough audience, into whose heads I am hammering the basics of contemporary physics with all my strength. This is even more difficult because my students have practically no preparation. I can, nevertheless, state that my words are not offered in vain, and that I have managed to arouse the audience's interest in my subject.»

«Simferopol', 3 December 1918.
... Recently I am staying at home (with Sergei, of course), when not at the lectures. The idle wandering to which I was doomed before finding the flat became

absolutely distasteful to me. I wish to get involved in study as one should. Unfortunately, the German scientific journals which I waited for so impatiently got hopelessly stuck in Khar'kov. So I have nothing to read in my field except for reading aloud my lectures, but there is nobody to listen to them.

... We are starting to arrange somewhat our Physics Cabinet; perhaps soon we will be able to organize the mechanical shop as well».

«Simferopol', [the beginning of December] 1918.
... I give 2 lectures on Mondays and Tuesdays (one in the morning, the other in the evening), and not 4, but a minimum of 6 hours go for them. For a change, I am absolutely free on Thursdays and Fridays, except for the damned examinations. Along with the lectures I attend with passion, I am engaged in the composition of my course, which is progressing with difficulty ....»

A large file of Sergei Frenkel's letters to his parents adds much to the picture of J.I. which his own letters present. The brothers were very attached to each other. J.I. wrote about his younger brother with invariable tenderness. Sergei supplemented the lyrical part of his letters about his elder brother with an ironic side. Here are some typical excerpts from his letter of 28 November 1918: «... I saw Jacov strongly and hopelessly rubbing a glass stick and trying to discover electricity on it with the help of corresponding devices. And he was swearing like a barbarian, since nothing came of his efforts (on account of the humidity), and he had solemnly promised to demonstrate experiments to his students that day; he had done this with his light-mindedness and passion, not even knowing if the things necessary for it existed in Simferopol'. ... I have not yet begun to attend lectures mainly because of the lack of the room and the disorder this causes. Yasha's lectures are in full swing. Even successfully, with the medical students of the first semester; at least he says they have become interested in physics. If this is the issue (and he is the man of passion), then it is a colossal service on his part, since one has to possess the wisdom of Solomon to accomplish this with the medical people. But the things are worse with the second semester: only 20 to 40 people out of 240 really attend the lectures (however, everybody is in the same situation, even Nikolay Ivanovich [Kuznetsov]; the second semester medical students consider and call themselves second year students and so do not want to study).»

Let us return to J.I.'s letters.

«Simferopol', 5 February 1919.
... The printing of my lectures has temporarily stopped. Thus I am not in a hurry with my course composition. Krylov advises me to deal with my dissertation[26], and it seems that I will follow his advice. I have to put myself on solid ground as soon as possible.»

---

26) The Master thesis is meant.

«Simferopol', 7 February 1919.
... Krylov today gave his first lecture at the university in front of ... 6 people! There are some grounds to believe that at least 3 of these 6 will stay. The third mathematician[27], Koshlyakov, has been trying in vain for two weeks to get rid of the flu. So, the Mathematics Department is kept mainly by M.A. Tikhomandritsky. This honourable scientist, who has recently celebrated his 75 jubilee, cannot, however, attract more than 3 to 4 people to his lectures.»

«Simferopol', 16 March 1919.
... Yesterday I received a letter from Ioffe dated December 20, in which he let me know that I had been elected a scientist and staff member of the Rentgenovsky Institute (of which he is president) and suggested that I come to Petrograd, if possible.»

Events in the Crimea have turned for the better by this time. After smashing Kolchak on the Eastern front the Red Army started to mount an offensive in the South and on 5 April, 1919 entered the Crimea, and on April 11, liberated Simferopol'.

In the very first days after the establishment of the Soviet power Frenkel began to work at the Crimean Narkompros where he was commissioned to lead the Department of Higher and Vocational Education (this work was mainly concentrated around the university). He also continued teaching.

Unfortunately, not all of Frenkel's letters from this brief Soviet period in the Crimea (April–June 1919) are extant. In the beginning of August, after Denikin came to power in the Crimea, he was arrested and put into Simferopol' prison. The first extant letter of his from this period was written there.

«From Simferopol' prison [to his mother], August 1919.
... Despite all the calming news from you, I do not believe that I will be liberated soon. I only hope that you — and all of you — worry and fuss less about me. I myself feel excellent and cannot complain of anything. I am not at all bored; I spend rather a lot of time reading Drude and Grave.[28] Have started writing my paper; now am expecting the delivery of my lectures which have still not reached me. From time to time I play chess with Rabinovich and Krasnopyorov [the cellmates of J.I. — V.F.]. This game is prohibited within the prison walls, and the day before yesterday the drunk Deputy Warden took the chessboard away from us, so we made another one to replace it. We usually talk about politics and private circumstances. So, I repeat, I am not bored at all. I am saved from ill temper by optimism and a philosophical spirit, of which I've already written. If one does not brood over what could be, etc., which I try to avoid, then one can live here splendidly, like in a sanatorium. The difference is, the rooms in a sanatorium are locked from the inside while in prison cells, from the outside.»

---

27) The second one was probably Professor L.A. Vishnevsky.

28) Apparently, «Optics» by P. Drude and «Elements of Higher Algebra» by D.A. Grave.

Frenkel was lucky: two months later he was set free under the care of a group of professors from Tavrichesky University. N.M. Krylov played an important role in his liberation; Professor A.N. Derevitsky also helped J.I. very much. Academician N.K. Gudziy, who then was also a professor at the same university, recalled that Derevitsky's attitude towards Soviet power and those who sympathized with it, was at first very moderate, but then he became much more of a «leftist», especially in the spring of 1919, when there was a «white» terror in the Crimea.

However, J.I. was removed from his teaching position and he lived either in Yalta or in Simferopol', making a living by giving private lessons and suffering from the situation.

Much later, in Leningrad, J.I. suddenly learned some details regarding this earlier time. Referring to Professor K.S. Shifrin, a student of J.I.'s, I.E. Tamm wrote: «The late T.P. Kravets, who had been then [in 1919 — V.F.] a professor at Khar'kov University, said that a paper arrived for him, via the Rector's office, in which some officer of the Denikin Counterintelligence Service inquired if the «Bolshevik agitator» J.I. Frenkel was really a physicist, as he pretended to be.» In reply to this Kravets wrote that J.I. Frenkel was the author of one of the most outstanding papers published in recent years at the ZhRFKhO (he meant the paper on the double electric layer). Kravets recalled this episode shortly before the war when he, J.I. and several other physicists were returning from one of the sessions of a conference dedicated to the nature of a latent photographic image. Kravets' lecture had raised some objections on the part of J.I., and Kravets jocularly said: «And do you know, J.I., that some time ago I was a much less severe critic of your work», and spoke of his response to the Denikin's investigation.»[29]

At the end of January 1920, after a half-year break, Frenkel returned to the University. He wrote to his parents in Yalta on 2 February:

«My dears! Today at 2 a.m. I arrived safely in Simferopol'. The whole way I trembled from cold and fear. I blessed you for the fur coat and, to warm up my feet, sat on them Turkish style. As for the fear, it proved unwarrented, since the robbers, obviously, were satisfied with the burglary of the day before yesterday. Robberies now occur exclusively in the daytime (why strain one's vision).

... So, I have completely established myself at the university and will start my hand organ in 2 or 3 days. Am feeling perfect. I am very glad that I'm finally returning to scientific research and teaching. I've become acquainted with one of our new assistant professors, Tamm, in whose room I am writing this letter and will spend the night. A very sympathetic fellow — it will be good knowing him!»

In Simferopol' Frenkel lived alone. His brother Sergei was forcibly made to serve in the «volunteer» Denikin army. J.I. said that in those years he happened several times to see the population ordered to «volunteer» for it.

---

29) UFN, v.76, # 3, p.405, 1962 (in Russian).

The authors of these orders probably did not notice the absurdity of the term «volunteer» when applied to their army.

In the fall of 1919 J.I. succeeded in acquiring an official form with the forged signature of some officer which ordered his brother Sergei to report to the Yalta municipal military command responsible for the electrical power plant's security, and Sergei was actually sent there. But J.I. carelessly mentioned this in front of some married couple previously unknown to him. Soon after this Frenkel was called before the investigating officer of his case. He ordered J.I. to make some additions to the file on him opened by the counterintelligence service. Looking through the documents there, J.I. noticed a statement, at the end of the file, of everything he had said that time. The statement was signed by the woman of the afore mentioned couple. This was, in Frenkel's opinion, the critical moment. He tore the statement out of the folder, knowing that if the officer knew about it and was proceeding against him, the action would make the situation worse. Everything turned out luckily: either the officer did not pay attention to the appearance and disappearance of the document — there were a lot of them then! — or he was a decent person, which could also easily have been the case.

But though J.I. managed to save his brother from being conscripted into the Denikin army, he could not prevent an accident. On 20 March 1920, Sergei Frenkel was accidentally wounded by a comrade soldier who took a shot at him in jest, not knowing that the rifle had been charged. The bullet went into the shoulder and cut an artery. Sergei did not guess how dangerous the wound was and asked to postpone informing his parents. He was driven to a hospital after a long delay and there died from blood loss.

J.I. was deeply shocked by his brother's death. Of the five Frenkel children only two remained alive: he himself and his brother Vladimir, then a medical student in Rostov. The letters from Simferopol' which Frenkel sent to his parents almost every day were full of pain.

«... I tell you the truth (but ask you not to announce it). To perpetuate Sergey's memory at Tavrichesky University, I want to write a physics book dedicated to him, and put all the profit from it into a stipend, named after him, for the most capable students in the Physics and Mathematics Department. I would like to start working immediately, but right now I'm too occupied with lectures and exams. Besides, I do not yet feel ready to start such serious work. Therefore I probably shall postpone it until my return to Yalta.»

In 1926 J.I. finished his book «Electrodynamics». In it he wrote: «To the memory of my brother Sergei, my first pupil and assistant.»

«Simferopol', 14 April 1920.
... Today I was elected Associate Professor by a majority of 14 to 6. The outcome is not especially honourable. Among the people who gave me the «black» marks were several new professors of a solid age and ancient spirit. On Friday the final elections at the Council will be held. I think everything will be all right there — there are no bad precedents — but anything is possible.

As you know, I wrote the review of my papers myself, and the commission members, who did not know a damn thing about them, just applied their signatures ...

We live peacefully with Nikolay Mitrofanovich [Krylov]; I am glad he is present. At times the loneliness becomes unbearable. I am going now to visit Frank who, with Tamm, expressed a wish to play chess with me.

My modus vivendi is very antisocial; I do not converse with anyone except my closest friends. Recently many people were arrested in Simferopol', among them university and even high school students. They are trying to eliminate the Union of Communist Youth. The counterintelligence agency is committing outrageous acts: rape, violence and even torture. Our professors have to guard students, boys and girls, who are absolutely uninvolved in politics but evidently related to those arrested earlier.

The food situation in Simferopol' is desperate but still seems to be better than in Yalta.»

«Simferopol', 24 April 1920.
... I spoke to Nikolay Mitrofanovich yesterday evening, with the result that our new comrade, Smirnov [Vladimir Ivanovich — V.F.], a very capable mathematician and exceptionally fine man, has just moved into our room.

Today I had to give a lecture to only 5 second-year students! Bunge and other professors have long since become satisfied with an even smaller audience! It's swinish, really!

My election to the Council is assigned for Wednesday. The outcome is difficult to guess. I think, anyway, I'll get elected, though perhaps by a small majority.

Today at 7 p.m. went with Frank to Professor Vernadsky's lecture «On the role of the life processes in the chemistry of the Earth's crust». The lecture had interesting ideas and was gracefully presented but too long. And generally it was difficult for me to listen, because my thoughts, not sufficiently captured by the lecturer, constantly wandered. I thought about Sergey, about those eight poor fellows hung today on the station square for belonging to the Union of Communist Youth. Really — «o tempora, o mores!»

«Simferopol', 1 May 1920.
... Obviously, I panicked my medical students, for only a few people came to the exam which tormented me for 2 1/2 hours, and three of five still failed. Yesterday morning, yielding to Smirnov's insistence and partly to my own inclinations, I bought a pound of salted fish in the market and 1 1/2 pounds more today. The fish was marvelous and vanished almost instantly. Because of the lack or, more correctly, shortage of knives, forks and plates, I am managing without them, tearing the fish with my hands and teeth. As for Smirnov, he, a specialist and connoisseur, ate it directly with tea, replacing sugar with it ... Lately, since our flat has become overpopulated, I often stay alone with my thoughts. Regular conversations with my colleagues and acquaintances cannot, of course, help entertaining me. I study

rather little; not enough concentration for serious reading. One concentrates more easily on his own work. I am not burdened by my many lectures and reports; on the contrary, they provide me with an impetus to study. The next presentation is for the Gurvich seminar. I'll talk about it later in more detail. The subject is «The opposing tendencies of animate and inanimate processes».»

In the Crimea Frenkel became close not only to mathematicians but also to biologists, especially Professor A.G. Gurvich, who gave the course of histology at Tavrichesky University, and his staff members A.A. Lyubistchev and Professor S.I. Metal'nikov, a zoologist.

We shall dwell more on the report for biologists mentioned by J.I. in the preceding letter; the manuscript exists in the Frenkel archives.

J.I. began with the known fact that for any system of inanimate bodies, a state of a stable equilibrium is normal. The stability of this state, its conservation, manifests itself in the fact that any action moving the body away from the equilibrium state gives rise to a force tending to reduce this motion. Frenkel called such a relationship of mutual weakening between the action and the system antagonistic.

On the other hand, in an unstable state of equilibrium the action (cause) produces the reaction (consequence) tending to enhance itself (this relationship of mutual amplification J.I. called a harmonic one; evidently, it is opposed to the antagonistic relationship).

Can inanimate systems remain for long at an unstable equilibrium? The simplest example of the situation is a vertically standing stick. The smallest deviation from the state of equilibrium causes it to fall down. However, the hand of a juggler keeps it from falling. Here the stabilizing factor which maintains the unstable equilibrium state is a live organism, a human body. Actually, the equilibrium of the human body is unstable. Standing on our two feet we, in Frenkel's words, «juggle with our bodies by means of a stabilizing mechanism of nerve pulses and muscle shortenings.» The malfunctioning of this mechanism leads to a fall.

So, the stabilizing system appears to be as follows. Cause A produces consequence B which tries to increase A. Thus at the first stage an animate system behaves like an inanimate one. But specific features of the former are revealed in the appearance of a stabilizing factor C which tends to change the sign of A (in modern terms, this is actually the famous damping feedback loop).

The stabilizing factor enters the game after some delay. A «normal» man balancing his body has a normal delay, while a drunk's delay is greater (and its owner falls), and the juggler's delay is smaller due to training, manifesting thus his art. The delay grows with any deviations from the normal life process. Therefore, for example, sick or old people have to support themselves with a stick while walking, etc.

«From the mechanical point of view,» stated Frenkel, «death is a transition from an unstable equilibrium state to a stable one, and life is reduced to support for this unstable equilibrium.»

Professor A. Gurvich, biologist. Oil portrait by Ya. Frenkel, Kazan, 1943

This definition, derived from the consideration of simple mechanical examples, proves correct in a number of other cases, and Frenkel turned it into a principle. He observed that, obviously, some effects similar to the mechanical delay must also take place when considering the life process from a chemical perspective. And indeed, the phenomenon of the delay is manifested in feeding (taking food is exactly the stabilizing factor) which occurs at discrete intervals. During the intervals between eating the live organism deviates from the state of the unstable equilibrium; if no food is taken for a long enough period, then death by starvation occurs. The smaller the intervals, the higher, usually, the organism is. Serpents, for instance, can live for months without food, and microbes for thousands of years. According to this viewpoint, lifeless systems have an infinite delay.

It is interesting to compare the preceding arguments with the assertions Norbert Wiener made when discussing the feedback problem and, generally, the analogy he saw between the operation of the human nervous system, on one hand, and the operation of the modern computers, on the other.[30]

Naturally, Frenkel's arguments and the trends he noticed did not form, in his opinion, any closed system of views. J.I. simply looked at nature with the eyes of an artist, rather than a scientist, and made a sketch I find a very vivid and bright study of the chosen original and an unexpected point.

Let us return to J.I.'s letters from the Crimea.

«Simferopol', 6 May 1920 [to his parents].
My dears!
This evening at the council meeting I was let down, that is, my election was turned down by 23 to 16. Evidently my enemies campaigned against me as actively as my friends did for me. Despite the bright and even maybe pathetic speech by Krylov, who was supported by Vishnevsky and Bunge, our opponents, obscurantists, without any substantial arguments against me, just referring to my little experience and the fact that I am provided with the lectures of the obligatory course and the rights and privileges this implies, as well as remembering my previous vile actions,[31] did not elect me, rather unanimously.

I am afraid, my dears, that you will be much more distressed by it than I am myself. Believe me, I am worried very little by this issue. And to this end, even election by the Council would not settle the issue, since the final approval has to be obtained from the «People's Chief Officer of Education» in Sebastopol. This highest officer would not, very likely, approve of me in the position of Associate Professor and moreover, would perhaps suggest that I lose my assistant professorship as well.»

«Simferopol', 15 May 1920, [to his parents]
... Yesterday evening I gave an examination; one of ten people passed and five

30) N. Wiener. «I am a Mathematician». New York: Doubleday & Co, 1956.

31) Frenkel meant here his work at the Crimea Narkompros under the Soviet power.

failed, the remaining scattered. Do not, please, think I was too demanding: our medical students do not meet the very minimal requirements. After the exam came the report by Smirnov[32]

... Today I feel well. In the morning visited the hospital (i.e., University) where I was shattered by the sight of several girls preparing to be victimized on the altar of physics ... Later went to the meeting of the Mathematical Society and for 3 1/2 hours suffered rather unpleasant feelings caused by improper spiritual nourishment.»

At the end of November 1920 the Red Army approached the Crimea; and even earlier in October the arrests had started. The previous few weeks Frenkel had spent hiding with his acquaintances; he was searched for in Simferopol' and Yalta.

On 11 November the Red Army broke the Perekop fortifications and entered the Crimea, and on the 14 November Simferopol' was liberated.

On 16 November the People's Education Division was reestablished, and Frenkel started immediately working in his former position as Head of the High and Vocational Education Department. Simultaneously he was a member of the Editorial Board of the Newspaper «Krasny Krym» («Red Crimea»).

«Simferopol', 8 December 1920, [to his mother]
... At the People's Education Division I work between 10–11 a.m. and 4 p.m.; however, once a week I manage to give over a couple of hours for lectures. Aside from a lot of small chores (accepting late students, issuing safe-conducts and passes, etc.), I would be satisfied with my work. I have already carried out reforms of university life and continue with them and, as far as I know, my colleagues not only have no reproaches with me, but would be distressed if I were to leave my position. As for the essence of the reforms, they as of now reduce to the following. First, the departments and the University Council have been somewhat reorganized (all the people giving lectures and lessons on their own have had their rights equalized, and representatives of other tutors and students are now admitted to the Council, one from each grade, department and division) ... As much social work for students as possible is done; right now this consists of organizing the canteen and issuing bread ...

Along with this, I am taking measures to open a worker's department at the university ... You can read about all of this in the newspaper clippings enclosed.

Besides the university, I also deal with vocational education which does not take much of my time at present. Finally, as for my participation in the newspaper, it long since declined to zero. I presented my resignation from the Editorial Board to the authorities, with no response as of yet.

Naturally, I cannot get involved in scientific studies now; not enough time for them; and we live too tightly, the four of us in one room ...

---

32) On the G. Cantor works in the set theory.

I have started working to return at least some part of the university to Yalta, and wrote about this in Moscow ...[33] I also wrote to Ioffe in Petrograd. I asked him, just in case, to let me know if I can get settled there at either the Polytechnic or the Rentgenovsky Institute. All the same I do not do science here, and my best years are passing in the meantime.»

In the memory of a great contribution of Frenkel's to the organization of Tavrichesky University his name is listed, among the names of other founders and outstanding persons involved, on a memorial plaque on the wall of the main building of the Crimea Pedagogical Institute.

Driven by the desire to turn back to scientific activity and contacts with colleagues, Frenkel filed an application for a request to be transferred to the disposal of the Narkompros in Moscow. He obtained permission to do so and on 5 January 1921, went to Moscow.

He went there with his wife, Sarra Isaakovna Frenkel, nee Gordina, then a 19-year-old student in the Biology Division of the Physics and Mathematics Department of Tavrichesky University; they married at the very end of December 1920.

Thirty years later, on 7 August 1949, J.I. wrote to his sons «My dears! Yesterday your mother and I celebrated 30 years of our acquaintance. True, our first meeting was fleeting, limited to one evening I spent at your mother's house, after which I returned to Yalta and was sent to prison. Then a whole year passed during which your mother hardly remembered my existence, and then we met again, in the same house in Miskhor, where I soon was imprisoned by her for my whole life.»

I.E. Tamm wrote about Frenkel's wife: «Sarra Isaakovna, a very clever, sensitive and responsive woman, played a major beneficial role in the life of J.I.»[34]

S.I. Frenkel was very talented in the humanities. She had a perfect command of English, French and German and did scientific translations from Russian and into Russian, working on contracts with our physical journals and, during the war, as a staff member of the Academy of Science Publishers.

In his wife, J.I. found a faithful friend and assistant for all his life. She helped him with his work, typing and, if need be, translating his papers, and assisted him in the implementation of all his good deeds.

Short, slim and petite when young, she was the little lady of J.I.'s large house, where there were always many people, where people came with their joys and troubles. For a man who worked as much as J.I., a well organized everyday life and the constant care of a close and tender human being were of the utmost importance, and Sarra Isaakovna ruled the house skillfully, providing her husband with relaxation and peace in his soul.

---

33) The university was not moved to Yalta.

34) UFN, v. 76, # 3. p. 406, 1962 (in Russian).

Sarra Frenkel (née Gordina), Ya. Frenkel's wife. Crimea, 1920.

In the 1930s she took an active part in the movement of the wives of scientists and engineers. J.I. wrote about it in a letter to Niels Bohr (on 13 May 1936): «My wife is in Moscow attending a most unusual congress — the All-Union congress of the wives of men holding leading positions in heavy industry (engineers, technicians, professors and so on). These women have recently started a campaign to improve living conditions by creating kindergartens for their children and those of all the workers employed in corresponding enterprises, improving the quality of the factory canteens, eliminating illiteracy among people of the older generation, raising the cultural level, etc., etc. This movement, which tens of thousands of women have already joined (there are 3000 of them present at the congress), is sponsored by the government and promises to be extremely effective in furthering the improvement of living conditions here — which have greatly improved in general during the last year.»

Mrs. Frenkel undoubtedly had a literary talent, to which her verses and letters testify. She had fine taste in literature, knew Russian poetry excellently, and could recite from memory for hours poems by Pushkin, Blok, Mayakovsky, Pasternak, Akhmatova.

She was of a light, easy-going character, and she lavished her small store of strength on her work, family, and friends, all people who turned to her for support — she just could not live otherwise. In October 1953 she wrote to her friends: «I gave my solemn promise to my sons that if I get well, I will take care of my health and change my light-minded attitude about it. But I myself do not believe I will manage it. Is it generally worthwhile to spoil the remainder of one's days by attempting to prolong them?»

She survived J.I. by several years and played an important part in the preparation of the posthumous edition of his selected works published 1956–59. Shortly before her death (in 1960) she said she wished to write a book about J.I.

When working on this book, I could not help thinking of this unaccomplished desire of hers.

Chapter 2
Back to Petrograd

On their way to Petrograd the Frenkels made a stop in Moscow. There J.I. dealt first of all with the matters associated with his recent life in the Crimea. At the end of December 1920, essential disagreement had occurred between J.I. and the Office of the Crimea Narobraz[1], of the People's Commissariat for Education (Narkompros) concerning the further reorganization of Tavrichesky University and the measures to be to taken.

In the letter to his brother of 22 January 1921 (already from Moscow) J.I. wrote that one of the reasons for this dissension had been «the excessive moderation in my relations with the students and professors of Tavrichesky University.» Having had no doubts about the necessity of reorganizing the University and making changes in both teachers and students[2], he believed that serious mistakes had been made in several cases, which had done only harm instead of being of some use. For instance, along with the Faculty of Social Sciences and the Faculty of Literature and Philosophy, on 11 January 1921 the teaching body of the Faculty of Physics and Mathematics was disbanded. Those were people whom J.I. had known very well, the majority of whom had progressive views.

In the other letter from Moscow, Frenkel mentioned his memorandum concerning the political situation in the Crimea which he had prepared. An attempt to find it several years ago at first failed. The difficulties, now overcome, were with how to get inside the Central Party Archives of the Institute of Marxism-Leninism at the Central CPSU Committee. This done, the rest was rather easy. The document was found in the Archives immediately, a small file # 31286 in the division # 461. On a scrap of graph paper, 6 by 6 centimeters approximately, which was torn, I think, from a university notebook, I read:

«Vladimir Il'ich [Lenin, — V.F.], Professor Frenkel who arrived from the Crimea ... gave me his report (describing the situation in the Crimea), the first part of which does not concern me, as I explained to him. Then he began to beseech me to pass it to you. I could not refuse. M. Pokrovsky.»

M.N. Pokrovsky, a famous historian, was the Deputy Director of Narkompros under Lunacharsky. The second half of the report touched upon the development of Tavrichesky University. It seemed to me so sharp that I doubted that censors would allow it to be printed. The danger was related to the fact that it was necessary to give a detailed reference to the archival source. Well, my imagination, heightened by fear, drew the following picture. The domestic political situation in our country might worsen, and my father would be declared a slanderer and counterrevolutionary, a ban would be placed not only on publication of his books,

---

1) Russian abbreviation for People's Education Department.

2) See J.I.'s letter of December 8, 1920.

or on him as a scientist, but also on his very name, and, as it seemed to me at times, a page would then be crossed out of our nation's history of physics. And even after these great waverings I still decided to include in my book an episode about the memorandum for Lenin.

Fortunately, the situation in our country changed radically and without the threat of returning to old times. I am glad to present my father's memorandum to the reader.

The Situation in the Crimea

The Political Situation

1. The decree of central authorities on terror in the Crimea is carried out by local organs (special divisions and emergency teams) with frantic zeal and unscrupulousness. This fanaticism transcends all limits and is turning the terror into brigandage, and into mass murder. Not only persons who had in some capacity participated in the counter-revolution are murdered, but also those who had absolutely no part in it. In Simferopol' the banishment of suspect elements to the north is also practiced, while in the country, especially on the south shore of the Crimea, the arrested are either freed or executed. For example, in Yalta two special divisions and two emergency teams are operating, and in three or four weeks they have executed a minimum of 700 people (more likely, 2000). Not only former officers, soldiers and other participants of Vrangel's army are shot, but mostly representatives of the bourgeoisie, and especially of the democratic intelligentsia, who had sought refuge in the Crimea, for the most part, from hunger. (The people from upper bourgeois levels had run abroad from the Soviet power in due course).The massacre takes place on the basis of special written forms taken from the citizens who had arrived in the Crimea after 1917, practically always without any kind of interrogation and explanation. The officers of special divisions and members of the emergency teams bathe in wine, of which there is so much on the South Shore of the Crimea; they perform executions with a drunken hand, not even reading the forms submitted (a fact which is firmly established and witnessed, in particular, with regard to Chernogorov, the leader of the Special Division of the Black and Azov Seas). Along with common citizens, entirely blameless as far as any politics is concerned, a lot of highly qualified specialists perished, such as Soviet workers, cooperators, medical doctors, and so on, that is, persons who knowingly sympathized with the Soviet power and who harboured communists and helped them during the reign of the White Guards. Totally, about 30,000 people were executed in the Crimea, and their number grows by the day.

Because of the fact that the Crimean Revolutionary Committee and, especially, the Regional Committee, took no measures to harness the special divisions, and that the central authorities did not pay enough attention to the Crimea, the Terror or, more precisely, the massacre, which is the arena of the Terror, has not shown to this date any signs of weakening ...

The most fervent enemies of the Soviet power have mostly left the Crimea. As terror continues, it turns neutral people and even those sympathetic to the Soviets into enemies and, accordingly, it does not destroy but, on the contrary, sows the seeds of counterrevolution. Terror must be immediately stopped, and the activity of the special divisions must be investigated to bring punishment to the guilty.

2. Referring to the «directives from the centre» (the existence of which is exceedingly doubtful), the Regional Committee began to banish from the Crimea to central Russia (or simply to the north) not only a group of Mensheviks who had established themselves by virtue of their selfless aid to the communists during the reign of the White Guards and who had received more or less responsible positions in the Crimea. But they also sent to exile some newly converted Bolsheviks who had attempted, in one way or another, to protest against the excessess of terror. Recognizing an intolerable Menshevik spirit in these protests, the Regional Committee «commands», in the line of party discipline, the aforesaid comrades at the disposal of the central party organs, depopulating thus the Crimea, which is now scarcely inhabited even without such actions. This depopulation threatens the Crimea with the saddest of consequences, for this part of the country, which has almost no knowledge of the Soviet power, needs particularly experienced and honest people for the building of the Soviet system.

The Regional Committee must be appropriately instructed on this matter and, generally, a tight control must be established over it.

3. In general, the centre must pay attention to the Crimea if it intends to convert it into a national health resort in the near future. The food crisis in the Crimea is rapidly becoming more and more acute, especially on the south shore, where the Soviet civil servants are given all of 1/4 pound of bread a day, and where the means for feeding the rest of the society are practically not organized at all.

The south shore is being relieved of the public that has arrived there. These people, for the most part, are on their way to relocating north. Removing them will free the place for sick workers and Red Army soldiers who need to be sent to the south. If accomplished by the special divisions, this relocation can hardly be conceived as being actually implemented, no matter from what point of view. Experienced party workers should be sent from the centre to the south who have the broadest possible investiture of power. Otherwise the «orthodox» communists operating in the Crimea will turn it not into a health resort, but into a wilderness bathed in blood.

Professor Ya. Frenkel
January 21, 1921
Moscow, 4th Sokol'nich'ya str. 16, Rudnev's apt.

There is no information about whether Lenin ever read Frenkel's memorandum. He did not send any telegrams or inquiries to the Crimea about the problems discussed in the report.

Soon after Frenkel's arrival in St. Petersburg, J.I. talked about the Crimean affairs with M. Gorky in the House of Scientists. Gorky was highly valued as a writer, but the conversation with him, as J.I. once briefly remarked, had disappointed my father.

J.I. had special talks about the university affairs with M.N. Pokrovsky and O.Yu. Schmidt, who was the head of the university branch of the Narkompros (the Faculty of Physics and Mathematics of the Crimean University was reopened in the spring of 1921).[3]

In a letter of 24 January 1921 from Moscow to his parents in the Crimea, J.I. wrote:

«Today I heard from Prof. Lazarev that Ioffe is going to visit Holland, retaining the title of President of the Rentgenovsky Institute, since this voyage is of a business nature. I have to hurry to Petrograd to find him there. However, Petrograd without Ioffe is not very attractive to me scientifically, and perhaps I'll settle temporarily in Moscow under these circumstances. But here they are trying to burden me with some administrative work at the Narkompros that doesn't particularly please me.»

Pyotr Petrovich Lazarev in 1921 headed the Institute of Physics and Biophysics he had organized in 1918. Frenkel told him about his plans and papers (one of them, related to electrodynamics, would appear a year later in the journal published by the Lazarev Institute). J.I. also shared with P.P. Lazarev his thoughts about the opposition between animate and inanimate processes. As a result, P.P. Lazarev invited J.I. to his institute. But J.I. met Abram Fyodorovich Ioffe in Moscow and decided finally to work in Petrograd. He wrote to his parents about this:

«Moscow, 26 January 1921
... Today I met Ioffe in Glavprofobr[4]. It turns out that he had written and cabled me to try to get me to come to Petrograd where the positions of a researcher at the Rentgenovsky Institute and a Professor of the Physics and Mechanics Dept. at the Polytechnic Institute had been prepared for me. So, we would have had to live in Lesnoe[5]. Abram Fyodorovich will go abroad to Holland in the near future, for about 3 months, to buy various scientific instruments. A pity we have to stay without him».

---

3) See: A.K. Synopalov, Transactions of the Crimean Pedagogical Inst., Book 1, 1927, p. 1 (in Russian).

4) The Central Office for Professional Education.

5) A part of Petrograd where a number of research institutes and the Polytechnic Institute were and continue to be situated.

The Rentgenovsky Institute depended on the Narkomvneshtorg[6] for the subsidy for purchasing the instruments and equipment. The Narkomvneshtorg was headed in those years by Andrey Matveevich Lezhava, who had known J.I.'s father well. Lezhava's deputy was E.A. Litkens, J.I.'s colleague at Narkompros in the Crimea. J.I. got an offer to go abroad with the delegation of scientists but refused it, since he wanted to settle down in Petrograd and wait for his parents to move there. However, he participated actively in the organization of the trip for the delegation, which included A.F. Ioffe, P.L. Kapitza and M.N. Kirpichova, as well as A.N. Krylov who was sent by the Academy of Sciences.

At the end of January 1921, Frenkel and his wife moved from Moscow to Petrograd.

During the two years of J.I.'s invariable stay in the Crimea, the scientific community in Petrograd had undergone essential changes.

More than a few pages have been dedicated to the history of the foundation of the Physical-Technical Institute (PTI) in Leningrad. Therefore, here we shall only briefly work out the main «junctions» in the road taken by Soviet physics from revolutionary Petrograd in the fall of 1918.

The contribution of Abram Fyodorovich Ioffe to the organization of the PTI is common knowledge.[7] Let us now discuss the important role played in it by Professor Mikhail Isaevich Nemenov (1890–1950), the head of the Soviet medical radiology school.

M.I. Nemenov, like many others, understood the great potential X-rays, discovered by Röntgen at the end of the 19th century, had for use in medicine. That is why as early as 1913, while working at the Women's Medical Institute in St. Petersburg, he came to the Ministry of People's Education with the suggestion to organize a radiological[8] institute. According to his idea, the institute should have invited both research and clinical institutions. The main objective of the institute Nemenov envisioned would have been the study of the influence of X-rays and radium rays on the functioning of organisms, especially humans. By then their effect on variations of cancer tumours was already known.

The Ministry even organized a special committee to evaluate Nemenov's suggestion. But matter progressed at a snail's pace, and it became clear very soon that there was practically no chance of success, especially once the war started.

After the October Socialist Revolution, Nemenov resumed his efforts to organize the institute. This time he met with complete support from the Soviet government in the form of the Science Department of Narkompros. And as early as the spring of 1918, the decree on the transfer to the future institute of some rooms not far from the Women's Medical Institute was signed. Together with its

---

6) The People's Commissariat for Foreign Trade.

7) M.S. Sominsky. Abram Fyodorovich Ioffe. Moscow-Leningrad. Acad. of Sci. of the USSR Publ., 1964 (in Russian).

8) «Rentgenovsky» in Russian.

director, Professor B.V. Verkhovsky, Nemenov at first led the organizational work. Soon, however, a disagreement arose between them over the problems in organizing the institute. Verkhovsky had envisioned it only as an extension of the Medical Institute; the activity of the Rentgenologichesky Institute was thus supposed to be limited to a very narrow range of auxiliary problems of a technical nature.

M.I. Nemenov's views on the subject were much broader. He wanted to concentrate research on the radiation effects on organization at the medical-biological department. Along with it, a technical physics department was also to be organized at the institute, where the work of manufacturing X-ray tubes and facilities should have been done in the first place. Studies in the physics of X-rays were also to be done there.

Academician Pyotr Petrovich Lazarev was asked to head the Physical-Technical Department, of the necessity of which Nemenov had convinced Verkhovsky at long last. Verkhovsky had chosen Lazarev, this brilliant pupil of P.N. Lebedev's, probably because Lazarev had had medical training in addition to education in physics[9]. As for M.I. Nemenov, he had regarded A.F. Ioffe as a more suitable candidate, since he was the most prominent X-ray specialist in Russia. Abram Fyodorovich had responded vigorously to Nemenov's proposal and had returned from the Crimea; in the beginning of September 1918, he started to work energetically on the organization of the Physical-Technical Department. But Ioffe understood that X-rays presented a tool not only for diagnostics of the human body, but for studying crystals as well. It was from this more general viewpoint that he considered the future activity of the department. To the same extent, the study of radium should have been carried out not only for medical applications, but for special physical research as well. Thus, the new institute should have represented the whole integrated plant with a very broad range of problems arising from both physics and medicine[10].

One might think that even then the core of the problems facing the institute had already shifted from medicine to physics. This change was reflected in the fact that the People's Commissariat for Education, on 24 October 1918, commissioned A.F. Ioffe personally to take over all the matters concerning the organization of the Rentgenovsky and Radiologichesky Institute from Professor B.V. Verkhovsky. And as early as 1 December 1918, A.F. Ioffe reported at the meeting of the Physics Branch of the Russian Society on Physics and Chemistry[11] on the organization of the State Rentgenovsky and Radiologichesky Institute in Petrograd and its present objectives[12]. A.F. Ioffe gave an extended version of this report before the Congress of the Russian Association of Physicists in February 1919 (at the same Congress

9) P.P. Lazarev, by the way, was a personal doctor of his teacher's.

10) See the paper by M.I. Nemenov in: «Gosudarstvenny rentgenovsky, radiologichesky i rakovy institut», Leningrad, 1928 (in Russian).

11) Russian abbreviation: RFHO.

12) ZhRFKhO, ser. phys., v.50, NN 7–9, 1919, p. 278 (in Russian).

D.S. Rozhdestvensky reported on the organization again in Petrograd State Optical Institute).

Thirty years later, on 22 June 1948, on the day of the 30th anniversary of the Central Rentgenovsky, Radiologichesky and Rakovyi Institute of the Ministry of Health, at the grand meeting of the Science Council of this institute, an address from PTI to the Director of the institute, Professor Mikhail Isaevich Nemenov, was read, among others. It stated: «Our institutes are, in fact, Siamese twins, who have existed separately only since the age of 3. For three years you and A.F. Ioffe have headed both institutes»[13].

The point was that, according to the Rules of the Institute which had been accepted at a joint meeting of the Physical-Technical and Medical-Biological Departments, there had to be an annual change of director. For the year 1919, A.F. Ioffe had been elected; in 1920 M.I. Nemenov became the director.

Nemenov probably was the first Soviet scientist who went abroad after the revolution. He had gone there back in 1920 and purchased in Germany a lot of equipment and books for the institute, using a substantial sum from the Golden Fund of the Republic which Narkompros had subsidized. This fact by itself testifies as to how highly the young Soviet government valued the development of science in general and one of its first-born in particular. One should bear in mind the extreme economic hardships caused to the country by economic destruction and the ongoing Civil War.

There was then the opinion abroad that science in Soviet Russia had almost perished and that scientists had practically no opportunity to work. Therefore the visit by a Soviet scientist on such a mission caused great interest among the scientific community in Germany. Nemenov's lectures attracted a large audience and were accompanied by many questions. Some of them were very peculiar and demonstrated the total lack of information, a presence of disinformation, in fact. Mikhail Isaevich liked to recollect how, after the first of these lectures, a well-known German medical doctor, Professor Miller, had asked him whether it was true that «a friend of his, famous physiologist I.P. Pavlov», was earning his bread by selling matches on the corner of Nevsky Avenue and Sadovaya Street (i.e., in the centre of Petrograd). Nemenov's mission contributed to a great extent to the elimination of these and similar misconceptions.

The Physics-Technical Department of the Rentgenologichesky Institute was then situated (or nestled, rather) in several rooms of the Laboratory of General Physics of the Polytechnic Institute, thanks to the hospitality of the Head of the Physics Chair, Professor V.V. Skobel'tsyn.

Lesnoe, which was then at the far outskirts of Petrograd, was very far from the apartment room in the centre of the city, where J.I. lived during the first months. Recalling his impressions of Petrograd, J.I. later said that he was much surprised by the absolutely clean and smoke-free city air: it was the air of a mountain resort.

---

13) Archives of the PTI, f. # 274, p. 9.

In the winter of 1921 practically all the plants and factories of Petrograd were out of work because of the fuel crisis. For the same reason, city transportation operated poorly. Trams ran to Lesnoe quite irregularly, and only until 11 a.m. Often J.I. had to travel to work on a buffer or between the tram cars . . .«Commuting by trams», wrote J.I., «is limping in both feet . . . Even in the morning the traffic is supported only by the «number 11», that is, by means of the lower limbs, and I have now gotten so accustomed to this trip that I read books on my way, to save valuable time».

Abram Fyodorovich Ioffe wrote regularly from abroad. Vera Andreevna Kravtsova, his wife, displayed his letters on the part on the information board concerning institute affairs so that people working at the institute would know of all the successes and failures of Ioffe's mission. In a letter of April 7, Abram Fyodorovich wrote: «Have Jacov Il'ich go to Moscow and get from Lesnoe or Grinberg the actual assignation of money, at least within the limits I had indicated, and then telegraph me . . . with the number of the document and the date. After the letter of credit arrives, I'll need at least two months to make a purchase, so I ask persistently that you hurry up.»[14] On 11 May 1921, after returning from Moscow, Frenkel wrote (to his brother, Vladimir Il'ich): «I've accomplished my official mission successfully. Due to the support of A.M. Lezhava and of one responsible official from Vneshtorg[15] (my Yalta acquaintance) I've managed to get the complete restoration of credit to the Rentgenovsky Institute (a total of 150,000 roubles), despite the cutting of all assignations down to 1/5 of their initial value by the Vneshtorg.»

In the fall of 1921, at the request of A.F. Ioffe and M.I. Nemenov, a special meeting called by Narkompros decided to organize 3 separate institutions on the basis of the Rentgenologichesky and Radiologichesky Institute. The Physics-Technical Department became the SPTRI, The State Physics-Technical Rentgenological Institute (with A.F. Ioffe as the head). The Medical-Biological Department became the X-ray and Radiological Medical Institute (director, M.I. Nemenov). The third was the Radium Institute, headed by Academician V.I. Vernadsky during the first years of its existence.

Practically all the participants in A.F. Ioffe's pre-revolutionary seminar, such as P.L. Kapitza, N.N. Semenov, P.I. Lukirskii, N.I. Dobronravov, Ya.R. Schmidt, M.V. Kirpichova, and electrotechnicians of early repute, V.A. Tolvinsky, A.A. Gorev and A.A. Chernyshov, who soon became A.F. Ioffe's deputy in SPTRI, were among the SPTRI researchers (about 50 people all in all). There were also several theoreticians, such as Victor Robertovich Bursian (Scientific Secretary to the Institute), Yuri Alexandrovich Krutkov, and the first person to graduate from the Physical and Mechanical Department of the Polytechnic Institute, Georgy Abramovich Grinberg. J.I. had known the first two before, from the university (from which

14) M.S. Sominsky. Op. cit., p. 253.

15) The Department for Foreign Trade.

they had graduated before Frenkel) and from the meetings of RFHO. According to the «Regulations» for the Institute elaborated by Ioffe, the scientists held the positions of «senior physicists» (heads of the studies), «physicists», «assistants» and «consultants».

In spite of the small number of people who worked at the institute, several directions of research were pursued. The most important of these were X-ray physics; atomic structure and spectra; and studies of the electrical and mechanical properties of crystals, to which the «X-ray method» suggested and developed by A.F. Ioffe was applied.

V.R. Bursian was in charge of the Theoretical Physics Cabinet: there were too few people at that time to call it a department. Bursian himself was involved then in the study of thermodynamics and developed the theory of electron motion in a vacuum. Yu.A. Krutkov continued the research in thermodynamics and adiabatic invariants that he had started previously (under Ehrenfest's influence).

Frenkel rapidly got used to the institute team and involved himself in its work. In 1923 he already became a senior physicist. We are going to dwell below only on a few of J.I.'s most interesting papers.

According to a tradition adopted by the institute, the majority of the papers were reported at the Science Council meetings, where everyone gave a survey of their scientific activities.

In those years the Council of the Institute met regularly once every two weeks. Along with the persons named above, the youngest members of the staff — Yuliy Borisovich Khariton, Viktor Nikolaevich Kondrat'ev, Alexandr Filippovich Valter, Nikolai Nikolaevich Mirolubov, etc. — who studied simultaneously at the Physics & Mechanics Department, took part in the council's work.

And at almost every session of the Council, minuted briefly by V.R. Bursian, J.I. reported on his studies. We give here a chronological list, based on the documents retained, of Frenkel's contributions:[16]

Year 1923
June 15: Report on the calculation of the interactive energy for electrified systems.
June 22: Note on the electrical nature of cohesive forces in metallic crystals.
August 13: Remark on the stability of crystal models.
August 17: Report on the structure of diamonds.
August 24: Report on the calculation of the crystal surface energy.
September 14: Report «On the theory of the electrical conductivity of dielectrics».
September 28: Remark on the theory of a hydrogen molecule.
October 5: Report on the development of the idea formulated at the previous meeting of the council.
October 25: Survey of the Griffith paper on the theory of the fracture of solids.
November 2: Contribution on the synthetic foundation of the principal vector operations.

16) Archives of A.F. Ioffe PTI, f. N10, pp. 1–20; f. N.14, pp. 1–26.

November 30: Report on the theory of the critical temperature of the metal molecule adhesion to a dielectric.

Year 1924
January 25: Contribution to the theory of the cohesive forces in heteropolar liquids.
February 24: Report on a possible explanation of the elasticity shear limit under melting temperature.
April 4: Report on the theory of thermal expansion of crystal lattices; report on the temperature dependence of the elasticity shear limit under melting temperature.
July 17: Report on the mechanism of the intrinsic friction in liquids.
August 1: Report on the foundations of the electron theory of metals.
September 26: Report on the theory of cohesive forces in dielectrics.

The consideration of this list allows one to reconstruct the main directions of J.I's work in the 1920s. Those were, in the beginning, studies in the molecular theory of solids and liquids, later in the electron theory of solids, and, finally, investigations of the mechanical properties of solids and the theory of the electrical conductivity of metals.

We should like to discuss in some detail J.I.'s work in molecular physics.

The first paper, «On the critical temperature of the metal molecule adhesion to a dielectric» (reported on 30 November 1923), was soon published (in 1924) in ZhRFKhO and the German «Zeitschrift für Physik».

By the example of this paper one may clearly see that there were beneficial and «mutually profitable» relations between theoreticians and experimentalists.

In January of 1954, at a meeting in memory of J.I. Frenkel, organized by A.F. Ioffe at the Semiconductor Laboratory headed by him at the time, I.E. Tamm said «that the possible classification of the treatise of physicists — theoreticians would assign each of them to one of the categories: the papers for theoreticians and the papers for experimentalists. The studies of the first group had the greatest effect on the development of theoretical physics, stimulating, at first, new theoretical investigations and calling to life experimental research only in the second approximation.» The theoretical papers of the second category developed out of particular experimental studies (often ones still being carried out), and gave rise at first to new experimental research and only then, «in the second approximation,» influenced the development of the theory.

Igor Evgen'evich himself, as well as many other participants in Abram Fyodorovich's seminar, observed that in J.I.'s vast scientific legacy there are papers of both types.

Frenkel's paper on the theory of condensation and absorption, which we will now discuss, belongs, at first sight, to the group of «papers for experimentalists». It appeared in J.I's collaborative effort with Nikolay Nikolaevich Semenov and Yuliy Borisovich Khariton. N.N. Semenov and Y.B. Khariton (his pupil, a student of the Physics and Mechanics Department at that time), studied in the early 1920s the behaviour of a molecular beam (it is more correct to speak about a neutral particle beam; the particles might be either atoms or molecules), i.e., a «jet» of molecules flying along a certain direction in a vacuum. Research into molecular beams had by then already attracted the attention of physicists, because they had won so many trophies with their help. Namely, using the molecular beam method, O. Stern, a member of Max Born's staff (in Frankfurt-am-Main, where Born worked in 1919), had measured molecule velocities in gas and their distribution, and later, approximately in 1921, together with W. Gerlach, the magnetic moments of atoms[17].

Semenov and Khariton, and a little later Alexandr Iosifovich Shal'nikov, a student who joined them, investigated the effect discovered by R. Wood in America (in 1915) and M. Knudsen in Germany (1916).

Schematically, the Knudsen experiment, carried out in a more «pure» way than Wood's, went as follows. A beam of mercury atoms formed by two collimating apertures was directed into a spherical glass retort placed in a Dewar vessel filled with some «cold» liquid. At the temperature of liquid air (−140° C) all the mercury atoms solidified at once, without reflection, on the spherical retort surface situated in the path of the beam. Knudsen managed to demonstrate with certainty that only 2 in every 10000 mercury atoms were reflected. By replacing the liquid air by some other coolant with the temperature −77° C, Knudsen observed the blackening of the whole sphere. This meant, in Knudsen's opinion, that, after having been reflected by the glass surface, a mercury atom was sedimented on it only after one of such multiple collisions. Thus, the reflection coefficient for metal (mercury) atoms by a dielectric (glass) changes from 0 to 1 within the temperature range from −140° up to −77° C. Knudsen tried to find the temperature variation of this coefficient. But, with only a discrete series of cooling liquids at his disposal, he had no way to «scan» the whole temperature range. Nevertheless, to his great surprise, he found that the reflection coefficient value changed by a leap which occurred somewhere between −140° and −130° C. Knudsen called the temperature at which this leap occurs a critical temperature.

Semenov and Khariton, by means of a simple and elegant method, found the way to study this phenomenon in every detail. In their experiments one end of a copper plate 1 cm wide and 13–15 cm long was placed into a cylindrical retort filled with liquid mercury. A Ni-Cr alloy wire was wound around the other end of the plate; it played the role of a furnace. The retort was lowered into a Dewar vessel with liquid air, the mercury froze and the copper plate became «frozen» in it. The frozen end of the plate and mercury took the liquid air temperature. «The furnace» heated the other end of the plate to a temperature of about 10° C.

---

17) It is interesting to note that similar experiments on the determination of atomic magnetic moments had been planned and prepared (but not implemented) in 1920 by N.N. Semenov and P.L. Kapitza (in 1920 the Physical-Technical Department of the Rentgenovsky and Radiologichesky Institute did not have the experimental basis required for these experiments). The publication of this paper was delayed until 1922. O. Stern and W. Gerlach knew nothing about the research by Kapitza and Semenov. In 1943 Stern won the Nobel Prize for his study of molecular beams and the discovery of the magnetic moment of a proton.

The gradient between 10° C and $-140°$ C was uniformly distributed along the plate (this fact was supported by special measurements made with the help of thermal couples situated on the plate). The source, a Ni-Cr alloy wire coated with an irradiated metal, zinc or kadmium, was mounted across the plate. The wire warmed up, and atoms flew in the direction of the plate. They were reflected by the part of the plate of which the temperature was above the critical one, and sedimented onto the other, colder part of it. The boundary between the parts was very sharp.

The results of Semenov and Khariton supported Knudsen's ideas and refined them in an essential way. We might add that, by depositing onto the plate various substances, such as picein, mica, or paraffin, or by replacing the copper plate by one made of some other material, it was possible to investigate the influence the substrate material has on atom absorption.

Another surprising effect was that, though the plate temperature dropping in the direction from the furnace to the mercury was constant at a given height, the boundary between the region of complete reflection and complete sedimentation was a straight line parallel to the surface of the frozen mercury, but had the shape of an arc.

Frenkel, who had heard about these experiments by Semenov and Khariton both at the Council meeting and from the conversations in their authors' laboratory, became interested in them.

As a first approximation to the solution of the problem, J.I. modified it in the following way. One should imagine, he said, that not the directed beam atoms, but rather the vapour from some substance collides with the solid (or liquid) surface of this same substance. Note that the principal issue of the chaos in the motion of gas or vapour particles is in the separation of all these particles into six molecular beams moving along each of the 3 Cartesian axes in two opposite directions. If the temperature, T, of a solid were below the critical value meant by Knudsen, all the vapour would condense on the solid surface, sticking to it. This scenario, however, contradicts both the theory and the experiment. The theory states that particle velocities of either solid or liquid are distributed according to Maxwell's law, their values clustering near the most probable one. Those particles which have much greater velocities break away from the surface. This phenomenon is called sublimation. Without sublimation, we would not, for instance, smell aromatic substances such as naphthalene, perfume and so on. Atoms come from vapour to the surface of a solid and transfer their extra energy (and momentum) to it. After these events the energy of an atom takes the value which is characteristic in the mean for the other «old residents». And already in this state these atoms have a definite probability of being vaporized again from the surface after some time. Thus, the concept that gas atoms colliding against a vessel wall simply bounce off it like billiard balls, proves wrong in reality (this concept can be easily obtained by way of the central formula of the kinetic theory of gases, namely $pV = RT$, where p is the gas pressure, V is the volume, T is its temperature and

R is the gas constant). Here is an example of the rather common paradox where correct conclusions are derived from incorrect assumptions. In this particular case the paradox is explained by the fact that both the gas and the solid are the same temperature, so the velocity distributions for particles of the vapour and of the solid surface are identical. Therefore «the delay» of particles on the surface makes no impression, since when they leave the surface, they carry away the same average momentum as they had when colliding with it.

The time between the instant of collision and the following instant of a particle breaking away from the surface (the sticking time) is determined by the surface temperature and depends on the properties of both the vapour and solid atoms determined by their mutual affinity. In other words, the sticking time depends on a cohesive force between the particles of the vapour and solid. For instance, the cohesive force between particles of a metal is greater than that between the particles of a metal and a dielectric. On the contrary, hydrogen particles have much greater affinity to the particles of a metal or a dielectric than to each other.

If the number of atoms in a molecular beam bombarding the surface is not large and the sticking time, or the atom life-time on the surface, is small, then only a monotonic (monomolecular) layer forms on the surface. The atom density of this layer does not depend on time, since a moment will come when the number of particles sticking to the surface per unit time equals the number of those vaporized from it.

So the density of the particles in a beam bombarding the surface of a solid or liquid is determined by three factors: the beam intensity, the degree of affinity, and the surface temperature.

Proceeding with these qualitative arguments (which conformed to the somewhat earlier ideas of Langmuir, as would become clear later on), J.I. constructed the quantitative theory of the phenomenon.

First of all, using results known from statistical mechanics, J.I. obtained a formula for the mean life-time, τ for an individual particle of vapour on the surface:

$$\tau = \tau_0 \exp^{U_0/kT},$$

where k is the Boltzmann constant, τ_0 is the oscillation period of the particle in the direction perpendicular to the surface, and U_0 is the energy required to remove a particle from the surface to «infinity». The quantities τ_0 and U_0 characterize the affinity between the beam particles and the surface. And a particle is an «individual» one when it is bonded only to the solid substrate and separated from similar particles also sticking to the surface by a distance large enough to negate their interaction.

Along with these individual particles, J.I. also considered «pairs», «triplets», etc., i.e. associations of particles on the surface, accounting also for the fact that the life-time, τ_i (i is a number of particles in the association), of each particle differs from τ.

Confining himself to only the «individuals» and «pairs» for the first approximation, Frenkel composed the balance equation for the particles on the surface. This equation describes, in the usual way, the following process: the rate of change of the number of sticking particles per unit area of the surface equals the difference between the number of particles on it due to the beam bombardment and the number of departing particles (from the amount of individual atoms and «pairs»). A simple analysis of the resulting equation demonstrates that at a given temperature of the surface it is always possible to choose such intensity of the beam at which the number of particles on the surface increases indefinitely. A steady state of the layer would become impossible and would face the sedimentation of particles from the beam onto the surface. J.I. called the corresponding value of the beam intensity a critical one. On the other hand, for any given value of the beam intensity there always exists the lowest temperature such that, when starting from it, a steady state is again not achieved; in this case sedimentation takes place. This temperature is the Knudsen critical temperature.

Let us now return to the experiments of Semenov and Knudsen. The boundary between the complete absorption and complete reflection was shaped like an arc, though, as mentioned, the temperature along this arc was neither constant nor equal to the critical one. This effect, unexplained by the Knudsen concept, is explained in a very simple and natural way by the theory J.I. developed. He noted that different densities of metal particles coming from nichrome scorching wire, that is, different intensities of the molecular beam, correspond to different points on the copper plate.

In his survey of molecular beam methods, N.N. Semenov wrote[18]: «At Frenkel's request we have carried out an experiment where the position of the wire (source) was determined as carefully as possible during the period of the experiment; then, by weighing the wire before and after the test, the number of vaporized atoms was determined, and «critical temperatures» near the end of the arc and near its apex were also measured. Using our data, J.I. Frenkel calculated this temperature difference by his formulae and compared the obtained value, which closely matched our experimental one (5° C).»

Summarizing all this, one can state that this study reveals one important property of J.I.'s scientific creativity, namely, the fact that experimentation was important to his developing a theory.

It is characteristic that, while analyzing a given problem, Frenkel always came to a more general one the particular case of which was the primary problem. Semenov's and Khariton's studies of molecular beams led Frenkel to elaborate the adsorption phenomenon theories a whole. In this paper Frenkel introduced a fruitful concept of a two-dimensional «gas» of particles absorbed on the surface and obtained a significant formula for their mean life-time,

$$\tau = \tau_0 \exp[U_0/kT].$$

18) N.N. Semenov. Uspekhi Fizicheskikh Nauk, v. 5, p. 57, 1925 (in Russian).

The life-time of this formula itself is very long, it is still valid[19].

Not confining himself to the problems of absorption, in the same paper J.I. also gives a plan of his further studies. Here is a passage very characteristic of his line of thought: «The surface layer of every solid may be considered, even in the absence of any alien adsorbed particles, as itself having been adsorbed with respect to the substrating mass»[20]. From this simple remark, a scientific metaphor of a kind, he passes to the estimate of τ_0, the oscillation period of a molecule on the surface of a solid in equilibrium with its vapour. In the case of cadmium, for which all the data necessary for such an estimate were available from the physics tables existing then, one has $\tau_0 = 10^{-13}$ sec.

Considering further the destiny of particles that fall out of the molecular beam onto the solid surface, Frenkel observed that some of them may diffuse into the interior of a solid; they are, so to say, «vaporized» in the intersite space of its crystal lattice. This consideration led to the investigation of questions associated with diffusion in solids and liquids and thermal motion in them.

A paper by Semenov, Khariton and Frenkel published in 1926 was also dedicated to the critical temperature of absorption. A further development of Frenkel's ideas that he had put forward two years before was contained in the theoretical part of this article. In particular, the paper took into account the fact that atoms of the surface layer were associated not only in «pairs», but also in «triplets», etc.

In the protocol of the Science Council meeting of 17 July 1924, V.R. Bursian wrote about the other report made by Frenkel: «The liquid state is supposed to be treated as an analogue of the solid, not gaseous, one.»[21] Perhaps this very report stimulated a whole series of Frenkel's studies in the kinetic theory of solids and liquids. The first of them was the paper written in 1925. It had a great influence on the subsequent development of molecular physics, so we will consider it here in detail.

According to the concepts which were commonly accepted in the 1920s, thermal motion in solid crystal bodies could be reduced to the atom oscillations near the sites of the crystal lattice. This picture corresponded to the notion of a perfect crystal; correctly describing a broad circle of phenomena, it led, however, to contradictions with experimental data for a number of gases. Within the framework of the perfect crystal picture it was absolutely impossible to understand, for instance, the reason for the mutual diffusion of solids. But G. Hevesy[22] in his experiments

---

19) We would like to note here, as a characteristic example, that J.I.'s paper on adsorption (1924) was translated into English in 1960 and published as a special issue by the California Radiation Laboratory in the USA.

20) J.I. Frenkel. «Collection of Selected Papers», v. II, Moscow-Leningrad, AN SSSR, Publ., 1958, p. 253 (in Russian).

21) Archives of the PTI, f. # 14, p. 19.

22) G. Hevesy studied the diffusion of a naturally radioactive lead isotope, the so-called radium D in the crystal of a normal (lacking the radioactive Pb isotope) PbCl . For the invention of the tagged atom method G. Hevesy won the Nobel Prize in 1943.

in 1920 demonstrated with a great clarity and accuracy using the method of tagged atoms that he invented, the effect of diffusion, or more correctly, of self-diffusion.

This effect could have been observed long before Hevesy; however, the mutual interpenetration of atoms from the two solids in contact would take an enormously long time at lower temperatures.

The other phenomenon not explained by the model of a «perfect crystal» was the electrical conductivity of ion crystals (cf., of the NaCl type). Hevesy tried to explain this phenomenon by a simultaneous displacement, in the direction of the applied field, of a gigantic chain of atoms piercing the crystal, whose beginning and end were situated on the boundary surfaces attached to electrodes.

A.F. Ioffe suggested quite a different explanation. In 1916, in a paper written with M.V. Kirpichova, he had managed to prove experimentally that some parts of the ions in a crystal lattice are not at sites, but in the so-called intersite space. This must be vast for such an ion to move in the direction specified by the applied electric field. Dissociated ions, after having entered the «intersite lattice» (a metaphorical notation of the intersite space suggested by J.I.), play the same role as free electrons in the conductivity of metals. Ioffe called the ratio of the number of dissociated ions, i.e., those in the intersite space, to the total number of ions the dissociation degree of a crystal.

The Ioffe ideas have found the qualitative formulation in the paper of J.I. whose name is strongly connected with the model of a real crystal.

The transition of an atom (or ion) from a lattice site to an intersite gap is treated by J.I. as its «vaporization» of some kind. The time spent by an atom in oscillations about the crystal lattice site is described by a formula similar to the one Frenkel had derived for corresponding oscillations of an atom adsorbed by a surface. The difference is that in the latter case oscillations in only the direction perpendicular to the surface had been taken into account, while in the former all three directions were of importance. Thus,

$$\tau = \tau_0 \exp[U_0/kT]$$

is as before, where τ_0 is the period of one oscillation. Having performed $\exp(U_0/kT)$ oscillations in a site of the lattice, an atom transits to the intersite lattice. The lower the temperature, the longer the period, τ, of the atom's «settled» life will be, and the life-time tends to infinity at $\tau \to 0$.

What is the further destiny of this atom?

It can either move permanently through the intersite space or get involved in oscillations about some new equilibrium position in the intersite lattice. The latter viewpoint on the mechanism of the atom or ion motion is supported by two experimental facts. The dissociated atoms (ions) have a very low mobility, and the elementary displacement time is much longer than the oscillation period.

A.F. Ioffe confined himself to indicating the role played by dissociated ions in the mechanism of electrical conductivity. Frenkel focused his attention on the destiny of a vacant site left in the lattice by the atoms departed from it. This vacant site J.I. called a hole, a little jargon term which, however, became absolutely natural and commonly used in the physics of solids and liquids. A hole can absorb an atom from the intersite gap, either «its own» atom — the one which just left this site or some other atom that has produced a hole at some other point of the crystal when leaving for the intersite space. In both cases the filling in of the vacant space, the «collapse» of a hole, is called association.

But an atom (or ion) from a neighbouring site can also jump over to the hole. This process may be required as a motion of a hole in the opposite direction. «Thus,» J.I. concluded, «thermal motion in a crystalline solid is constituted by: 1) atom oscillations about their (proper and improper[23]) equilibrium positions; 2) dissociation of bonded atoms; 3) wandering of dissociated atoms (by means of elementary shifts along intersite gaps); 4) association of dissociated atoms; 5) wandering of vacant sites (holes).»[24]

All these summands of the total process of atomic motion in a real crystal are also described, in the main (processes 2, 3 and 5), by the same experimental formula with the appropriate values of the oscillation period τ_0 and energy U_0.

Constructing again the difference balance equation for the particle number in intersite gaps (and the number of holes equal to it), and considering the steady state, when the number does not change with time, J.I. obtained an expression for the dissociation degree in a real crystal. This value thus relates to the time scale of the establishment of the given steady state and the diffusion coefficient, D, for atoms in the intersite space and holes. The dependence of the diffusion coefficient on temperature proves exponential, $D \sim \exp(-U_0/kT)$.

When an electric field is not applied to the crystal, the ion motion in the intersite space is chaotic (in the absence of temperature and density gradients). The presence of the field marks a preferable direction for the ion and hole motion. Frenkel thus derived a formula for the electrical conductivity of a crystal relating it to the temperature; the degree of dissociation; the periods of the ion oscillation near a site and in an intersite gap, τ_0 and τ_0'; and the corresponding values of energy, U_0 and U_0'. Holes did contribute to the conductivity and were treated as ions of the opposite charge.

Frenkel then took the next step. Indeed, what makes the difference between an amorphous body, regarded as an overheated liquid, and a crystalline one? Obviously, it is the lack of a crystal lattice with the difference between the sites and intersite gaps (proper and improper equilibrium positions). An amorphous body is a completely dissociated crystal. Dissociated atoms oscillate about their (temporary) positions in a real crystal. Similarly, in an amorphous body thermal motion

23) A proper equilibrium of an atom occurs at a site on the crystal lattice, an improper one at an intersite gap.

24) J.I. Frenkel. «Collection of Selected Papers», v. II. Moscow-Leningrad, AN SSSR, Publ., 1958, p. 256 (in Russian).

Leningrad, September 1924. From left to right: I. Obreimov, P. Ehrenfest, N. Semenov, A. Ioffe, A. Chernyshov.

of atoms has to occur likewise: several oscillations near some equilibrium position (and the lower the temperature, the more oscillations), then a leap into a new, also improper equilibrium, then several more oscillations and so on. And it is very easy to pass from an amorphous body to a liquid.

So, the formula for the diffusion coefficient derived for a real crystal remains valid also for atoms of a liquid, and it reads:

$$D = [\delta^2/(6\tau_0)] \exp[-U_0/kT]$$

where δ is the distance between two neighbouring equilibrium positions (the length of one step in the wandering of an atom over a liquid or solid).

The expression for D should be inserted in Einstein's well-known formula $f = kT/D$, where f is the friction coefficient of a particle, equal to the ratio of the force applied to its velocity, v ($1/f$ is the particle mobility). This coefficient is known to be simply related to the viscosity of a liquid. From the Stokes formula it is known that

$$F/v = 6\pi\eta a,$$

with a generally being the radius of a ball moving with velocity v in a liquid, to which the friction force F is applied. In the experiments where a ball falls into a viscous liquid, when the gravity force, mg, is balanced by the friction force, viscosity is found by measuring the velocity (this is one of the first laboratory studies in any university course in physics). The Stokes formula holds only for very small values of the radius. Einstein, in his time, had bravely applied it to the measurement of the radius of a sugar molecule dissolved in water, treating it as a ball and using the known viscosity value for water. Frenkel went even further in this direction. He used the Stokes formula for atoms of the liquid, meaning the atomic radius is a and the viscosity of the studied liquid is η. As a result, Frenkel derived the following formula for the viscosity of liquids:

$$\eta = [\tau_0 kT/(\pi a\delta^2)] \exp[U_0'/kT],$$

which is widely known now. It determines the character of the dependence of η on temperature. The formula allows one to determine, by means of a comparison of a theoretical value of η with an experimental one, the molecular parameters (U_0' and τ_0') involved in it, producing the correct orders of magnitude.

When assessing the given simple derivation of this formula for viscosity, a question may arise about the validity of applying the macroscopic Stokes formula to microscopic subjects like atoms. Frenkel dealt especially with this question and demonstrated why such an approach is possible. The point is the following. The atoms of liquid have to come into motion, that is, to restrict themselves, to make way for atoms to move from one equilibrium position to another, and the laws of viscous fluid hydrodynamics take these collective effects into account. Frenkel also obtained the resulting formula by other methods, which is characteristic of him[25].

The set of an atom in an intersite gap and the vacancy (hole) left by it in a crystal lattice is known in literature as «the Frenkel defect». This concept, introduced by J.I. in 1926, gained the broadest spread in solid state physics. What is the destiny of an atom in an intersite gap? Wandering over the intersite lattice, it can «come up» on the surface of a crystalline body and there take a place in a new layer under construction, in a site on the normal crystal lattice.

Thus, inside a crystalline body, an isolated defect, a hole, uncompensated by the intersite atom, would be left. These kinds of defects are called «the Schottky» ones after the German scientist who paid special attention to them in 1930. So we see that «Schottky» defects are particular cases of the «Frenkel» defects. The transition of atoms from the bulk of a body to its surface J.I. called, in an imaginative way, absorption of atoms of emptiness.

As experiments with ion conductivity demonstrate, in ion crystals, for instance, both Frenkel defects (in silver halogenides) and Schottky defects (in alkaline metal halogenides)[26] may appear. Note that one of the distinctions between

---

25) See, for instance: J.I. Frenkel. «Collection of Selected Papers», v. III, Moscow-Leningrad, AN SSSR Publ., 1959, p. 202 (in Russian).

26) Ch. Kittel. «Introduction to Solid State Physics», Ch. 17. N.Y.: Wiley, Z. Chapman and Hall, 1954.

these types of defects is that a crystal with a Schottky defect has a smaller density, since its volume increases under the constant total mass (atoms or ions from the bulk drift to the surface).

J.I.'s papers on the theory of critical adsorption (1924) and on thermal motion in solids and liquids (1926) were so closely related to the experimental studies carried out at SPTRI that they seem to serve as an example of theoretical papers written for experimentalists. However, their later destiny is such that one is on even sounder ground to consider them theoretical papers for theoreticians. This work suggests that the physical origin of dielectric and semiconductor crystals, their electrical conductivity, optical, thermal and mechanical properties, is determined by not only geometry and «chemistry» (composition), but by the Frenkel and Schottky defects as well (and also admixtures and dislocations).

Mobile holes (vacancies) and the concept of their motion were successfully transferred to the electron theory of solids: a hole in a populated band and an electron in a conductivity band are analogues of a hole in the crystal lattice and an atom in the intersite space.

In 1926 Frenkel wrote: «Taking the holes' mobility into account, it is possible to treat them as «negative atoms» of a kind which, in contact with the usual atoms, cannot, however, undergo dissociation.»[27] It is interesting to compare this quotation, and the whole theory described above, to the Dirac theory of the positron which had made such a great impression on the development of quantum mechanics. In 1933, in his speech at the first conference on the physics of atomic nuclei, Dirac said: «A vacant state with negative energy, i.e., «a hole» in the distribution of electrons with negative energy, would be recognized by us as a particle with positive energy, since the lack of negative kinetic energy is equivalent to the presence of positive energy.»[28] The motion of a positron thus may be regarded as being similar to that of a hole: in a magnetic field, for instance, this motion corresponds to the motion which, in the case of a hole, leads to the anomalous Hall effect. The similarity continues: lattice dissociation with the origin of a pair a hole — an atom in the intersite space (such a pair is sometimes called a Frenkel pair)[29] corresponds to the birth of an electron-positron pair, and the process of association, that is the «shut-down» of a hole by its own or some other atom from the intersite space is similar to the annihilation of electron and positron.

The correspondence between the concept of a positron as a hole in the ocean of electrons with negative energy (introduced by Dirac), and the concept of a hole in a crystal lattice (introduced by Frenkel) seems evident.

The two papers of Frenkel considered above are written in a very simple manner. To understand the new and profound ideas developed in them, it is enough

---

27) J.I. Frenkel. «Collection of Selected Papers», p. 257 (in Russian).

28) P.A.M. Dirac. Theory of Position. In: «Atomic Nucleus», Moscow-Leningrad, GTTI, 1934, p. 133 (in Russian).

29) See c.f. V.S. Vavilov. Uspekhi Fizicheskih Nauk, 1964, vol. 84, N3, p. 431 (in Russian).

to know second year university physics. To explain purely scientific questions, J.I. widely used an imaginative language of metaphors and comparisons which not only made the exposition easier to understand, but also fixed the physical picture in the memory of the reader (just as we remember a poetic description of nature more easily than a prosaic one). And to see and show his model better, J.I., as an artist, looked at it from various points of view, and every new view enriched the model with more details and made it more profound.

In parallel with the SPTRI, J.I., as has been stated, also worked intensively at the Physics and Mechanics Department (PhysMech) of the Polytechnic Institute. The department was the youngest faculty at the young Institute (founded 20 years earlier), a faculty of a new type.

Its organization had started back in 1918 and was completed by the spring of 1919. In March of that year the Council of the Polytechnic Institute decided to found a new department. The departmental rules were worked out by a group of professors joined around A.F. Ioffe (F.Yu. Levinson-Lessing, I.V. Mestchersky, V.V. Skobel'tsyn, A.N. Krylov and others) who had started all the talk about the PhysMech[30]. Ioffe and other PhysMech founders, in particular P.L. Kapitza, who was then a senior student in the Electricity and Mechanics Department of the same institute, thought that the course at the PhysMech should last 4 years. In accordance with its name, the department was to prepare specialists in physics (general physics, thermal technology, electrical technology) and mechanics (theoretical construction mechanics, hydro- and aerodynamics). These specializations became, of course, more and more diversified over time. The first-year programme for the PhysMech students coincided more or less with those of the institute's other faculties. But starting with the second year of studies, special attention was paid to physics, particularly theoretical physics, and mathematics. In the third year, students usually became involved in their own scientific work, done mostly within the walls of SPTRI or in the laboratories of the Polytechnic Institute itself (often some students began scientific research in the 2nd or even 1st year). The graduates had to join new research centres at the national design bureaus and plant laboratories. So the teaching aimed to provide the students with skills in some narrow field, but their knowledge and culture in physics would help them to become quickly oriented toward any new problem or question that might arise in their work after graduation. It is not by chance that, starting in 1922, the faculty graduates earned the title of engineer-physicist. Later the department itself changed its name to the Engineering and Physics Department, under which it celebrated its 20th anniversary. During World War II, the earlier name was restored.

From the first day of its existence, PhysMech was headed by the Dean, A.F. Ioffe, and the Praesidium of the Department, which included Alexey Nikolaevich Krylov and Pyotr Leonidovich Kapitza. Theoretical physics courses were given by V.R. Bursian, Yu.A. Krutkov and Vsevolod Konstantinovich Frederiks (all three

---

30) M.S. Sominsky. «Abram Fyodorovich Ioffe». Moscow-Leningrad, AN SSSR Publ., 1964, p. 284 (in Russian).

of them were at the same time SPTRI staff members). Experimental physics was covered, besides by A.F. Ioffe, by Professor V.V. Skobel'tsyn and his assistants — P.L. Kapitza, Ya.R. Shmidt, M.V. Kirpichova and others, who were mostly participants in the Ioffe seminar. A part of the courses in physics and mathematics was also given to students of other departments of the Polytechnic Institute; that these should be more «physical» and less technical was clear both to the PhysMech professors and to all progressive scientists working at the Institute.

About 40 students were admitted to the PhysMech the first year, but a large number of them soon went away. Nikolay Nikolaevich Mirolubov, a PhysMech student of the first year, recollects that 20 people of the 40 accepted with him were dismissed or left on their own within 6 months, and only 5 graduated. By the beginning of 1928, 24 people had graduated from the PhysMech. But one should note that six of them became academicians; three, corresponding members of the Academy of Sciences; and ten, professors!

At the PhysMech J.I. gave a course in theoretical physics which included thermodynamics, statistical physics, electrodynamics (in turns with V.R. Bursian), and a number of special courses, first of all, the theory of relativity and the theory of quanta, mathematical physics. He also led exercises, both in his own course and in mathematics and molecular physics courses, and headed seminars in theoretical physics and the structure of matter.

Rather small groups of students attended Frenkel's special courses. G.I. Aksyonov, A.F. Valter, V.N. Kondrat'ev, N.N. Mirolubov and Yu.B. Khariton were in one of these groups. So small an audience promoted the establishment of equal and friendly relations between the students and the lecturer, the more so, because J.I. was not much older than his listeners. And this implied, as N.N. Mirolubov reflects, a unique atmosphere that distinguished the lectures (it is not by chance that Mirolubov, Kondrat'ev and Khariton later became Frenkel's friends). Sometimes someone made fun of Frenkel. For instance, once J.I. observed at the beginning of his lecture that the size of his audience had increased by 20% due to the addition of one more student. «So, there are six of you now,» said Frenkel and started the lecture. Only an hour later J.I. discovered, to everybody's amusement and his own surprise, that it was a stuffed student dummy which Kondrat'ev had made from his coat and hat! One should not be surprised that the «new student» sat in a coat and hat without eliciting any reaction from the lecturer; on the contrary, it was extremely surprising that Kondrat'ev managed to survive for a whole hour in the cold auditorium!

Frenkel very likely succumbed, under the spirit of the moment, to exaggeration when he wrote in one of his letters from the Crimea that the years he had spent there had given nothing to him as a scientist. Of course, had they been lived through in Petrograd, communication with colleagues at the university and Ioffe's seminar would have been even more profitable. But the lecturing in general physics and the contact with mathematicians and biologists of Tavrichesky University expanded J.I.'s scientific horizon. Scientific problems about which he learned from these two groups of colleagues were of great interest and suited his tastes. He had gained teaching skills. Teaching new courses at the PhysMech, communication with students, and the search for the most gifted of them was vigorous creative work that J.I. was very fond of.

By the fall of 1921, Frenkel had gotten an apartment with 4 large rooms in a wooden house in Yashumov Lane in Lesnoe, near the Polytechnic Institute (Yashumov Lane is now Kurchatov street). In those early years one and two-storey wooden houses with glazed verandas were put up all over Yashumov Lane.

The winter of 1921/22 was very cold, there was no running water or firewood. Ya.G. Dorfman recollected: «Night. No street lamps in Sosnovka[31]. In Priyutskaya Street, at the very place where the former experimental plant of the PTI now stands, there was a pine forest. Three suspicious shadows were prowling among the tall trunks ... A technician from one of the laboratories of the Polytechnic Institute with an ax in his hand was in front and the two people toting a saw after him were Professor Jakov Il'ich Frenkel and myself. We were going silently, looking around every now and then ... Silence. Somewhere far away somebody was sawing a tree. We went to work. After our great efforts, a pine tree fell. Fortunately, nobody saw us. With tremendous difficulty we cut the log and carried part of it away as our trophy ... Only one part remained to be moved home. Hark! Some people were coming toward us. We left the log in a hurry and retreated. But for some reason the people turned and went in the opposite direction. We didn't risk coming back for the log, cursing these idlers in our hearts ... The next day we found out that the «idlers» were students, my fellow student V.N. Kondrat'ev among them. They had also gotten worried when they met us, because they also had gone firewood hunting illegally»[32].

Frenkels' life during their first years in Petrograd was hard enough: the parents of both spouses were their dependents. On 26 June 1922, J.I. wrote to his mother: «We assume you'll stay with us for some time to get well. True, our «resort» would suit Father better than you, as it's time for him to lose weight. Aside from that, we, of course, have not been starving yet. A crisis which has many times hung over our heads (or better to say, stomachs) was countered by the mercy of Providence. Now there's a light at the end of the tunnel: issues of products at the House of Scientists are improving, and money has reappeared in the cashier windows at the Polytechnic and Rentgenovsky Institutes.»

Not only food and fuel shortages occurred in Russia at that time, but there was a real hunger for textbooks which virtually had not been published during the war and revolution. And yet during these very years Rutherford's papers on a nucleus appeared, as well as Bohr's new quantum physics. Moreover, an army of students had greatly increased because thousands of young people, children of

---

31) A sector in Lesnoe.

32) In: «Twentieth Anniversary of the Engineering and Physics Department of the Leningrad Polytechnic Institute», Leningrad, 1939, p. 41 (in Russian).

workers, peasants, intelligentsia, who could not have even considered a higher education in tsarist Russia, had gained access to the universities.

It was proposed that J.I. publish his lecture courses in the form of textbooks. His writing, the ability to explicate freshly and simply not only the material of his original papers, but also the surveys of various branches of physics as well, were well known. Frenkel responded eagerly to those proposals and, without any slowdown of his pace at the SPTRI and the Polytechnic Institute, wrote the books.

In a small study in Frenkel's apartment the light was on from the evening until the late night: it was then, in the quiet hours, that he worked on the books. While writing down his lectures, J.I. refined their composition. In one of the letters of this period Frenkel wrote: «My book got stuck on some shallows, but this is beneficial, because in clearing the channel I'm clarifying my own thoughts. This has resulted in some new ideas which would make the foundation for a good paper.»

The first of Frenkel's books was «The Structure of Matter», written on the basis of his lecture course.

This treatise of his was published in 1922–1924 in three issues. Of all of Frenkel's books this one became the most obsolete. But it remains of interest not only because it shows its author's creative features (which already reveal themselves in it, though maybe not so brightly yet), but also because it presents a rather exhaustive picture of physics as understood in the first part of the 1920s. It was a time when all the known natural forces (except gravitation) were being reduced to the electromagnetic ones, when the list of elementary particles consisted of the electron and proton and no surprises and novelties were envisioned in this direction, when physics, having had refrained from elastic deformation of ether, was making, as Frenkel wrote, fast progress towards the «materialization of electricity and electrification of matter.»

When explaining the history of the subject, J.I. did not stick to chronology, since it often contradicted the logical sequence. But what was contemporary in 1922–24 is only history now. So, for instance, the hypothesis Jean Perrin suggested in 1919 was rather widely discussed (by J.I., in particular) and considered quite reasonable in those years. It stated that the radioactivity of heavy elements, which has been called natural for a long time now, is actually stimulated by some penetrating ultra-X-ray radiation coming from the centre of the Earth, the Sun, or even stars, and which was not observable because of the instruments' imperfection. To «rehabilitate» Perrin, in case he needs any rehabilitation, one should add that it was he who first indicated the intra-atomic origin of a star's energy.

Frenkel put into the last part of «The Structure of Matter» his own original papers on the electrical conductivity of metals and the theory of contact phenomena.

Frenkel's book was well received by readers and critics. O.D. Khvol'son was among them, and he wrote in his review of the first issue: «The author is well-known as a profound connoisseur of physics and the author of his own clever studies in the field of theoretical physics. The book is right for anyone who, having the necessary preparation, desires to get acquainted with the most recent achievements of science associated with the fundamental questions about the structure of matter».[33] Soon after all three issues of the book were published, J.I. was asked to write a simplified version aimed at a much larger circle of comparatively unprepared readers. As a result, a new book, «Electricity and Matter», was published in 1925.

In the same year that Frenkel wrote «The Structure of Matter», he also worked on «The Theory of Relativity», which was the first non-popular study on this subject in Russian and included an explanation of not only the special, but also the general theory of relativity (the latter had been proposed by Einstein shortly before).

In 1923 «The Theory of Relativity» was published, and in the same year J.I. submitted for publication yet another book written at Ioffe's suggestion and primarily associated not with his work at the SPTRI, but with teaching at the PhysMech. It was entitled «The Electrical Theory of Solids» and dedicated mostly to the electrostatics of crystals, particularly of metals.

The mutual penetration and kinetic energy of a particle in a lattice are of fundamental importance for the calculation of all the macroscopic characteristics of solids: electrical parameters (conductivity in the first place), elastic one (compression modulus), the dilatation strength limit, thermal parameters (heat conductivity and specific heat, latent heat of sublimation). Frenkel paid the most attention to the calculation of just this energy for both heteropolar (of the NaCl type) and homopolar (of the diamond type) lattices, basing his work mainly on the papers of M. Born and A. Landé and sharpening their results (the latter is basically associated with Landé's study of homopolar lattices). As it is known, the value of the strength limit of solids observed experimentally is essentially smaller than the theoretical one.

In the very first years of SPTRI's existence the research of A.F. Ioffe and his co-workers was dedicated to the explanation of this difference (these studies in particular had inspired J.I. to deal with the theory of crystal lattices). The reason for the deviation lay not in the fallacy of the basic theoretical assumption but in the deficiency of the samples studied. On the sample surfaces there were a lot of microcracks which, under the action of the applied load, grew catastrophically and led to the fracturing of the sample. However, if a sample (NaCl crystal in Ioffe's experiments) had been put in a liquid, then the surface layer, pock-marked by cracks and scratches, would be dissolved and the measured strength of the crystal would have been of the same order of magnitude as the theoretical one.

In «The Electrical Theory of Solids» Frenkel's papers were reflected, but also considerably expanded. It is interesting to note that it was in this book that «the first approximation» of Frenkel's concepts of thermal motion in solids and

---

33) O.D. Khvol'son. Pedagogicheskaya mysl', # 1, 1923, p. 69 (in Russian).

the electrical conductivity of metals were presented. He developed these ideas in detail in journal articles one and a half years later. The other remarkable feature of the book was its close relation with experimental studies, especially with those of the Ioffe school, which were fairly completely described by Frenkel.

Lastly, one more book of Frenkel's, «The Book of Vector Calculus with Applications to Mechanics», was published in 1925.

Lecturing in mechanics and electrodynamics had showed J.I. that students mastered the problems associated with the use of vector calculus and even vector algebra only with great difficulty. This might be explained by the fact that in the middle of the 1920s there was no vector calculus in higher mathematics courses. Even the best Russian textbooks on theoretical mechanics of that period, such as the books by I.V. Mestchersky and E.L. Nikolai, tended to manage without it. And if one makes an excursion to the middle of the 1910s, that is, a decade before Frenkel's book was published, it is possible to judge the situation of vector calculus from an example from the practices of Moscow University as testified by Igor Evgen'evich Tamm. He wrote[34] that Professor Bachinsky had given him the top grade for the electrodynamics exam simply because Tamm knew the notion of the vector product. And this was in his 5th year of study!

Twenty years after this episode, in 1935, Professor Ya.N. Spielrein, the author of one of the first vector calculus books published in Russia, noted[35] that he had recently heard that one of the most outstanding scientists thought that the articles on mechanics in the Encyclopedia of Technology had been «spoiled» by the use of vector notations.

Under such circumstances Frenkel's book was of the utmost value. On 19 September 1924 he wrote his father to say: «I have news, and all of it is good. First, my book on vector calculus not only safely passed the GUS[36], but was honoured by being commended as a sample guide to the subject. By the way, its first referee happened to be Spielrein. As for the second one, upon whose reference a commendation, rather than a simple approval, depended, I had an opportunity to speak to him several times in Leningrad. It was Kagan «himself», the head of the Science Department of the Gosizdat (The State Publishing House (Russian abbreviation)), a well-recognized mathematician. He reproached me only for having given the book not to him, i.e., not to the Moscow branch of the Gosizdat, but to the Leningrad one. As a compensation he proposed that I write a course on electrodynamics, but a much shorter one than that I had suggested to him in the spring.»

---

34) I.E. Tamm. UFN, v. 76, N3, 1962, p. 401 (in Russian).

35) Ya.N. Spielrein. «Vector Calculus for Electrical Engineering and Physics», p. 1. Moscow-Leningrad, ONTI, 1936, p. 3 (in Russian).

36) Chief Scientific Council of Narkompros (Russian abbreviation).

So Frenkel wrote and published 5 books in 4 years, and it was done without a break in either his research or teaching activity. This kind of work required genuinely titanic efforts, and one can hardly believe it was done by one man!

One can see from more than the reviewers' responses how popular Frenkel's books were. For instance, N.A. Rubakin included «The Electrical Theory of Solids» in the list of books published in Russian in 1924. Such lists were issued by the «Institut Bibliopsikhologii» [Institute for the study of the influence of books on readers] situated in Switzerland and headed by N.A. Rubakin. Rubakin sent a copy of this book of Frenkel's to A.M. Gorky in Sorrento, at the latter's request[37]. The famous Russian writer Andrey Belyi praised Frenkel's books in his recently re-issued memoirs «Na rubezhe dvukh stoletiy.»[38]

As a result, Frenkel's name became known, by the middle of the 1920s, to a very wide circle of scientific intelligentsia and students in the Soviet Union. And this recognition, along with his original papers, as well as his numerous reports at the Congress of Russian Physicists called on regularly already in those years, put him in the front lines of the leading theoreticians of the country.

At that time, the papers of Soviet physicists were published, as a rule, in the physics section of the ZhRFKhO and in the German «Zeitschrift für Physik». Approximately one-third of the space in the latter, which was then and until the Nazis came to power in Germany one of the most popular physics magazines abroad, was allotted to the papers of Soviet physicists, especially SPRTI staff members. This promoted the growth of the influence of Soviet physics on the development of this science in general and testified to the high reputation which it had gained even then. One of the reasons for the popularity of the «Zeitschrift für Physik» was the fact that articles were published in it approximately 2 months after they had been submitted to the Editorial Board.

By the end of 1925, 6 papers by Frenkel (all already described here) had appeared in «Zeitschrift für Physik». These papers received a positive response from physicists abroad.

In September of 1924, the IV Congress of Russian Physicists was held in Leningrad, to which scientists from Leningrad and Moscow contributed most intensively. The Congress was very well organized and took place in a very ceremonial atmosphere. Here is a small detail which may now seem amusing: two trams were specially provided so that the participants could attend one session at the Polytechnic Institute in Lesnoe.

The presence of Pavel Sigizmundovich Ehrenfest promoted, to a considerable extent, animated discussions at the congress.

J.I. Frenkel took an active part in the Congress and delivered five reports there. He wrote to his father in September of 1924 that he had had a great success at the

37) S. Belov. «Knigoizdateli Sabashnikovy» («The Sabashnikovs Publishers»). Moscow, 1974, pages 134 and 146 (in Russian).

38) Meaning: «At the Border of the Centuries».

Congress and generally had been lifted to a height that he, in his own opinion, had not deserved. And further on he said: «However, Ehrenfest contributed significantly to this; he tried by all possible means to push me forward. Three days ago he spoke to me about a trip abroad and it became clear that the condition that I leave my family in Russia or some other place, of which condition of the Rockefeller committee Ioffe had informed me, had actually been imposed by him. «You can,» he told me, «send all the money to your family and live on black bread; it's your own business. But you have to spend these six months without family to use your time properly.» I could, according to Ehrenfest, also get very well settled abroad without his help, but he is ready to help really only on the condition stated. And I think that he is quite right.

«I have to tell you that Ehrenfest has conquered the hearts of all our youth, and maybe of the elderly as well. This man combines the simplicity and directness of a child with an extraordinary love of people, inexhaustible wittiness and the brain of a great and ingenious researcher. In his performance inanimate objects, such as molecules, atoms, and electrons, start to speak a Russian rather broken in the sense of affixes, case and genders but otherwise very delicate; they love, hate and generally exhibit life, transforming themselves into microscopic inhabitants of the animate universe. For Ehrenfest physics is not so much precise science as it is a drama or comedy of the life of atoms and electrons. You may see from the previous characterization the extent to which I'm influenced by the charisma of Ehrenfest's personality. I don't know whether this charisma is related in any way to his good feelings toward me; nevertheless, I guess there are no egoistical motives at all here, since neither I nor anyone else can help smiling with pleasure when looking at and, especially, listening to Ehrenfest. This smile reflects the good humor and goodwill he radiates.»

Paul Ehrenfest, or as he was called in Russia, Pavel Sigizmundovich Ehrenfest, the student of Hilbert and Boltzmann, is known for his studies in thermodynamics and quantum mechanics. Analyzing the papers of Planck which had initiated the onset of new working quantum physics, he demonstrated that the concept of energy quanta was a postulate, not an implication of the theory Planck developed. In many textbooks one finds an elegant theorem credited to him stating the average values of quantum-mechanical variables which satisfy the formulae and a commonly accepted classification of phase transitions. Ehrenfest had developed the notion of adiabatic invariants introduced by Boltzmann and showed, within the framework of Bohr's quantum mechanics, that under the slow variation of an external field acting on a system, its state does not change and the transitions between quantum states are absent. One could easily continue the list of his scientific achievements. But Ehrenfest has entered the history of science of the first third of this century as a person with the exceptional gift for criticism in physics and a very fine physics intuition. He was indispensable in any discussion of the most exciting questions of physics. A broad erudition, a clear understanding of physics phenomena and high moral standards were the basic features which attracted young people to Ehrenfest. He encouraged young scientists in feeling that their investigations were on the right track. One of the best examples of this is the discovery of the electron spin by Uhlenbeck and Goudsmit. This paper was presented at a seminar of Ehrenfest's that gathered each Tuesday in Leiden. Ehrenfest supported the results of the two physicists, and revealed and clarified to a great extent, the physical essence, despite the fact that authorities such as Bohr and Pauli had cast a doubt on these results. It is not for nothing that many years later, in the 1950s, Uhlenbeck, on the occasion of being awarded with the Oersted Medal for achievements in teaching physics, dedicated his speech completely to Ehrenfest. According to B.M. Pontekorvo's testimony,[39] Ehrenfest's influence on the outstanding Italian physicist Enrico Fermi during the latter's scientific trips to Germany and Holland in the mid-twenties was as great as his influence on Uhlenbeck. Fermi regarded his communication with Ehrenfest as the most valuable experience he had made abroad.

These abilities of Ehrenfest's, despite the tremendous recognition his contribution had garnered, left him dissatisfied deep down inside. Einstein discussed this in the paper dedicated to the memory of his friend[40]:

«His stature lay in his unusually developed ability to grasp the essence of a theoretical notion, to strip a theory of its mathematical accouterments until the simple basic idea emerged with clarity ... It was on this account that he was invited to scientific congresses, for he always brought clarity and acuteness into any discussion. He fought against fuzziness and circumlocution ... Some of his utterances could have been interpreted almost as arrogant, yet his tragedy lay precisely in an almost morbid lack of self-confidence. He suffered incessantly from the fact that his critical faculties transcended his constructive capacities. In a manner of speaking, his critical sense robbed him of his own ideas even before they were born ... In truth he felt unhappier than anyone else who was close to me».[41].

P.S. Ehrenfest, an honourary member of the Russian Physics and Chemistry Society (RFHO) and a corresponding member of the USSR Academy of Sciences, made great contributions to Russian and Soviet Physics. As of 1907 Ehrenfest lived in St. Petersburg. The same year, supported by I.I. Borgman, O.D. Khvol'son and K.K. Baumgart, he was elected a member of the RFKhO and took the most active part in its work, being the reporter at many meetings of the society and the author of many papers and reviews published in the society's journal (ZhRFKhO).

However, under the conditions in tsarist Russia, Ehrenfest, a Jew by nationality, could not find a proper job and in 1912 had to go abroad again. At the Polytechnic Institute, P.S. Ehrenfest gave in 1908 a brilliant lecture course on the

---

39) Uspekhi Fizicheskih Nauk, v. 57, N3, 1955, p. 350 (in Russian).

40) In his letter to Ehrenfest in 1922 Einstein wrote: «I seldom meet people in whose company I'm in good spirits. I need our frienship even more than you do.» (Quoted by: K. Zeelig. «Albert Einstein».)

41) A. Einstein. «Out of My Later Years». Phil.libr., New York, 1950, p. 236.

theory of differential equations illustrated with the examples from physics (he was invited by the Institute Council, an invitation issued by V.L. Kirpichov and A.F. Ioffe). Though not working at St. Petersburg University, he was closely involved with the university physics department; he organized a colloquium at his place where St. Petersburg physicists gathered twice a week. In particular, Ehrenfest greatly influenced V.R. Bursian, A.A. Friedmann and Yu.A. Krutkov, university students then, by directing their first steps in theoretical physics.

Tat'yana Alekseevna Afanas'eva-Ehrenfest, the wife of Pavel Sigizmundovich, played a significant role in the growth of Ehrenfest's sympathy for Russia and his love of Russian culture.

Great friendship existed between P.S. Ehrenfest and A.F. Ioffe (it is said that Ehrenfest was the only one of Ioffe's friends with whom Abram Fyodorovich was on a first name basis). Ioffe left, unfortunately, only very short notes about Ehrenfest in his book «Vstrechi s fizikami»[42].

P.S. Ehrenfest promoted in every way the success of Ioffe's mission abroad[43].

Ehrenfest did much for J.I. Frenkel as well, with whom he had gotten acquainted in 1924 at the IV Congress of Russian Physicists in the USSR. At the Congress, according to A.F. Ioffe's memoirs, Ehrenfest met with the same popularity that he had enjoyed in 1909 at the Physics section of the XII Congress of Naturalists and Medical Doctors (it was then that Ehrenfest met the head of the Russian school of physics, P.N. Lebedev; Ehrenfest made a great impression on Lebedev).

In Leningrad Ehrenfest contributed a paper devoted to the attempt to constuct a corpuscular theory of diffraction. There was no quantum (wave) mechanics then, and one could only speak, perhaps, about the difficulties physics faced. The report was followed by a discussion, and one of the speakers was J.I. It is impossible to reconstruct his argumentation and the way he suggested to soften the corpuscular-wave contradictions described so brilliantly by Ehrenfest. Professor J.L. Zel'manov merely recalls Ehrenfest's reaction to J.I.'s remarks: «Frenkel», he said in a jocular attempt at pathos, «Why are you spoiling such a perfect picture of our misunderstanding of physics!» This sentence is very characteristic of Ehrenfest, who was an extremely witty man.

And one more episode. J.I. was saying something at a blackboard. Ehrenfest said, smiling, after he had listened to J.I.: «So you have dealt with this also? Tell me, Frenkel, is there at least one region in physics you have not left a trace in? Or, better to say, where you have not left dirty footmarks?» Both the teller of the joke and its butt were very satisfied by it and laughed heartily together.

---

42) «Meeting with Physicists» (in Russian).

43) Literature about P.S. Ehrenfest that we would especially recommend: M. Klein. «Paul Ehrenfest, Vol. 1. The Making of a Theoretical Physicist». Amsterdam-London, 1970. See also: V.Ya. Frenkel. «Paul Ehrenfest». Moscow: Atomizdat, 1977; «Ehrenfest-Ioffe. Scientific Correspondence». Ed. by V.Ya. Frenkel. Leningrad, Nauka, 1990 (in Russian).

Here Ehrenfest turned a pun in Russian, one which is not possible to express in English. The Russian words indicating «leaving a trace» and «leaving dirty foot tracks» sound very similar (one is «ostavit' sled», the other is «nasledit'»). In addition, this little witticism shows that, in spite of Ehrenfest's «very specific» twisted Russian, he expertly used different idiomatic puns.

Frenkel's success at the Congress was primarily due to his paper on the theory of metals[44]. By then there existed the theory of metals elaborated by Drude and improved by Lorentz. Drude, using the concept that free electrons existing in a metal are responsible for its electrical conductivity, had developed a simple theory wherein these electrons are treated as a perfect gas with the metal's temperature, and enclosed in metal. Using this model he had been able to find a theoretical foundation for the Wiedemann-Franz law of proportionality between the electrical and thermal conductivity of metals (as for the ratio of these coefficients, it was dependent only on temperature the same way for any metal). This result had given life to the Drude theory. Soon the theory was refined by Lorentz who took into account the fact that electrons have the Maxwell velocity distribution. This led to the correction of the numerical factor in the formula describing the Wiedemann-Franz law and, aside from that, allowed a very simple explanation of the phenomenon of thermoelectron emission. Namely that electrons in the high-energy «tail» of the Maxwell distribution left the metal.

But both the Lorentz and the Drude theories remained internally inconsistent.

The issue was that electrons taking part in thermal conductivity seemingly should also contribute to the thermal capacity of metal. Experiments demonstrated that both metals and dielectrics (there was not even a word for semiconductors then) have an equal thermal capacity specified by the known law of Dulong and Petit to be 6 calories per gram-atom. Thus the free electrons characteristics of metal have no effect on the thermal capacity. However, according to the classical perfect gas theory, free electrons have to have thermal capacity and to contribute to the metal's total thermal capacity.

The contradiction seemed so obviously pronounced, and multiple attempts to overcome it had been so fruitless, that Ehrenfest awarded it the title of the «thermal capacity catastrophe».

This catastrophic situation was resolved by Frenkel with the help of the Bohr theory of atoms applied to the description of a crystal body. His line of arguments was extremely simple.

Consider metal vapour. It is an insulator, as is any gas. One specific feature of metals is a relatively low ionazation potential, i.e., a small bond between the

---

44) We read in the Leningrad «Vechernyaya Krasnaya Gazeta» of 17 September 1924: «A vigorous discussion followed Professor J.I. Frenkel's report «Foundations of the Electron Theory of Metals». Professor Ehrenfest of Leiden University participated very actively in it.» According to the same newspaper (of September 18), Ehrenfest also spoke about Frenkel's other paper on the electron theory of solids.

nucleus and the electron rotating at the atom's outermost orbit[45]. Such an orbit, a prolate ellipse, was called «comet-like» by J.I. Thus, the atoms of metals may be characterized from the geometrical viewpoint by having a large distance between the valence electron and its nucleus. In the aphelion, according to the Bohr-Sommerfeld theory, the valence electron of a sodium atom is 150 times further from the nucleus than in the perihelion; the orbit of the valence electron is stretched so strongly!

In a sodium crystal lattice the distance between atoms is 3.72 A, while the afore mentioned aphelion distance between the valence electron and the nucleous is 2.56 A. Therefore it is easy to see from purely geometrical considerations that this valence electron spends spends part of its time nearer to the nucleus of the neighbouring atom than to the master one. The situation here is similar to that appropriate for the formation of a molecule according to Bohr. A molecule of hydrogen, for instance, consists of two hydrogen nuclei, protons, which are surrounded and simultaneously united by two electrons. Each of them belongs to the molecule as a whole and does not have its own master nucleus.

Thus, valence electrons in metal belong to the whole lattice, are not connected with any particular atom and, under the influence of the lattice's atoms, transit from one of them to the other, wandering through the metal. This is a picture of Frenkel's wandering electrons which replaced free electrons of Drude and Lorentz. Wandering electrons acquire, in J.I.'s imaginative words, only one kind of freedom, the freedom to change the master, and, passing from one atom of the lattice to the next, they become «the slaves of the collective».

To characterize Frenkel's imaginative language, let us give here yet one more example also related to the electrical conductivity of metals. In a later paper J.I. wrote:[46] «The motion of an electron [in a perfect crystal lattice of metal — V.F.] may be likened to the motion of a point of a wheel rolling along a smooth road. The rotation of an electron about some particular atom corresponds to the rotational motion of the wheel, and the transition from one atom to the other is like the progressive motion of wheel. In reality, the arrangement of atoms even in absolutely pure metal is not absolutely perfect because of the thermal motion. The latter creates «the roughness» of the three-dimensional field along which «electron wheels» «roll». (Note that the electrical resistance of metals is due to this «roughness»; it grows with temperature, because the number of «holes» and «bumps» on the road, as well as their depth and height, increases.)»

The wandering of electrons in metal is disordered (similar to self-diffusion) without an electrical field and is oriented when one is present (or when the temperature gradient is present). By proving the dependence of thermal and electrical conductivity on temperature, Frenkel conquered the same territority which had

45) J.I.'s arguments were in the spirit of the Rutherford-Bohr atomic model.

46) J.I. Frenkel. Theory of Metals. Moscow, AN SSSR Publ., 1933, p. 8 (in Russian).

provided the success of the Drude-Lorentz theory, that is, he gave a theoretical explanation of the Wiedemann-Franz law.

But now it became clear why wandering electrons make no contribution to the thermal capacity. Their velocities are on an atomic scale, and their kinetic energy is 2 orders of magnitude greater than the kinetic energy corresponding to room temperature and has nothing in common with this usual temperature. Only at temperatures of several tens of thousands of degrees does electron motion acquire the classical character described by the kinetic energy formula: $(1/2)mv^2 = (3/2)kT$. However, it is impossible to compare the heat capacities of dielectrics and metals at such temperatures, since they melt and vaporize at much lower ones.

The paper by Frenkel is very simple in its assumptions (simplicity is a historical concept; for instance, what could be «simpler» than Newton's law!). «Geometrical» consideration of the atom arrangement in the lattice led Frenkel to classify crystals as metallic ones (with comet-like valence electrons) and dielectric ones (with planet-like valence electrons), as well as to construct a theory of metals which eliminated the thermal capacity catastrophe.

After several years J.I. would turn again to the theory of metals armed much more powerfully with de Broglie-Heisenberg-Schrödinger quantum mechanics, instead of the Bohr atomic model.

Ehrenfest was a member of the committee of the Rockefeller Foundation which sponsored scholarships for the most talented young scientists. In particular, Fermi and Heisenberg had Rockefeller scholarships, and among Soviet theoreticians receiving them were Yu.A. Krutkov, V. A. Fock, G.A. Gamov, and L.D. Landau.

Documents about the Rockefeller Foundation and its scholarship-holders are stored in New York, in the Rockefeller Archives Center. J.I. Frenkel's file is there as well. It was P.S. Ehrenfest who suggested Frenkel as a candidate for the Rockefeller Foundation Scholarship. One can trace the stages of pertinent organizational efforts and negotiations by Ehrenfest and Ioffe through the correspondence and the letters about J.I. Let us give here additional material from the Rockefeller Archives[47]. Professor A. Trowbridge, Director of the Scientific Department of the Rockefeller Foundation in Europe, wrote to Professor W. Rose, President of the Foundation, on 8 April 1925:

«My dear Doctor Rose:
On the occasion of my visit about a month ago to Leiden, I talked with Professor Ehrenfest on the question of possible candidates for fellowships in physics or mathematics from Russia. You will remember that Professor Ehrenfest, while himself not a Russian, has been many years in that country as a teacher and has been in correspondence with his Russian colleagues throughout the war and revolution and up to the present time. Through his efforts, already several promising

47) Corresponding materials are published in the paper by V.Ya. Frenkel, P. Josephson, Uspekhi Fizicheskikh Nauk, v. 160, N11, 1990, pp. 103–134 (in Russian).

young Russians have been enabled to leave their country for short study abroad. Professor Ehrenfest, I believe, has had some conversation with you on the subject of getting a few of the more gifted young Russians into Western Europe at the time he visited you in New York. He now makes application for two fellows: Jacob Frenkel and George A. Krutkov. This letter is with references to his first choice among these:

JACOB FRENKEL
Jacob Frenkel, age 31, married, with one child, is at present a lecturer in theoretical physics at the Polytechnical Institute in Leningrad; he is also a senior physicist at the Physical-Technical Institute at the same place. Since 1917, he has published 14 articles on varied topics in physics, mostly connected with electron-quantum theory. He has also published five books.

His application for a fellowship for study abroad is supported by Professor M. Born, who wrote: «He shows in his work a very complete mastery of mathematical methods of physics, a deep insight into physical relationships».

Professor Ehrenfest states as his opinion that if one can help Frenkel and two other young men at present in Russia, one will be doing the most one can for this branch of science in Russia at the present time, since the men will go back with new ideas, gained by residence for a year in the West.

Professor Ehrenfest applies for the fellowship of Jacob Frenkel to allow him to begin his studies at Göttingen with Born for three or four months, and possibly later to spend four or five months at Cambridge, England. He would wish to begin his work at Göttingen in the autumn of 1925.

With regards to the stipend, I think that the usual $ 185 a month for a married man will probably do in this case. Yours very sincerely
A. Trowbridge.»

There are two more documents concerning J.I. Frenkel. They are also kept in his file in the archives and illustrate the Rockefeller Foundation's system of thoroughly checking the work of its scholarship holders.

During the visit to Göttingen of Professor W.E. Tisdale, assistant to Professor Trowbridge, he contacted Professor R. Courant, and in a memorandum containing the account of this meeting he stated: «According to Courant, Frenkel is extremely good.»

Being in Leiden on 13 October 1926, Tisdale met Ehrenfest and directed the following message regarding Frenkel to the Rockefeller Center: «Ehrenfest is the most pleasing scientist I have met in Europe. I discussed with him the fellows with whom he has been in contact and he said of FRENKEL that he was a very good appointment, that he uses his time well and remarks that it is extraordinary for a man to be able to receive the new impressions which Frenkel has received and at the same time get so many good results in a short time. He assured me that Frenkel is going back to Russia and that his appointment has been very much worthwhile.»[48].

Thus, in accordance with the programme confirmed by the Rockefeller Foundation, J.I., his wife and his son went abroad at the end of 1925. So in a way Frenkel acted contrary to Ehrenfest's requirement. Sarra Frenkel reflected later that Ehrenfest's reaction was contained in the following sentence: «Frenkel, you make it so difficult to love you sometimes!» Ehrenfest's anger changed, however, to mercy when he got to know that Frenkel's wife and son would not live with J.I. but would stay in Nice. (There they lived for about a year at the home of the Frenkel brothers' tutor, M-lle Robaglia [see Chapter 1]. She later returned to her motherland from Russia in 1923.)

In the conclusion of this chapter we would like to balance Ehrenfest's praise of Frenkel. His relation to J.I., as one may see from Ehrenfest's letters to Ioffe of the beginning of the 1930s, grew somewhat rocky: they were too different in their style and temperament. As early as 14 August 1924, Ehrenfest wrote to Ioffe: «Frenkel's pattern of thought differs so greatly from my own that I have no hope for any mutually helpful influence: to him, »results» are infinitely more important than «understanding»: thus far, almost nothing has come of our conversations. Only Pauli might be able to tame him, since he thinks clearly on his feet.»

Needless to say, the word «understanding» in this phrase of Ehrenfest should not be taken in its literal sense. New and general principles of developing theories are very often built on rather boggy ground. Only much later are these theories shored up with a strong foundation containing a minimal number of non-contradictionary axioms. For a scientist with Ehrenfest's critical acumen and talent, it is these searches for fundamental axioms and rigorous proofs of intuitively hazarded theorems and conceptions that constitute the essence of understanding. Without this understanding and elucidation, Ehrenfest obviously did not consider it possible to work on applications of new theories for solving specific problems. Frenkel, on the other hand, believed that the most important criterion for the correctness of new theories should be nothing other than the successful solution of old problems and the statement of new ones with their aid. This is why these «results» were, in Ehrenfest's (probably correct) opinion, more important to Frenkel than Ehrenfest's «understanding».

As an example, let us cite the concept of the de Broglie waves, which was advanced in 1924. Unexpected and paradoxical, it at first appeared invalid to many: how could the notation of electrons (whose «granularity» no one doubted) and waves be combined in a single form? Commenting at Leningrad on de Broglie's paper, which had been read at Frenkel's theoretical seminar at the Physical-Technical Institute, Ehrenfest said half-seriously that «de Broglie and I — either one or the

---

48) The assurance of Ehrenfest that Frenkel would return to Russia was caused by the corresponding demands of the Rockefeller Foundation whose aim was to support scientists of various nationalities.

other of us must be be cracked.»[49] Frenkel, who was the first in our country (as observed by I.E. Tamm) to recognize the fruitfulness of de Broglie's idea, applied this relation not just to free electrons, but also to a description of the behaviour of electrons in a metal, and used this conception as an aid in deriving all of the positive results of the Drude-Lorentz electron-gas theory.

An attitude similar to Frenkel's in approaching the «understanding/results» problem was also characteristic of Enrico Fermi, and he also came under well-meaning criticism from his colleagues for it. The American physicist Ph. Morrison recalls that Fermi usually discussed advances that had already been made, while Bohr focused his attention on what had not yet been done and was presenting difficulty. V. Telegdi remarked on the same subject that «I think Fermi, if we speak of philosophy rather than style, was always a pragmatist. A theory was all right as long as it explained facts — and to the devil with philosophical organization». This prompted Pauli to call Fermi the «engineer of quantum mechanics.»[50]

Frenkel liked to repeat to his students, who sometimes found it difficult to accept quantum-mechanical conceptions, which conflicted with their so-called common sense (something our subconscious learns through everyday experience in dealing with the «understandable» laws of classical physics), that «to understand something is to get used to it.» His quick and agile theoretical physicist's mind and his surprising intuition reduced to a minimum the time he needed to accustom himself to new ideas.

In giving their due to his quickness to respond and to the tolerance of new ideas that is so typical of the scientist with a romantic bent, we do not by any means intend to disparage the efforts of scientists of the other persuasions, who devote their talent to casting light on these new ideas, which at first seem to be confused and groping for the truth. The efforts of scientists of these two types are not so much contradictory as complementary, illustrating the validity of Bohr's complementarity principle.

It is interesting to note that Ehrenfest's view of Frenkel's approach to physics problems can be contrasted to the following judgment of Blokhintsev:[51] «Frenkel placed the understanding of a physical phenomenon ahead of everything else. I have never been able to help the impression that once having understood a phenomenon, he lost interest in it. The later fate of an idea that he had set loose was of little interest to him. He was even less concerned with extracting any gain from his achievement. The pleasure of discovery was apparently reward enough for him.»

---

49) Ehrenfest was not alone in reacting like this to de Broglie's paper; the majority of western theoretical physicists shared his view. Ironically, E. Schrödinger said that he did «not want to talk about such nonsense» when P. Debye suggested that he review this paper at a seminar. And it is while reading this paper that Schrödinger came to the equation now figuring under his name. (See P.L. Kapitza's paper in «Voprosy Istorii Estestvoznaniya i Tekhniki» # 18, Moscow, 1965, p. 25 (in Russian).)

50) «Exploring the History of Nuclear Physics» (Proc. of AIP Conf. #7). Ed. Ch. Weiner. N.Y, 1972.

51) In a private letter to the author of this book.

Chapter 3
In Germany

In November, 1925, Frenkel goes to Berlin and spends almost a year in Germany, France and England. His letters to his parents[1] form the basis of this chapter.

«Berlin, 17 November 1925.
... I have seen none of the Berlin scientific world so far. Tonight I'm going to meet Einstein's assistant, Grommer, and via him I shall maybe get to Einstein. Tomorrow I'm going to visit two scientific institutions, and the day after I'll attend the meeting of the German Physics Society, where I have been invited in writing by Schell, the editor of the journal where I published my papers [«Zeitschrift für Physik». — V.F.]. Next week, at the university physics seminar, one of my papers will be reviewed. Generally, I seem to have a rather good reputation among Berlin physicists. I have to apply all my efforts to justify and strengthen this impression. Until now I have done nothing in this direction. I'll start to work in the proper way after my arrival in Hamburg.»

«Berlin, 20 November 1925.
... This noon I spent 2 hours at Einstein's. With a sort of excitement I climbed up the staircase in the house where Einstein lived. But my uneasiness vanished as soon as I saw Einstein «in the flesh». He appeared an extraordinarily nice man who combines Ioffe's gentleness with the directness and sincerity of Ehrenfest. I spoke to him exclusively of physics, mainly expressing my own ideas. Those views he quite approved of, particularly those regarding my theory of metals which I am going to lecture on at the university seminar Wednesday evening. We touched as well on some other questions concerning the theory of quanta, and Einstein summed up his relations to the latter in the following way: the situation is desperate, it's impossible to understand anything!

The meeting took place in Einstein's study; the host made a rather proletarian appearance in a knitted waistcoat, worn-out trousers and sandal shoes, which are so popular in our Leningrad. However, the apartment seemed well-appointed.

Last night, I had been at a place of the young German physicist (R. Bekker) — the one who will review my paper. He and his wife were extremely hospitable and treated me to a supper; after that we retired to his study and sat down to physics. Then it became clear that he did not agree with me, and the final word in our argument would rest with Einstein. Today the latter decided in my favour.

I spoke to him in German without any special eloquence but well enough to be understood. I hope to master colloquial German soon, since I will be dealing almost exclusively with Germans. And that's that, it seems. I could complete the epics of my letter with some lyrics. However, I am too shy to put my feelings

---

1) His correspondence with his wife seems to be lost.

A. Einstein

down on paper; you know this, and hardly need to have them expressed in words. Please, do not miss us too much, yes? It is too early to count the days before we meet again. There will be plenty of time for doing so. For my part, I will write to you frequently if not always extensively.»

«Berlin, 26 November 1925.
... If not for Einstein's occasional departure to Leiden, I might have stayed for another month and worked with him (not on the relativity theory) ... As a result of my first discussion with Einstein, he expressed his consent, (if not desire) to take me under his patronage.

The day before yesterday, at the university, in the presence of Einstein, Planck, Nernst, Laue and other coryphaei (Ioffe among them), my theory of electrical conductivity of metals was reported. The reporter, Bekker, disagreed with me on a certain essential point. But Einstein came out with quite energetic a reproof and stated that he considered Frenkel's arguments perfectly correct, and his results highly remarkable.[2] I must with no false modesty admit that I had by no means expected such appreciation of my work which I had regarded as either in a qualitative or quantitative respect. At the colloquium there were debates, and the questions were put by Herr Geheimrat Nernst himself (whom I took the liberty to address as Herr Professor, thus horrifying every one present). I gave explanations in German. My German, however, in the opinion of A.F. Ioffe, only few people could understand. I will try to master spoken German as soon as possible. It seems to me that I have already made some progress in this field, in spite of the fact that I still have not started any systematic studies.»

«Berlin, 30 November 1925.
... Tomorrow I will go to Hamburg. I will stay there probably till the 26th, and then go to France for ten days or a fortnight, if there is enough money. From the middle or end of January until the middle of March, I'll work in Berlin with Einstein. We already have a firm arrangement about it. (I was at his house again the day before yesterday and spoke to him not only about physics, but about philosophy and politics as well.) I have just been to Springer, the owner of one of the largest German scientific publishing houses. He already knew everything about me (from my Göttingen friends) and immediately agreed to accept my «Electrodynamics» (whatever the volume). I promised to present to him the 1st book by 1 June, i.e. the corrected and extended translation of what I have already written in Russian. And the second part will follow the first. However, I shall not be in a hurry, bearing in mind that research and not writing books is my main goal.»

J.I. had kept the lifelong memory of his meetings with Einstein, the impression of his personality, and admired him as a remarkably open-hearted and human personality. Einstein's portrait over his desk decorated Frenkel's study.

---

2) A.F. Ioffe wrote from Berlin: «Yesterday I was here at a colloquium, where Frenkel's paper on metals was presented and very much praised (especially by Einstein)» (cit. by: M.S. Sominsky «A.F. Ioffe», p. 470 (in Russian)).

Much later, in the late 1940s, in a paper contributed to the memorial collection in honour of Einstein's 70th anniversary, J.I. mentioned one of the discussions he had had with Einstein. Then they touched on the question about gravitational waves. J.I. wrote, «A. Einstein was perhaps the first to indicate the relation between gravitational waves and corresponding particles (when speaking to the author in 1925). A detailed mathematical study of this problem was published in our country by M. Bronstein in 1936. His results have been recently developed by D.D. Ivanenko.»

J.I. reflected that in the conversation with Einstein mentioned in the letter of 30 November, 1926, he asked, knowing about Einstein progressive views and his sympathy toward the USSR, whether Einstein was going to visit the Soviet Union. Einstein replied with a joke, «I am afraid as a result of such a visit I would be considered as a bolshevik here and a bourgeois there, in your country.»

Here is another small detail of everyday life associated with Frenkel's meetings with Einstein. J.I., when seeing his guests off, always helped them put on their coats, making no difference as to their age or sex. One of his students, while putting on the coat held by J.I. and saying goodbye remarked that he became convinced «the Frenkel effect» really existed. He explained that this was the title under which the ceremony of dressing accepted in Frenkel's home was known to the public.

«In this case, it would be more correct to speak about «the Einstein effect»,» J.I. answered, «because I acquired this habit after I had visited him».

Ya.I. Perel'man, a prominent writer of popular science books, with whom J.I. had become acquainted in Petrograd in the early 1920s, became aware of Frenkel's contacts with Einstein. Under the agreement with Perel'man, J.I. in 1925–1926 wrote small articles for Leningrad's «Vechernyaya Krasnaya Gazeta»[3] about his impressions abroad, approximately one article monthly. Perel'man asked J.I. to describe his meetings with Einstein as well. Answering him in April, 1926, Frenkel stated: «I presumed to accept your offer to write not only popular scientific, but also, let me say so, «columnist» articles for «Krasnaya Gazeta». However, I would feel uneasy writing about Einstein.»[4]

Frenkel's letter of November 20, 1925, where he noted about how shy he was to express his feelings in writing, may help in understanding the last sentence.

In 1931 Frenkel wrote a small note on Einstein for the «Granat» encyclopaedic dictionary. It is invigorated by his recollection of personal meetings with Einstein. In the conclusion, J.I. wrote:

«. . . During the last decade Einstein was looking for the development of his general relativity theory in order to let it incorporate, in a natural way, not only gravitational field theory but electromagnetism as well. In 1929 he managed to

3) «Evening Red Gazette.»

4) J.I. Frenkel. Memoirs, Letters, Documents. Leningrad: Nauka, 1986, p. 336 (in Russian).

obtain some results following this trend, i.e., as to elaboration of «the unified field theory». However, the validity of these results as yet remains in some doubt. It is interesting to note the persistence with which Einstein continues his work, ignoring all the other problems of physics until the question he has formulated and considered as a fundamental one still is resolved.»

«If I do not manage to solve this problem», Einstein once said privately,[5] «I will be virtually lost for physics.»

«Einstein has no pupils and has never had any, in spite of the fact that he is always ready to share his thoughts and is rather active as an opponent in various scientific gatherings. It's even more surprising since Einstein is an extremely charming person.»[6]

In the middle of the 1930s Frenkel gave a lecture in Odessa about the achievements of modern physics. Among the questions asked afterwards there was also one about what Einstein was working on, why he stopped publishing his works, and J.I. answered something like: «Einstein's activity in the first quarter of the century was practically volcanic. Today I regard him as a sleeping volcano.» He made a pause here, then raised his right indexfinger and added: «Sleeping, but still a volcano.»

Starting in early December 1925, Frenkel spent a period of time in Hamburg during his sojourn in Germany. Four months spent in that city, as his letters witness, proved to be very fruitful for him. Frenkel's communications with Wolfgang Pauli promoted this to some extent. Pauli studied physics under the supervision of so famous a master as Arnold Sommerfeld. At the age of 25, Pauli was an absolutely mature scientist; he wrote the survey of relativity theory in the «Mathematical Encyclopaedia», 1921, which was recognized as a classical work. He was also the author of a famous paper which allowed him to be considered one of the founders of quantum mechanics. We imply the fundamental law discovered by him in 1925, which was at first referred to as «the Pauli exclusion», and later «the Pauli principle».

The other physicist whose name is found in Frenkel's letters from Germany was Otto Stern, professor of physical chemistry and the director of the laboratory of the Physical Institute at the Hamburg University. J.I. knew Stern very well from his studies of molecular beams mentioned previously; they had a lot of common interests in the field of molecular physics.

The letters from Hamburg presented below were addressed to the parents and need no comments.

«Hamburg, 2 December 1925.
... Yesterday on the arrival in Hamburg, I went to the university where in the Studentenhilfe [the bureau of student assistance] I got the addresses of several

5) Probably, one of the conversations of Einstein with J.I. himself is implied.

6) Encyclopedic dictionary «Granat», v. 51, p. 151.

inexpensive rooms. I chose a little room on the upper floor (attic) of a certain boarding house. The choice seems to be correct, since I feel comfortable enough in my quiet dwelling not so far from His Highness the Lord. I have a small sofa and a soft arm-chair at my disposal, a little table in front of the sofa, a closet, a chest of drawers, a huge bed and a wash-basin of matching dimensions. What else do I need?

This morning I went to the university, where I made the acquaintance of Prof. Stern (physical chemist, semi-theoretician — semi-experimentalist) and with Pauli (associate professor, one of the most gifted theoreticians in Germany and «the Universe» as a whole, since there are practically no theoreticians outside of Germany).

What I shall be doing and how, I still don't exactly know. But in any case, I'll work hard.»

«Hamburg, 7 December 1925.
... These days I study quite a lot, trying not so much to expand my knowledge as to make it more profound. Every day, approximately from 11 or 12 in the morning till 5 p.m., I spend my time in the library of the Physical Institute. In the evening, after 7–8 p.m., I usually stay at home and work until late at night, about 2 a.m. I decided to leave this nasty habit starting today, because as a result, I wake up at 10 a.m. without feeling vigorous enough. My night vigils were caused by a paper I was writing on three previous evenings and at last completed yesterday. It is related to my study of fluid viscosity which was done last year but still remains unpublished. At the request of Professor Stern, I will present this paper in the nearest future at the local seminar on theoretical physics.»[7]

«Hamburg, 11 December 1925.
... Everything is O.K. with me. My mood has improved and is steady. As before, I continue working only on physics and a little on music (an hour every day). Yesterday I attended a colloquium and then had supper in a restaurant together with several participants. Thus I somewhat broke my seclusion. Today after a scientific discussion with Pauli I went to dine with him and Stern at a restaurant, contrary to the routine.»

«Hamburg, 13 December 1925.
My scientific work until now has mainly been in reading. Yesterday I came across a rather interesting question which may perhaps be easily answered. This would amount to another small paper and thus prove my stay in Hamburg was of use. As I put it in a letter to you the day before yesterday, I'm going to leave for Nice for the Christmas holidays around the 22nd–23rd. Both Stern and Pauli are leaving anyway for this period; whether I shall again return to Hamburg or not, I don't know at the moment. I asked Ehrenfest about it and am going to follow his

7) We have described this article earlier (see Ch. 2).

advice. In any case, I shall return to Berlin no later than mid-February to work there with Einstein for some time.

As I wrote to you many times, I'm doing only physics here and a little music. I read almost no papers. So I practically have no encounters with German living conditions, trends, hopes, the German economy and ideology in general, because I have no acquaintances besides Pauli and Stern. With the latter, I speak little and almost only about scientific subjects. I have an impression that German professors are not interested in politics. Political movements are broadly developed among students, and the right are dominating. For instance, recently the students of some department of the Hamburg University decided not to elect a Jew Dean ... Ideologically, Germany is too diverse for it to be encapsuled within it inside one formula. One day I saw, for instance, a signboard «... supplier of the Court of His Emperor's and King's Majesty» in one street and a poster «vote for communists» in the other; neither this nor that is torn off or painted over.»

«Hamburg, 17 December 1925.
Today I received a letter from Ehrenfest and Einstein, who is visiting him now. It happens that they had a discussion about where it was better for me to work and came to the conclusion that the best place is Hamburg (until Born comes). Einstein does not at all refuse to talk with me about the subjects I'm interested in but considers me as being able to visit him in Berlin while working in Hamburg. I find this issue to be correct, because in Berlin I had nobody except Einstein, while here I have just established good relations with Pauli, Stern and G. Wentzel,[8] and I think that I'll find my further stay in Hamburg very satisfying. Meanwhile, encouraged by Pauli, I have accomplished one study which seems to be rather good and which I'm going to convert into a new paper in Nice.»[9]

«Nice, 26 December 1925.
... Two years ago I would have hardly believed that I would have the chance to visit Nice. However, by a whim of destiny, I am here. Arrived the day before yesterday. It is so warm here that I can go walking without a coat. It is just as well, since my coat is quite indecent, especially for Nice. By all means, I'll buy a new one some day now. As I've told you in my letter from Paris, prices in France are fantastically low. From the minute I crossed the border my status changed from that of a poor student (as in Germany) into that of an American billionaire, and on the basis of this the first evening I visited various cafes where I successively drank tea, chocolate and coffee. During this short period (I mean my trip from Hamburg to Nice) I had so many different impressions which cannot be explained in a single letter. I'll probably be a most interesting person to talk to back in Petrograd.»

---

8) G. Wentzel later became distinguished through his studies in quantum electrodynamics.

9) Frenkel speaks here about the paper on the electron spin («Electrodynamics of a rotating electron»). It was finished in Nice in April, 1926, and ended with acknowledgements to Pauli and Langevin.

«Hamburg, 11 January 1926.
... I spent 3 days in Paris. Visited Prof. Trowbridge, one of the most prominent members of Rockefeller's committee, whose scholarship I now hold, and the director of the Paris office of this institution. I arrived exactly on time, because just at that moment Trowbridge was trying to find my address. He made me very welcome and in about an hour the talk was reflecting his interest in the situation of science and scientists in Russia. He expressed his regret concerning this situation in France and Italy, where governments care very little about science. In conclusion he asked whether my scholarship was sufficient, to which I replied in the affirmative. In the evening I went to the College de France to see Brillouin[10], with whom I had established a very good relationship, and there met Ehrenfest's daughter (last year she lived in Leningrad). Together with her and a friend, we went first to a restaurant and then to the cinema. Finally, yesterday morning, I visited one young French physicist, de Broglie, (whose papers were of great interest to me) and then looked over the Luxemburg Museum and liked it very much. After that I went again to see Brillouin, and then on to see Metal'nikov.»[11]

«Hamburg, 17 January 1926.
... In the morning I usually go to the Institute of Physics and return home at 3 p.m. where I stay until the next morning, with the exception of Tuesdays and Thursdays when there are evening meetings (5 to 7 p.m.) at the institute, like at our Röntgen Institute. I am not very satisfied with the productivity of my work. Yesterday I started to translate into German my «Electrodynamics», however, I'm not going to spend more than 2–3 hours a day on this matter, so as not to hinder my scientific studies. Every day after evening tea I play music for about an hour and a half but, to the distress of my kind hosts, I am learning pieces and not just playing melodies.

If one adds to this that every day I receive letters from my wife and, for my part, also write to her, one can get a full enough impression of my uncomplicated Hamburg life.

In Paris, as I informed you in the previous letter, I visited the Louvre (enroute to Nice) and the Luxemburg Palace (on the way back). The grandiosity of the Louvre is striking: it is several times larger than the Hermitage. As for my impressions of the Louvre, the most vivid is that of the extreme disenchantment with Leonardo da Vinci. His paintings — and there are quite a number of them in the Louvre — are worse than their reproductions: they are monotonous and not at all picturesque. In particular, the impression made by the «Gioconda» is as if it is painted in only two colours, black and yellow. Rubens is exhibited abundantly

10) Leon Brillouin, French physicist, theoretician, who worked in the field of solid state quantum physics (the bands in the energy spectrum of solids are known under his name), information theory, etc. In 1941 he emigrated to the USA.

11) S.I. Metal'nikov was a biologist, lecturing in Tavricheskiy University (in the Crimea) from 1918–20, and then emigrated to France. He worked at the Pasteur Institute.

enough in the Louvre, but I don't like him. Generally, my impression of the Louvre is similar to that of the Hermitage. The Luxemburg Museum fully corresponds to our Russian Museum; there I saw a series of very good paintings, already partly known through reproductions, and it would give me greater pleasure to return there than to the Louvre. There are virtually no «ultramodern» paintings in the Luxemburg Museum . . .

I've decided to put off my musical programme until the hosts go to bed. I play this «music» daily, in a most precise way spending one and a half hours on it at a time. (By the way, I am learning the Glazunov concerto.) I seem to be having more success in this field than in physics. Incidentally, I use one and the same method in both fields, and pursue the same goal: everything should be precise, accurate and clear. I am now most decisively eliminating a habit being satisfied by a superficial and hazy knowledge I recently acquired, which is due to either perhaps fatigue or some other reasons.»

«Hamburg, 24 January 1926.
My dears! Thank you very much for the letters I received yesterday and this morning. It gave me great pleasure to read them. In particular and with a sense of satisfaction, I absorbed the information about my confirmation to the rank of professor . . .[12]

I've absolutely plunged into physics and only written communications and music distract me from this. I should, however, confess that such an ascetic way of life is the result not only of a thirst for knowledge, but, in part, of the lack of financial resources as well.

Yesterday I started work on my new article which appears to be interesting. The corresponding study has just been successfully completed. My stay in Hamburg, i.e., scientific communication with Pauli and partly with Stern, as well as my enhanced studies have began to influence and increase the growth rate of my knowledge and broaden my horizons.

I give rather little time to my «Electrodynamics», or rather, to its translation and updating. No wish to spend too much time and energy on it.

I've recently started to exchange anecdotes with Pauli (he is also very fond of this genre). Thus I plan to widen my knowledge in this area as well.»

«Hamburg, 1 February 1926.
. . . All through the week, a little longer in fact, I've been sitting over my paper; «giving birth» was hard, but then «the child» proved to be of a tremendous weight (45 pages!); however, qualitatively speaking, there is nothing extraordinary about it.[13] For me though, the important thing is that this work justifies my stay in Hamburg to some extent. Before I leave, that is, within the next month and a

12) An extremely flattering presentation marking the title of professor was composed by a committee of A.F. Ioffe, Yu.A. Krutkov and V.R. Bursian after J.I. had departed to Germany (See Archives of the A.F. Ioffe Physical-Technical Institute, personal file # 548, pp. 6–12 (in Russian)).

half, I hope to produce at least one more paper and then to help advance on the translation (with modifications) of my «Electrodynamics».

This morning I found only 3 pfennigs in my pocket, with which sum I survived until 3 p.m. when I was expecting Rockefeller's money. But it hasn't arrived, despite Trowbridge's promise that it wouldn't be late. Meanwhile my stomach began to cry «help» so loudly that I was incapable of absorbing any spiritual nourishment. So I went to a colleague of mine and borrowed from him 50 marks, and, with a sense of joy and also to recompense myself for the distress which one feels after having had only 3 pfennigs, I went to a good restaurant and had a solid dinner. My limited resources do not affect my studies at all; on the contrary, it's a conducive factor since nothing distracts me from my work. The only thing I regret is the impossibility, under the described circumstances, of getting closer to my comrades in science. All of them — Stern, Pauli, Wentzel, Minkowski — are bachelors (very nice fellows) and besides the institute, meet only in restaurants, cafes, cinemas and similar places which I do not visit. Well, I'll write «Electrodynamics» and then will go off into such a drinking-bout that one's heart will ache (whose?).

Hamburg as a free city, along with the federal institutions and the corresponding black-red-yellow flag of the republic, also decorates its own municipal institutions with a ... red flag, though not with a hammer and sickle, but with a building symbolizing the Hamburg harbour. By the way, I'd like to go there again (by a small steamship); the harbour looks majestic and very tempting to a person who is inclined towards vagrancy (such as me, who also has inclinations of the opposite nature; incidentally everyone proves to be a Tartarin from a Tarascon of his own kind).

I spent an hour and a half daily «strumming» the violin (i.e., I play Bach sonatas and also some other pieces, not excluding etudes) ... Due to a good school I feel in no need of any guidance.»

«Hamburg, 7 February 1926.
... Aside from the expectation of money (also a pastime), I spent last week clarifying some difficult questions related to my studies, rummaging through literature and grounding the milestones of the next paper, but all rather slowly. Hope to make up for lost time during the coming week.

Here in Hamburg time is slipping past undetected due to the extreme monotony of my life. Never had time to look around, and already a whole month has passed since I returned. I have used this time not so badly, but could have done much better. However, a lot of time and energy went into my gigantic paper.[14] By the way, it proved to be almost perfect from a literary point of view, so

13) The paper about the Faraday effect (i.e., about the rotation of the light polarization plane in the magnetic field).

14) See the letter of January 24, 1926.

despite everything I'm making some progress in German, though I have very little conversational practice.

Once during dinner at a restaurant, after the regular Tuesday colloquium, I asked Stern why all — or almost all — Hamburg physicists were bachelors. And he answered that not only Hamburg physicists but also other specialists at Hamburg University did not enjoy family life. And that the reason lay in the large quantity and high quality of Hamburg restaurants. There existed a theorem established by Ehrenfest which read: «every man dines in restaurants until it bores him; then he marries», and as the menus at Hamburg restaurants did not bore, then ... and so on. Taking into account my infrequent visits to restaurants, one has to assume that with such a life in Hamburg as I have now, I would remain a bachelor till the end of the century.

Two days ago I found out that among other scientific disciplines in Hamburg there also exists astrology. Namely, walking along my Grindellallee (it is not a garden path but just an ordinary and very noisy street) I noticed on a neighbour's house a marble plaque with the inscription: «A. Dauhrer. Astrologischwissenschaftliche Beratung».»[15]

«Hamburg, 15 February 1926.
My dears! I am returning today at 5 p.m. to Hamburg after a refreshing two-day trip to Braunschweig, a small but peculiar town with a variety of gnarled two- and three-story buildings constructed back in the XVII century and which have not been in a good state since then. I've seen a lot of new people whom I had known before only by name. I won't name the names since even to you they mean almost nothing. The congress was not large, approximately 75–100 people. There were several very interesting presentations. I spent both days very pleasantly. On the first day a friendly get-together of congressmen took place in the grand hall of one of the hotels. I met, by the way, Kondrat'ev and his wife[16] who came from Göttingen. We were very pleased to meet each other. I also met some Berlin acquaintances, among them the 1000-pound Geheimer Regierungsrat Doctor Karl Schell, the editor of the same journal («Zeitschrift für Physik») where I'm publishing my German articles. He lost a quarter of his original weight after reading and correcting my last paper.

The next day the congress sat until 2 p.m. Each presentation took 15–20 minutes plus discussion. But more than 20 reports were given in $1\,{}^1\!/_2$ days.»

«Hamburg, 20 February 1926.
... P. Debye from Zürich,[17] one of the most prominent theoretical physicists will soon be coming to Hamburg. He will give a lecture at the local colloquium on

---

15) Scientific-astrological advice.

16) Viktor Nikolaevich and Yelena Ivanovna.

17) P. Debye is a famous physicist, the author of classical papers on the thermal capacity theory, electrolyte theory, etc.

Thursday on the occasion of its anniversary. On Wednesday at the apartment of Prof. Koch, director of the Hamburg Institute of Physics a party to mark this occasion, i.e., in honour of P. Debye will be given, and Herr Professor Doctor J. Frenkel is also invited. «Now you have a chance to really make use of your dinner-jacket!» — you, dear father, will probably exclaim while reading this. Nothing of the kind in fact! In the invitation cards it is stated that a regular suit is acceptable. (Gefl. Strassenanzug).»

«Hamburg, 25 February 1926.
... Yesterday I was at Prof. Koch's party, along with 25 other invited physicists, chemists and mathematicians. Everything was very nice, simple and pleasant (and also tasty; but due to the lack of either thirst or appetite I paid very little attention to these things). I got to know Debye, who was the hero of the evening and is one of the few stars of contemporary theoretical physics. I liked him very much for his spontaneity and simple manners. He is approximately the same age as Abram Fyodorovich but he appears much younger. And his face is rather plain, as, I imagine, goes for all Dutch people.»

«Hamburg, 28 February 1926.
... Last week I began to work more and in a more productive way than in the previous two weeks. Some new ideas even started to toss and turn in my head, though they have not yet been implemented in anything specific.

Yesterday I wrote a letter to Ehrenfest in Leiden in which I asked him to specify the date of my visit to Leiden ... To tell you the truth, I think that my sojourn in Leiden with Ehrenfest as a constant companion and tutor would be much more useful for me than sitting here in Hamburg where I'm living in almost complete seclusion, meeting only with Pauli a couple of times a week. But I wouldn't like to contradict Ehrenfest anyway, letting him arrange my life as he deems it proper. One may ask — why stay in Hamburg until Easter? Partly to enhance the work on «Electrodynamics» (Ehrenfest will not let me do it). And then it is necessary, at least before the end, to get to know my Hamburg colleagues more closely. When I have more money, I'll be able to meet with them more often, for instance while dining at a restaurant.

I have visited the local Trade Mission a couple of times [The Soviet one — V.F.]. I'm made very welcome there; they suggested I visit their club and listen to the radio concerts from Moscow and Leningrad. I think I'll use this invitation some day soon. Otherwise I'll turn completely sour in my seclusion. Besides, I'm going to get acquainted with the local cinemas. This acquaintance could not be made until now due to reasons beyond my control.

I wake up late as before; usually I'm ready by 10 a.m. And later — later I have no definite programme. Sometimes I go to the institute, sometimes I cope with electrodynamics or write letters, or go for a walk musing over some question or other. Maybe my studies would be more successful if I introduced some sequentiality or systematic approach to them. But this is probably beyond my capabilities.

Usually after 7 p.m. I sit at home and, after tea, play music.»

«Hamburg, 8 March 1926.
... On Monday, though completely unplanned I went to Milstein's concert (there was, as it happened, my favourite Grieg sonata G-dur in the programme) organized by our Trade Mission for the local Russian community. It was a great pleasure. They asked me to give a lecture in my field at the Trade Mission Club, which I will do on Monday evening. However, I have still not chosen any definite topic. I have been working rather intensively in recent days; I am zealously pushing forward with «Electrodynamics».»

«Hamburg, 8 March 1926.
... Today I received Ehrenfest's reply to my letter. He writes he feels very tired and unable to work, so he advised me, since I want to use our meeting for scientific work, to postpone my trip to Leiden by several months. Thus, I shall go from Nice directly to Göttingen and from there, at the end of the summer semester, shall move on to Leiden and England. Incidentally, all this is still far away!»

«Hamburg, 11 March 1926.
... Yesterday I gave a «religious-philosophical» lecture at the Club of our Trade Mission: «On the Ether, God and the Theory of Relativity». One of the listeners asked me: «Why do you deny the existence of ether? It is actually on sale in any pharmacy.» Judging by the harmonious laugh I heard, I can conclude that the audience was more prepared than this listener. I think the public was interested in my lecture.

I am expecting Krutkov from Göttingen any day. To my mind it will be both a pleasant and useful experience for us, living together for several days. There are no theoreticians in Göttingen at the moment, so there is nothing to do there for Krutkov. Recently I have been meeting with Pauli more frequently. I would like to write one more paper before leaving Hamburg. Whether I shall manage to I don't know.»

«Hamburg, 14 March 1926.
My dears! Since yesterday I have had a guest, Yu.A. Krutkov, whom my hosts have put up in the room next to mine. I am very glad he is here; we are spending our time together very pleasantly and usefully. Till 8 p.m. yesterday we read a new paper of Heisenberg's on quantum mechanics, commented on it and partly developed it further. This morning, as it was of Sunday and the weather was good, we went to visit Hagenbeck. There's no sense in describing the Hagenbeck Tiergarten[18] to you ...

18) A zoo in Hamburg founded by K. Hagenbeck.

Krutkov is a very nice and pleasant fellow; in one day during our get-together in Hamburg I have got to know him better than in 5 years in Leningrad. Pity he had not thought of coming here earlier. However, I could have gone to Hamburg just as successfully. From the middle of April until the beginning of August, we shall both work in Göttingen anyway; probably together a great deal; we both are a little bit lazy, and collectivism is the best antidote in such a case. Besides, a joint discussion of various subjects often is an absolutely necessary addition to personal thoughts on them.

... I am glad to be able to inform you that yesterday I resolved one of the «mysteries of nature», as the male part of my parents likes to say, namely, the question about the so-called «elasticity limit» of solids. Abram Fyodorovich was very interested in it, studied it experimentally and managed to get 500 roubles from the Science and Technology Society to pay for my theoretical investigations. But then, 2 years ago, when nothing came out of it, I, having already taken 100 roubles, refused to accept the remaining sum. Today, that is tonight, after speaking to one physicist who has also held experiments in this field, I seem to have solved this problem, moreover, not seem, but have solved. I have already had a talk about this with Stern who found my arguments quite correct (but was frightened by such a large amount of formulae).»

«Hamburg, 17 April 1926.
My dears! In two hours I'm leaving Hamburg. Yesterday I finally got a French visa (and a Belgian transit one); tonight I will be in Berlin, will definitely spend three days there, then I'll leave for Paris, shall be there not more than a couple of days, and finally come to Nice. As regards my scientific studies I can say that my stay in Hamburg was completed brilliantly.

I've written two papers in the last one and a half weeks; I have already written to you about one of them;[19] the second I more or less finished (with Pauli's help) only yesterday, and today I sent a preliminary communication about it to the English journal «Nature».[20] I definitely consider both articles very good.

We have enjoyed very good weather in Hamburg for two weeks now and the last two days it has been just marvelous. Due to the good weather, good papers and good relations between myself and Pauli and other colleagues, I'm rather sorry to be leaving Hamburg. Will Göttingen be better? — We'll see!»

«Nice, 17 April 1926.
... My departure from Nice is planned for Wednesday. There is no possibility for me to stay here any more. In Paris I'm going to meet Langevin and discuss one of my recent papers in more detail than last time.[21] I have just received the proofs of the preliminary communication concerning this, and three days ago I

19) «On the theory of the elasticity limit and the strength of crystal bodies» (Zeitschrift für Physik, Bd. 37, S. 572, 1926).

20) About a rotating electron.

discovered that the theory presented there does not exhaust the problem despite Pauli's opinion and my original viewpoint. However, I won't go into detail. I last saw Langevin very briefly. He has just come back from a small tour of France devoted to antifascist propaganda. He is an extremely nice man and one of the most enlightened representatives of the contemporary French intelligentsia.

Mme Curie's Institute did not impress me much. Generally, in France the matter of scientific institutions is quite sour. However, this year some trend may be observed of the increase of financial support for scientific research and institutions (at the universities; others are nonexistent).»

«Paris, 23 April 1926.
... Again I am in Paris, for the fourth time. Today I met one French physicist, E. Cotton, previously unknown to me, and spent two hours with Brillouin at the laboratory of the College de France where we talked science and drank tea.»

«Paris, 26 April 1926.
... The day before yesterday I breakfasted at Langevin's, and visited him again yesterday to speak about one of my articles which he rather approves of. Langevin and his household made me very welcome. Today I'm going to the plant where a device for underwater signaling, designed by Langevin, is manufactured. In the evening I shall be at the Sorbonne with F. Perrin for the regular meeting of Paris physicists. In a word, as you see, I am not wasting time but using it to see people and to show myself off.»

An outstanding French physicist, Paul Langevin (1872–1946), about whom J.I. writes, belonged in the mid-twenties to a senior generation of physicists. A student of Pierre Curie, Marcel Brillouin and J.J. Thomson, he entered the history of science due to his classical studies on dia- and paramagnetism, gas discharge, piezoelectricity and hydroacoustics. He was a recognized head of French physics and F. Joliot-Curie among others, was his direct pupil. For Langevin-the-teacher, whose motto was: «better to understand than to know», his exceptional generosity was characterized by his sharing of ideas with students and coworkers. This is a mark of broad and really gifted personalities. «Such authorities as Einstein and Sommerfeld», testifies O.A. Starosel'skaya-Nikitina in her book on Langevin,[22] «have indicated more than once Langevin's absolute lack of interest in establishing his priority as an author.»

Langevin was also a very pleasant and charming person; involving the people who surrounded him into his circle was an additional source of attraction for him.

At the grand meeting dedicated to the election of Langevin to the French Academy of Sciences (1934), his student, a member of the Academy George

21) In a postcard from Paris, of March 20, 1926, J.I. wrote: «This morning I was honoured to be promoted into the company of Einstein, Langevin, and Mme Curie at her institute, and even gave them a brief account of my latest paper».

22) O.A. Starosel'skaya-Nikitina. Paul Langevin. Fizmatgiz, Moscow, 1962, p. 169, (in Russian).

Claude, addressed a welcoming speech to him. Having touched on the personal features of his teacher, G. Claude noted that the idealization of people was Langevin's drawback, that Langevin pictured mankind as it should have been rather than it regretfully was.

If one thinks over these words of Claude,[23] his reproach turns into high praise which faith in Humanity and the Good deserves. Langevin loved people, he always came to the defense of the persecuted and the oppressed, against the oppressors. His love for people incited his hate for fascism and war.

In the summer of 1928 Langevin visited the USSR, staying in Leningrad at the Physical-Technical Institute. On the agenda were visits to the Hermitage and the Russian Museum. The inscriptions made by Langevin and Frenkel have remained in the Book of Honored Visitors of the Russian Museum.

In 1945, F. Joliot-Curie flew to the jubilee session of the Academy of Science of the USSR. Langevin, who was to have been among the most respected guests, was unable to make use of the invitation. Years spent in prison in France during the Occupation had undermined his health. Via Joliot-Curie, Langevin sent Frenkel and several other Soviet physicists a book published in honour of his 73rd birthday which coincided with the Day of Victory over fascism (the birthday was celebrated in Paris). Langevin made an inscription in the book (June 11, 1946): «To my dear Frenkel in memory of our last meeting in Leningrad and hoping that we will meet and talk again soon.» Along with the book, Joliot-Curie brought Frenkel Langevin's photograph which since then has hung over J.I.'s desk, alongside with Einstein's portrait.

On 27 February, 1947, at the joint meeting of the Physics and Mathematics Department of the USSR Academy of Sciences, the Institute of Physics of the Moscow State University and the P.N. Lebedev Physical Institute, J.I. gave the speech «Paul Langevin as a theoretical physicist» and related his impressions of him.[24]

From France J.I. returned to Göttingen, which was probably the headquarters of physics in the mid-twenties; the development of new quantum mechanics in these years is not called «the Göttingen period in the history of physics» for nothing. A lot has been written about this period; the Göttingen letters of J.I. all add to the general picture.

Max Born was the central figure on the Göttingen physics horizon at that time, and Einstein and Ehrenfest had made a good choice when they advised J.I. to continue working specifically with him. Pauli, Heisenberg, Fermi, Dirac, Wigner, Weisskopf, Oppenheimer, Heitler, Jordan, Hund, Maria Goeppert-Mayer were Born's co-workers, — all these names mean a lot to any physicist, and almost all are known to non-physicists as well.

---

23) This excerpt from the G. Claude speech may be found in the op. cit. of O.A. Starosel'skaya-Nikitina.

24) Vestnik Akad. Nauk SSSR, # 5, pp. 44–45, 1947 (in Russian).

When one ponders why certain scientists, leaders of schools, exert a strong attraction on other young scientists, one inevitably arrives at one and the same combination of characteristics: talent and high personal qualities. Such, among others, were Rutherford and Bohr, Sommerfeld, Born and Langevin abroad, and in Russia Pyotr Nikolaevich Lebedev, Pyotr Petrovich Lazarev, Abram Fyodorovich Ioffe, Dmitriy Sergeievich Rozhdestvensky, Leonid Isaakovich Mandelstam, Sergei Ivanovich Vavilov, Jakov Il'ich Frenkel, Igor Evgen'evich Tamm and Lev Davidovich Landau.

This is why, when one starts to characterize such a scientist, trying to «pack» everything into a set number of lines, a depressingly monotonous pattern emerges. It is possible, though, to console oneself with the fact that the whole infinity of nature is also «packed» into a set number of dry and invariable laws, ordering and systematizing it to a certain extent. But it cannot be said that these laws impoverish and rob nature!

This standard scheme is brilliantly illustrated by Max Born. «The secret of his success», states Soviet physicist and theoretician Yu. Rumer, who also worked with Born, «was in the extraordinary broadness of his personality, in the combination of the talent of a great scientist with the warm heart of a very good man»[25] — it is a description with which J.I. and all those who knew Max Born would doubtlessly be in agreement.

Max Born started as a student of the Breslau University where he studied, in particular, matrix calculus under mathematician Rosanes. A profound knowledge of the matrix theory later gave him a hand in the creation of the matrix form of Heisenberg's quantum mechanics. Heisenberg, being ignorant of matrices, had tried to develop his own corresponding theory. At the turn of the century, in Germany students often wandered from one university to another, enticed by the glory of this or that lecturer. In Zürich Born studied mathematics under the famous Hurwitz, in Göttingen — under Hilbert and Minkowski, and astronomy under Schwarzschild.

Born became Professor at Göttingen in 1921. Yu. Rumer reflects that Born used to speak about his method of work: «first start calculating, then think». This statement of Born should, of course, not be taken seriously, characteristic though it is for his method of working: he did not like to speculate «on fingers» (in contrast, as it happens, to Ehrenfest), and «mathematics was always the via regia[26] that led him to the revelation of the mysteries of nature.»[27]

In Göttingen, where Born spent 12 years until the fascists came to power, he wrote all of his most well known memoirs on quantum mechanics, and it was there that he invented the concept of probability waves relating to the square amplitude of the Schrödinger wave function ψ. This work allows one to consider

25) Yu. Rumer. Uspekhi Fizicheskikh Nauk, v. 78, 1962, p. 695 (in Russian).

26) «King's road» (Lat).

27) See footnote 25.

its author, together with de Broglie, Heisenberg, Schrödinger and Dirac, as one of the founders of quantum mechanics.

The years of exile Born spent in England as the Head of the Theoretical Physics Department at the Edinburgh University until 1953, when he returned to West Germany. He settled in a small resort area, Bad Piermont, not far from Göttingen where he was made honorary citizen in 1953. Born lived there until the end of his days (1970). In his final years he deliberately played no part in the active work of theoretical physics and gave all his attention to the history of science. We can enjoy the fruits of this remarkable work by reading his numerous articles published in Russia in the «Soviet Physics Successes» and in special collections.

As an epigraph to the Göttingen letters of J.I., in which Born is mentioned frequently and with invariable warmth, one may take the words of a small note (dated October 10, 1961) sent by Born to the editorial board of the «Soviet Physics Successes» for the issue dedicated to the memory of J.I.[28] They read: «Frenkel worked with me in Göttingen in the twenties, and I rated him highly. He was not only filled with ideas but also had a perfect command of all the techniques of theoretical physics. Though his ideas were sometimes rather strange, they were always fruitful. I always read his papers and books with great interest. I am proud to be able to count him among my friends and scientific colleagues.»

Let us proceed now with the J.I. letters (to his parents).

«Göttingen, 28 April 1926.
... So I am in Göttingen. I like this town very much. Five minutes from my home there begins a wonderful huge park or maybe I should say forest, rising up over a gentle slope on one of the hills surrounding Göttingen. The central part of the town is surrounded by a bank where at one time the town wall was, and splendid trees now grow here which turn this earthen wall into a pure garden path.

I have not seen Born today because he is not well. But nevertheless I have been to the Institute of Physics where I received the papers and a postcard from you, Dad, and also a letter from Pauli about my paper. The latter seems in order, but virtually the same has been done simultaneously and independently by one of Bohr's coworkers (Pauli is there at the moment). Because of this I'm going to finish my article first thing tomorrow and forward it for publication (pity I hadn't done it earlier).

Today I only met with one of the Göttingen physicists, J. Franck, whom I had gotten to know at the Congress in Braunschweig. The summer semester has not yet begun but is starting any day now. By the way, Born is going to give a very interesting special course (on quantum mechanics) which I am going to attend together with Krutkov.

In Paris on Monday (the day of my departure) I visited Perrin and met there a number of young French physicists. Perrin himself (Jean Perrin) is a golden oldie

---

28) M. Born. Jacov Frenkel. Uspekhi Fizicheskikh Nauk, 1962, v. 76, # 3, p. 431 (in Russian). See also: J.I. Frenkel. Memoirs, Letters, Documents. Leningrad: Nauka, 1986, p. 76 (in Russian).

with curls as white as snow, and radiates goodwill all around. The atmosphere in his laboratory is a real family one. They made me very welcome.»

«Göttingen, 1 May 1926.
... I'm still enraptured by Göttingen. I haven't lived in such a small town for a long time (since Zoppot),[29] and the contrast with big cities is brightly pronounced in a very attractive light. Besides that, the Göttingen environs are absolutely fascinating. A half-hour walk, and even less sometimes, and you find yourself outside the town, on the hills covered with linden trees and other which fill the air with the delicious fragrance of spring blossom. From the hills, on which the inevitable Bismarkturm or Bismarkstein[30] tower, a wonderful view of the town opens up, abundant with greenery (except the small central part). In the town itself there are neither trams nor automobiles (almost), only cyclists and pedestrians, at least half of them students in gaily coloured caps of the various societies and «Burschenschaften».

I have settled in here magnificently. I'm going to get a bicycle and, along with physics, take up cycling. You need not worry, nobody will run into me here. Though maybe I'll run into somebody.»

«Göttingen, 4 May 1926.
... I'm getting on well. Yesterday I finished off the second paper and am now able to start on new studies. Today I finally met Born. He is over 40 but very young in appearance, not very tall, lean, clean shaven, with greying hair and sky-blue eyes. He is a very simple man and cheerful in conversation. In a word, I like my patron as much as I like Göttingen.»

«Göttingen, 9 May 1926.
... The day before yesterday I again turned to my «Electrodynamics». It is always difficult to start work on it and after a month and a half's break it was a struggle to begin threading new pages over the completed hundred and fifty. When it will be finished, it's difficult to say. I'll be glad if it is completed by July, because I do not plan to let electrodynamics interfere with my other studies.

I've already told you about my impression of Born. Lately I have been able to characterize him more precisely: in my (not so bright) imagination I saw Born as quite a serious man; in reality he happened to be not at all «serious». I visited him at home two days ago and spoke with him about various scientific subjects. I met his wife, still quite a young woman; I have also seen his son, a boy of some seven years in an Indian outfit obviously imported from America. Krutkov and I had been invited by the Borns to visit them as guests tomorrow evening; but due to a concert this invitation has been postponed until a so far unspecified date. Born and his wife seem to be very musical, since in an enormous room which serves both as Born's study and a guest-room there are two grand pianos side by side.

---

29) In Zoppot (Sopot), a resort town on the North Sea coast, J.I. lived in 1913.

30) Meaning: «Towers of Bismark» or «Stones of Bismark».

The most talented of Born's assistants, Heisenberg, has moved to Copenhagen, to Bohr. In Göttingen there remains Jordan, also a very gifted fellow, and Herr Doctor Hund[31] (what a dreadful name! No prefixes can soften it). There are also several Americans in Göttingen who would like to be enlightened by the new revelations irradiated by Born.[32] Unfortunately, he irradiates them at such a speed that they elude the majority of the audience. Despite this, though, the audience consisting mainly of German students stamp their feet (such stamping replaces applause in Germany), but it is still thinning.

The relations between students and professors, even Geheimrats,[33] are usually much simpler than I had imagined.

I am still almost ignorant regarding the intrinsic life of the local studentship. It is completely divided into societies and «Burschenschaften»[34] whose principal objective is, as it was long long ago, the collective drinking of beer, singing and, of course, Mensur[35], i.e., the decoration of faces by scars.»

«Göttingen, 15 May 1926.
... Yesterday at the theoretical seminar chaired by Born and the famous mathematician Hilbert I presented a small report of one of my Hamburg papers. It was revealed that I can speak fluent German if necessary, though with a pronunciation which differs drastically from the usual one, and with a systematic confusion of genders. I consider myself innocent in the latter and lay the responsibility for an unreasonable appropriation of masculine and female genders by inanimate subjects on the German themselves.

On Wednesday I visited Born for the second time. I found him as pleasant as before. I informed him of the programme of my studies (including also electrodynamics) and gave him some «promissory notes» associated with at least the thinking over, if not the resolution, of some new questions. Theoretical physics is now going through an extraordinarily violent development. The topic of the day is an attempt to bring closer mechanical phenomena, the motion of material particles in atoms, with the optical ones. In the doctrine of light one usually operates with the notion of *waves* which may be substituted, in a sense, by the equivalent concept of light *rays* (i.e., the lines of propagation of the waves). It happens that the *paths* of material particles may be compared with light rays; one is able to consider, respectively, «material particles» only as a schematic representation to which waves of a special kind correspond in reality. The latter differ from light waves but are closely related to them.

---

31) Hund is German for «dog».

32) Among them, in particular, was R. Oppenheimer.

33) A privy councillor in Germany.

34) Fraternities (German).

35) Duel (German).

I started to write a paper with Krutkov, a rather extended one-percent popular article concerning new quantum mechanics developed recently in Göttingen and the associated new theories mentioned above. The paper is aimed at the Moscow journal «Uspekhi Fizicheskikh Nauk.»[36]

I am living very peacefully with Krutkov, though bickering regularly in a lighthearted manner. He is a very good fellow and reminds me somewhat of Nikolay Mitrofanovich [Krylov. — V.F.] as a person to speak to, though he differs from him in his «quietness». S. Vavilov, whom I've mentioned at the beginning of the letter, is also a very pleasant man.»

«Göttingen, 18 May 1926.
... Yesterday evening Kapitza arrived; tomorrow he leaves for Cambridge. We have arranged that I will visit him there at the end of July to attend the congress of the British association which opens at the beginning of August in Oxford. I am quite well and studying very intensively today — all day long. Had a battle in chess with Kapitza. We are all invited to Franck's in the evening. I will write to you about my impressions on Thursday. My chums do not anticipate any special fun there and therefore suggest we leave earlier to have a good time at some of the local cafes. The suggestion was accepted. Recently my main problem has been the lack of money. However, I hope it will arrive at the end of the month so I can meet «the day of reckoning» (of payment!) with honour and dignity.»

«Göttingen, 20 May 1926.
... The day before yesterday we, that is all the Russian company, went to Franck's. By the way, I have to note that scientists here live in remarkably comfortable and nicely furnished apartments. However, as Franck says, this is not by far the case with everybody. We spent the evening rather quietly but towards the end the public became excited and joyful over the solution of various puzzles suggested by Kapitza and me. At midnight we, i.e., our company, left for a cafe and drank 4 bottles of wine between the five of us, after which I broke my glass and started displaying a trend to repeat this experiment with other subjects, both animate and inanimate.[37] Then we sensibly went, or rather «retreated», home.

Yesterday Krutkov, Vavilov and myself visited Prof. R. Pohl, and today went to the Borns. Born, Franck and Pohl live together very comfortably, are on first name terms and generally constitute very pleasant company. The photograph attached, which was taken the day before yesterday, before Kapitza's departure will help you to get a more complete impression of the people I have mentioned.»[38]

---

36) «Soviet Physics Successes».

37) One such animate subject, says Kapitza who remembers this evening very well, was a policeman met by the company.

38) This photo is reproduced in: «A.F. Ioffe. Meetings with physicists». Moscow, Fizmatgiz, 1960, p. 76 (in Russian).

«Göttingen, 23 May 1926.
... Worked harder on «Electrodynamics»; have started a new study, perhaps something will come of it. As for the violin, during my stay in Göttingen — and the first month will be over in 5 days — I have played it only twice. Born once suggested we play music together (he plays the piano very well). I have to resume my exercises. Several days ago Krutkov and I were at Born's, the conversation was in English and German, and due to the intervention of two languages I spoke each of them equally badly. Yet sometimes I can speak very well, everyone is quite surprised. From the point of view of the study of German my stay in the Russian community is not so useful. I suggested to Krutkov that he speak German, but nothing came of it.

Due to the fact that I read Russian papers daily I don't feel isolated from Russia in general or from you in particular, as I did in Hamburg.»

«Göttingen, 30 May 1926.
... These days I'm dealing a lot with my «Electrodynamics» to finish it in the beginning of July, or in the middle at the latest. As for new ideas, they are starting to form little by little. I am involved in the pondering over new questions mainly in the course of my evening walks which, one may state, have entered my schedule last week. At 11 p.m. the streets of Göttingen are almost empty, and it is a pure pleasure to walk them. The earth enbankment around the central part of the town is especially beautiful; there was the city wall there at one time, and now it is a high park path with beautiful branchy trees on both sides of it. So, well, when walking along the bank alone, as well as other Göttingen «paths», I think about quanta, the theory of relativity and the ways to unify these presently yet disjunct fields. It seems to me I'm on the right track, and I'm not losing my hope to achieve my goal. However, it is too early to write about this ...

Some day soon Bursian will come to Göttingen, and P. Lukirskii and N. Semenov with his wife will follow. In one word, a Russian or, better to say, Röntgen[39] occupation of this purely German town lies ahead ... Besides, in the middle of June Ehrenfest will arrive with the retinue of his colleagues, a Ceylonese parrot among them which has been taught by Ehrenfest to pronounce the phrase: «Aber, meine Herren, das ist keine Physik».[40] This parrot is recommended by Ehrenfest as a chairman of forthcoming discussions of the new quantum mechanics.»

«Göttingen, 18 June 1926 [to his parents].
From my reflection on the theory of quanta and other — as you, my genial father, used to say — «mysteries of nature», some particular results start to separate. Today I made Born happy with one of them. My apprehensions in regard of his

39) All the persons mentioned worked at the State Physical-Technical Röntgen Institute (later simply Physical-Technical Institute) in Leningrad.

40) «But, gentlemen, that is not physics».

Göttingen, May 1926. From left to right: Yu. Krutkov, Ya. Frenkel, S. Vavilov, M. Born, V. Kondrat'ev, P. Jordan, J. Franck, P. Kapitza

possible complaints to me [because of the work on «Electrodynamics» — V.F.] proved tenuous. The book is coming along successfully. Next Sunday it will be ready.

Today all the [Soviet — V.F.] community goes to a movie and I, motivated by sociability and against the orders of duty («Electrodynamics»!), shall go with them.»

«Göttingen, 23 June 1926.
... I'm still absorbed in my book. Hope to finish it on Saturday. Till then I am not a man but a writing machine. Am writing the book if not from the very morning, then after dinner anyway, and I don't leave it until I go to bed, with small breaks, of course, because one has to freshen up. I put away the manuscript for an hour or two in the hope of recovering my intellectual ability in this period. But soon the book is ready: and then ... Then I'll direct all my energy to the struggle with quanta. I'll rest after my arrival in England, and even better in Nice.»

«Göttingen, 29 June 1926.
My dears! In the end of ends I can say my book is finished. Some remaining trifles don't count. I really may be proud of my work. First of all, so to say, quantitatively. To those 150 pages which had been written I added here 400 more during a month and a half, 300 of them in the last month. As for a qualitative side, I've no doubts that my book has great advantages and small drawbacks. And that is all one could desire. Today I brought my manuscript (weight: 8 kilograms) to Born and explained to him in a very friendly and extended talk first the motives which

had encouraged my literary activity, then the structure and contents of the book. Tomorrow I'll get from him the letter of reference for Springer and the day after I'll move to Berlin for a couple of days. I had feared that a whole month of extremely intense work would tire me greatly. Nothing of the kind! I feel in excellent spirits, in spite of the fact that I have been going to bed very late and waking up rather early the last few days. This gives me hope that during the remaining month of my Göttingen sojourn I'll find time enough to do something appropriate. I've told Born about this. By the way, Born is now developing a remarkably interesting and witty theory which elucidates a lot of questions remaining still unclear.[41] And would you believe the very simple and modest way he speaks about what he is doing! I'm very fond of him, not only as a physicist, but as a person as well. He obviously has a good opinion of me, which I will try to live up to.»

«Berlin, 2 July 1926.
... Springer met me rather genially and, having looked through Born's letter, immediately took hold of the manuscript. And in Born's letter was written: «Having got acquainted with Prof. Frenkel's book, I find it a very good and original work which also appears to be just to the point, since there are no worthy handbooks on electrodynamics at present, except for the obsolete Abraham's monograph. I am sure the treatise of Prof. Frenkel will spread widely. I myself can learn a lot from it.»

So my book is going directly to press, and in September, that is, before I go home, it will be published. Very pleasant!»

«Göttingen, 10 July 1926.
By the way, Ehrenfest's wife[42] has recently arrived in Göttingen. Today, at her request I started to teach her new quantum theories. Though she is a physicist herself and the wife of Ehrenfest at that, she did not follow the development of these theories, the more so since she had spent half a year in Russia from where she returned only recently. The first lesson was a success. Hope that the following ones will be just as good. For me these lessons are also useful because, while explaining these subtle things, I'm forced to pay attention to the questions I had not managed to think over before. T.A. Ehrenfest is a very sweet woman, and it's a pure pleasure to teach her. As for her husband, the possibility of his arrival in the near future is not unreal; however, he is still hesitating. I would like to see him very much. I requested him to arrange a Rockefeller allowance for Tamm and obtained an affirmative reply (after quite a long wait). I'm very glad for Tamm; I instantly wrote to him about it today.

As I've reported, I managed presumably to catch «a big crucian by the tail», some mystery of nature, but, despite all my efforts, I cannot pull him out. I haven't yet lost hope of navigating the matter to a happy ending.»

---

41) J.I. seems to be speaking here about the probability interpretation of the wavefunction.

42) Tat'yana Alekseevna Afanas'eva-Ehrenfest.

«Göttingen, 22 July 1926.
... Yesterday, quite unexpectedly, Ehrenfest arrived to stay for some five days. I was very pleased to meet him, and personally have received a number of compliments from him. Two days ago I completed my private course of «quantum mechanics» for T.A. Ehrenfest. — These days I'm studying very little: too many friends and acquaintances distracting me from work. The proofs of my book have started to arrive in large quantities. I look through them immediately and pass them on to Krutkov and Bursian for further stylistic improvements.

The congress is over. Regarding the number of persons present and the presentations, it was an exact copy of the Braunschweig one. It was not very interesting there, but very hot, for a change. The most pleasant thing was that I met up with many acquaintances, in particular Pauli, Stern and other Hamburg physicists. In the evening of the 17th, after a two-hour session, all the public gathered in a restaurant located on a very picturesque hill, where all drank beer, ate and chatted. Curiously, all the married local physicists and mathematicians came with their wives; it's very characteristic from the point of view of the Göttingen traditions. I reported nothing, since it's impossible, with extremely limited time (15 minutes per «nose»), to say much. And besides, I had nothing new completed and unpublished.»

«Göttingen, 25 July 1926.
My dears! This morning I got an English visa, or, more exactly, a notice from the English Consul in Berlin that I'm permitted to enter the United Kingdom for the period from July 31 to August 31. I immedately sent off my passport to apply for the visa. On Thursday or Friday I plan to be on my way via Hamburg to Bremen.

Almost every day I receive proofs of my «Electrodynamics». Unfortunately, I will have no time to complete the first, most vital, portion of proofs before my departure from Göttingen.

Ehrenfest is here now. He has informed me that he is very pleased with me and that I may rest on my laurels, that is, I can peacefully investigate the problems I'm interested in and have no need to rush to publish my articles, since my reputation has become solid and steady anyway. He will stay in Göttingen for several days and then also go to England.

Yesterday after a long break I took my violin and played a little. In the evening I attended a concert of new music, with the Ehrenfests and Born. This evening I'm going to play with Natasha Semenova[43] Grieg's Second Sonata.»

«Göttingen, 28 July 1926.
... For the last few days I have been working a lot in view of Ehrenfest's arrival; he asked me to calculate something for him and to tell him about some new theories. Ehrenfest seems to promote me widely, perhaps even more than I deserve. He issued for me a type of diploma of the first theoretical physicist of the USSR. —

---

43) Nataliya Nikolaevna Semenova, the wife of N.N. Semenov.

Uspenskii, whom I wrote to you about, is the same mathematician (academician at present) who had read my first mathematical paper of 1910–1911 (The finite difference calculus).»

«Hamburg, 31 July 1926.
My dear parents! I'm writing this on board the steamship «Hermina» which leaves for England in several hours. It's a cargo steamship, and a very small one. There will be constant rolling and pitching. Well, no trouble! I'll get used to it in two days (though it's cold comfort because in two days I'll be in London). There are only 5 or 6 passengers. Probably out of respect for my rank, the captain has provided me with his berth, so I'm set up in the best possible way. I spent half of the day with Pauli — first in a restaurant, then in a canteen and finally in a cafe; we talked exclusively about scientific subjects. I'm brooding on whether to take a holiday from this during the trip or, on the contrary, to deal with it intensively. Probably this will depend on the weather rather than me. No wish to write more. I'm going to walk around the harbour: under the usual circumstances I wouldn't be there, since only those who are departing are allowed onto harbour territory by the police.»

«Oxford, August 1926 [to his parents].
... Have got some time to see London a little. The city is beautiful (except the East End which is grey and dirty); reminds one of Paris rather than Berlin. And Englishmen are more like Frenchmen, not Germans. I've walked the vast expanses of Hyde Park, where people sit or lie on the grass (in Germany grass, i.e., a lawn, is an untouchable, sacred thing); I looked over some streets; sent telegrams both to you and my wife and then, after a shower and relaxation at the hotel, went again for a walk around the city until the rain drove me back. — At midnight London sleeps, as if it weren't a capital but a kind of Göttingen. In Oxford life comes to a standstill even earlier. Restaurants, and they are comparatively scarce, are closed before 10 p.m. Generally, England seems to be mainly a country of family comforts ...

Yesterday the congress was opened by welcoming speeches and the presidential address of His Royal Highness the Prince of Wales who has been, for services unknown, elected to the position of President of the British Association for the Advancement of Science for the current period.

Today I had the pleasure of meeting and shaking hands with a whole group of celebrities, including E. Rutherford. Have also seen old friends — Ehrenfest, Born, Perrin. Then Semenov, Kapitza, (they are both here) and I drove to the suburbs, which are a special pleasure in view of the parquet smoothness of English highways and the picturesque English landscape, a mixture of meadows and groves. All in all, I'm very satisfied with my visit to England.»

«Oxford, August 10, 1926.
... Yesterday after lunch I went, with Kapitza, Semenov, N. Selyakov and several other persons, for a walk along a very small, picturesque river running through

Drawing of N. Semenov by Ya. Frenkel

Oxford. The same day I had the pleasure to behold Bohr himself and exchange several words with him. Generally, as far as concerns beholding, I manage to get very much in these days. But that is all. In fact, the congress lacks any particular interest. Maybe — and even certainly — it would be more interesting to listen to the presentations at the other sections, for instance, the biology, psychology, or anthropology ones. But it's too late now: the congress is practically over.»

«Cambridge, 15 August 1926.
... I'm now dealing with my proofs and tomorrow will send them to Springer. Besides this I did nothing in Cambridge, apart from a lecture given in the circle of local physicists and the ensuing discussion with two of them, as well as the discussion with Ehrenfest (who is also here now), Semenov and Kapitza. I am thinking of describing my impressions (in Nice) in several notes for «Krasnaya Gazeta». England is a very peculiar country. — Cambridge is a small and charming town, like Oxford. At 10 p.m. all life here comes to a complete halt; most of the streets even have no lights. Yesterday at midnight I saw some fellow climbing up a wall to a second floor window. According to Kapitza it was a student gone on a spree; students have to be at home by 10 o'clock in the evening, and to conceal his being late he had chosen a rather inconvenient direct route.»

From J.I.'s letters one can see the results of a year of his intensive work on this scientific journey.

We will dwell especially upon three points related to his scientific activity.

1. Starting as early as in approximately 1923, J.I. began to work on «Electrodynamics». The plan of this book was developed in Leningrad, through a course of lectures at the Physics and Mechanics Department, as well as during direct scientific work in this classical branch of physics which had yet to be touched by quantum mechanics.

As a result, J.I. wrote a short book and, as we know from his letter of September 19, 1924, presented it to GIZ.[44] But the publishing house declined the book at the time, by which J.I. was extremely distressed. V.F. Kagan[45] explained his position by citing the lack of paper in the country for printing handbooks in physics and mathematics, so the publishers could not sign contracts on books to be printed in small quantities. Along with this, the course by J.I., if one judges from its German version since the Russian manuscript is now not available, was difficult and more like a monograph rather than a handbook.

One is able to understand the contemporary situation in electrodynamics in world scientific literature, for instance, through Born's evaluation referred to by J.I. in his letters.

The book was published in 1926 and received a positive response by the critics. «This book,» I.E. Tamm writes about the 1st volume of «Electrodynamics»,

---

44) The State Publishing House (abbreviation in Russian).

45) Scientific Editor of the State Publishing House.

dedicated to the microscopic theory of electricity, «along with the second volume written later, was published both in German and Russian and became one of the basic monographs on the contemporary theory of electricity in world literature. It is enough to say that it was accepted as a standard handbook in a number of European and North American Universities.»[46]

In his «Electrodynamics» J.I. refused to represent the subject in its historical aspect, because it was often contradictory to the logical one, and invented a new plan of the material exposition. Even such a «superclassical» section of the electricity theory as electro- and magnetostatics was marked by this novelty.

J.I.'s course was the first one in which the concept of fictitious magnetic poles was eliminated, and the principle of equivalence between elementary dipoles and electric currents was accepted as a foundation, instead of the Coulomb law. According to this principle, the interaction of dipoles is identified by the interaction of elementary currents (in the limiting case both the former and the latter are treated by J.I. as points characterized by either a dipole moment or a current vector). This representation brought logical grace and beauty to the whole exposition.

The second illusion banished by Frenkel from his handbook was ether. «Ether», J.I. wrote, «has twice played an important role in the development of physics as a fruitful concept. The first time was with Huygens, when it formed the grounds for the theory of optical phenomena; the second time, Maxwell and Faraday exploited it as a bridge between optics and electromagnetism. The time has now come to understand that the historical role of ether is played down to the end, and that it deserves the right to a worthy place in the history of physics. Its introduction even as a supporting construction in the exposition of the modern theory of electricity does not help to understand the subject, but, on the contrary, only helps to obscure it.»[47]

Such a categorical demand of an honourary retirement of ether now, 65 years after it had been written, appears to be quite unnecessary. But in fact J.I. was one of the pioneers who promoted the formulation of this view. It was not for nothing that Philipp Frank wrote: «Professor Frenkel from Leningrad was probably the first to eliminate, in his «Electrodynamics», not only the theory of ether, but, in absolute succession, the whole method of exposition in which the word «ether» is generally used.»[48] But if Ph. Franck welcomed this, a critic from «Nature»[49] wrote about the matter in a sarcastic tone: «The author, as is natural in the atmosphere of Leningrad, seeks to break with the historical development of electricity, and as

---

46) I.E. Tamm. Uspekhi Fizicheskikh Nauk, 1962, v. 76., # 3, p. 411 (in Russian).

47) J.I. Frenkel. Lehrbuch der Electrodynamik, Bd. I, S. VII. Berlin, 1926. (in German).

48) Ph. Frank. Scientia, v. 25, p. 192, 1931. Philipp Frank, Einstein's successor at the Chair of Prague University, a famous physicist-encyclopaedician who, along with R. Mises, published a book which was well-known in the thirties: «Differential and Integral Equations of Mathematical Physics».

49) «Nature», v. 19, # 3006, p. 851, 1927.

a first step discards the conception of the ether, which he regards as obsolete. It is an interesting attempt, but whether it will be fruitful, is another matter.» Time has answered this reviewer's question.

Vital information was also the order of the material presentation chosen by J.I. Before him, macroscopic electrodynamics was explained in a phenomenological way without any relation to the microscopic one. J.I. begins his course with the microscopic theory of electricity and constructs the macroscopic theory on it's basis.

Finally, J.I. saturated his book with the results of the most recent studies of microscopic electrodynamics which had never been published in textbooks before. This is also true regarding his own papers on the electrodynamics of a point wise electron and rotating electron. Therefore this book has retained its meaning until today. I.E. Tamm in the introduction to the posthumous edition of «Electrodynamics» cited above, remarks that the concept of the belated far-range action formulated there holds a certain effect on the development of quantum electrodynamics as well. It concerns, in particular, the famous works of R. Feinman (1954) who refers directly to J.I. Frenkel.[50]

The book received a lively response from critics. Among those who reviewed the 1st volume, one may name Ph. Frank and L. Brillouin, and for the 2nd, which appeared in 1928, G. Breit and P. Jordan.

I.K. Kikoin's reflections concerning «Electrodynamics» are interesting. In 1928 he was in Germany, in Munich. The following quotation is from his article dedicated to J.I.: «When I had worked for some time at the Munich University, where Sommerfeld chaired the Department of Theoretical Physics, I got to know some young doctoral students, i.e., people who had graduated from the university, and found out they were studying Frenkel's «Electrodynamics.» To my question as to what were they doing they replied:

- We are preparing for our exam in electrodynamics held by Sommerfeld.
- But doesn't Sommerfeld have his own handbook of electrodynamics?
- Yes, but Professor Sommerfeld considers the Frenkel book to be presently the best in the world. And he will take the examination by this book.»[51]

2. In his letters J.I. often mentions his two papers on electrodynamics of a rotating electron accomplished in Hamburg and Nice. Let's examine them in more detail.

---

50) I.E. Tamm. In: J.I. Frenkel. Electrodynamics. Collection of Selected Papers, v. I. Acad. of Sci. of the USSR Publ. Moscow-Leningrad, 1956, p. 3 (in Russian). See also: B.G. Kuznetsov. The Evolution of Electrodynamics. Acad. of Sci. of the USSR Publ. Moscow, 1963, pp. 129–132 (in Russian).

51) I.K. Kikoin. In: J.I. Frenkel. Memories, Letters, Documents. Leningrad: Nauka, 1986, p. 66 (in Russian).

The notion of a rotating electron has a very peculiar history.[52]

In the first half of the twenties an energetic comprehension of atomic spectra continued. This branch was jokingly called «zoology» by Einstein. But the «zoological spectroscopy» formed the foundation on which Niels Bohr constructed his quantum theory of atoms, when he combined the spectrum analysis data with the Rutherford model of an atom. The Pauli principle, playing such an enormous role in the whole of modern physics, was also invented by his author in the process of the study of spectroscopic data acquired by a whole army of scientists who specialized in the classification of spectral terms.

Research into the anomalous Zeeman effect led them to the understanding, or more precisely, systematization of a large number of empiric rules (a long row of selection rules, such as the Runge rule, the Preston rule, the Paschen-Back effect). But a complete scheme which would incorporate all the seemingly whimsical and disorderly behaviour of atomic spectral lines in a magnetic field did not yet exist.

The other difficult problem of spectroscopy was the question of the alkaline metal spectra (Li, Na, K, ...) similar in many respects to the hydrogen atom spectrum. The lines in the spectra of these metals were split in doublets — and one could not explain this fact within the framework of the existing concepts. According to the Landé conjecture (this viewpoint was shared also by Bohr), the reason for this phenomenon comes down to the following. Since the positively charged ion of an alkali metal, including the nucleus and all the electrons except the «optical» one situated at the most outer orbit, possesses electrical and magnetic momenta, its interaction with the optical electron leads to the appearance of doublets: the electron energy level happens to depend not only on the so-called principal quantum number, but also on the magnetic quantum number. This effect is called the degeneracy removal. However, experiments demonstrated that the alkali metal ion, coinciding with its «architecture» with the complete structure of the electron shell of inert gas, possesses a high degree of spherical symmetry. This excludes the existence of the momenta which exactly reflect the symmetry break. Following Sommerfeld, Pauli therefore assumes the magnetic (and mechanical) momentum of the ion to equal zero. But then only one possibility remains: an optical electron itself is responsible for the spectral doublets. In his paper in 1924 Pauli assigns to the optical electron «a characteristic quantum ambiguity of quantum properties which cannot be described classically». To provide this ambiguity with a description, Pauli introduces formally one other quantum number for which he does not suggest any «classical model».

In 1925 two Dutch physicists working in Leiden, G. Uhlenbeck and S. Goudsmit, made a conjecture that this new quantum number, called a spin is related to the electron rotation around its axis. Such a concept demanded that its authors overcome enormous resistance, because physicists were used to the point

52) Theoretical Physics of the XXth Century. IL, Moscow, 1962 (in Russian). See also: V.J. Frenkel. Paul Ehrenfest (2nd edition). Moscow: Atomizdat, 1977, pp. 137–145 (in Russian).

model of electron, and this proved very difficult to abandon. As we mentioned earlier, these scientists received great help from Ehrenfest, with whom they worked, and from Einstein, who happened to be in Leiden then (just at this time Ehrenfest and Einstein wrote a joint letter to J.I. which Frenkel mentions in his letter of December 17, 1925).

Uhlenbeck reflects that Ehrenfest became interested in his and Goudsmit's work, after being attracted by a clear origin of the model it rested upon. He sent his students' paper to «Naturwissenschaften» commenting on it with the words: «Both of you are young enough to allow yourselves this stupidity.» (This «stupidity», one should note, immediately made their names widely known!) The paper came out in print in the end of November, 1925.

Ehrenfest talked and wrote about the article by Uhlenbeck and Goudsmit to his colleagues, Lorentz, Heisenberg, and Bohr; they also expressed some skepticism which in the spirit of diplomatic formulation one would call «reserved pessimism».

It is interesting to note that in 1924 the same notion of a rotating electron was put forward by a young American physicist R. Kronig. While in Germany, in Tübingen, he showed it to Pauli, who happened to be there; he did not, however, get any support from the latter and gave up the publication.

If one digs a little «deeper», though, it appears that another American physicist, Arthur Compton, had spoken about a rotating electron even earlier, back in 1921. Compton considered that specific features of some electrons' tracks in Wilson's camera could possibly be explained by ascribing a rotation to an electron, i.e., by treating it as a bullet coming out of the barrel with a thread. (These tracks were observed in the research of a phenomenon called the Compton effect.) Compton further assumed that such a rotation is characteristic not only for free electrons but for the electrons in atoms as well. However, he did not expand on this point. His idea was recalled after the paper of Uhlenbeck and Goudsmit had been published.

The story of the proper electron momentum demonstrates that an idea needs to acquire the rank of scientific revelation to be not only true but also timely.

So, the classical model for the fourth quantum number introduced by Pauli was formulated. According to this, an electron is a sphere of radius r with, uniformly distributed over its surface, a charge equal to the electrical charge «quantum», namely, to the electron charge e, i.e., with surface density $\eta = \frac{e}{4\pi r^2}$. It is assumed that the sphere rotates about its axis with angular velocity ω.

It is quite easy to calculate mechanical and magnetic momenta of such a sphere. True though, to find magnetic momentum, one had to split the electron charge down to «subelectron» parts, which contradicted the very concept of the *elementary* electron charge. This was nothing new for physicists: a similar procedure had been used previously to calculate the electron mass on the grounds of its «electromagnetic origin». Magnetic momentum of a rotating electron proved to be

$$\mu = -\frac{er^2}{3c}\omega,$$

and the mechanical one (determined by the electromagnetic momentum with the density

$$\vec{G} = \frac{1}{4\pi c}\vec{E} \cdot \vec{H}$$

with $\vec{E}$ and $\vec{H}$ being the electric and magnetic fields caused by the charged sphere rotation) was

$$\vec{i} = -\frac{2}{3}\frac{e}{cr}\vec{\mu}.$$

Substituting into this formulae the value of radius r obtained from the formulae for the electromagnetic electron mass,

$$m = \frac{2}{3}\frac{e^2}{r},$$

one finds that

$$\frac{\mu}{i} = \frac{e}{cm}$$

in the agreement with spectroscopic data (a value two times larger than the corresponding figure for the orbit motion of the electron).

However, this coincidence which played, perhaps, a decisive role in the recognition of a rotating electron, holds only for the uniform charge distribution over the electron surface. If one assumes a uniform volume distribution, the coincidence expires. Moreover, if one turns to the known and, multiply checked in experiments, value for the electron charge and mass ($m \approx 9 \cdot 10^{-28}g$ and $e \approx 4.8 \cdot 10^{-10}$ CGSE), and takes for μ one Bohr magneton,

$$\frac{eh}{4\pi mc} \approx 10^{-20}$$

(as required by spectroscopic data), then radius r becomes 10^{-13} cm, while the angular velocity $\omega = 10^{25}$ sec^{-1}. So the linear velocity at the electron equator would be

$$v = r\omega = 10^{12} \text{ cm/sec},$$

i.e., 2 orders of magnitude larger that the light velocity. An even more essential difficulty of the rotating electron model is that the electron has to possess an additional mass appearing due to rotation. The order of magnitude of this mass may be assessed by dividing, according to the Einstein relation

$$\epsilon = mc^2,$$

magnetic energy —

$$\frac{1}{8\pi}\int H^2 dV,$$

associated with the magnetic field H created by the rotating electron in the surrounding space, by the square of the light velocity. This additional mass is

$$\Delta m \approx \frac{\mu^2}{r^3c^2}.$$

Thus, introducing the values of μ and r, one finds

$$\Delta m \approx 10^{-21}g,$$

which is a million times greater than the basic electromagnetic mass of an electron. If otherwise one exploits the same formula to estimate the electron radius, assuming that its mass 10^{-27} g is of a purely magnetic origin, then

$$r = 10^{-11}\ \text{cm},$$

an extremely large value from the point of view of the «electron scale».

These relations and calculations were carried out by Lorentz just after he had been informed about the Uhlenbeck and Goudsmit idea (Ehrenfest wrote meaningfully that «all around the world the best physicists, having completed their studies, are always eager to know: «What does Lorentz say about it?».»)[53]

Judging by Frenkel's letter of December 17, 1925, (and basing it on what father told me at the end of the 1940s), he, after reading the Goudsmit and Uhlenbeck paper in «Naturwissenschaften», made the corresponding work stimulated at the first stage by Pauli and was going to publish it. But very soon Pauli advised him to refrain from publication — and Frenkel followed his advice.

The situation changed in the end of February, 1926. At this time Pauli had become acquainted with the manuscript of the English physicist L. Thomas, who had worked, at the time when it was written, at the Niels Bohr Institute in Copenhagen. Thomas's paper was dedicated to the relativistic theory of a rotating electron (the corresponding note was published in the «Nature» issue of April 10, 1926). Pauli showed Thomas's manuscript to Frenkel. As a result, J.I. returned to his prepared earlier work (we refer one more time to the letter of December 17, 1925).

One should note here that the calculation of the field generated by a uniformly charged rotating sphere had been already performed by the father of electrodynamics, J. Maxwell. The same problem was treated also by Abraham, the author of that very handbook on electrodynamics which was mentioned by Born in his report on J.I.'s book. From the book by Abraham, J.I. had taken the results about

53) P. Ehrenfest. Collected Scientific Papers. Amsterdam, 1959, p. 478.

the rotating sphere and applied them even in his first (unpublished) physics paper. He also returned to this calculation later, in the mentioned article published in ZhRFKhO in 1919. In the conclusion to this paper, discussing its results on a superlight velocity of point on the electron equator, J.I. wrote: «All the issues and formulae of the previous sections are derived on the assumption of a uniform charge distribution over the sphere ... They are inapplicable to real bodies just because electricity has a discrete (atomistic) structure and, hence, one cannot even speak about its steady flow.[54] With the acceleration, individual properties of each elementary charge, either a positive nucleus or an electron, have to be described more and more accurately (for instance, in the increase of the mass and like), and the light velocity has to be a limited one for any of them. But if electrons or positive nuclei were spheres or balls with uniform charge distribution, the above results could be applied, to a full extent, to them as well. However, one can hardly liken positive nuclei which in the case, say, of radioactive substances have, undoubtedly, a complicated structure, to smooth balls. As for the electrons, experiments have yet to give any indication regarding magnetic fields caused by their rotation.»[55]

We see thus that the notion of a rotating electron was very close to J.I.[56] One could only regret that he did not try to find, at an appropriate time, experimental facts concerning the rotating electron magnetic field in spectroscopic data.

So, as a result of reading Thomas' manuscript and after the discussion with Pauli of his old and new findings in the theory of a rotating electron, Frenkel sent on March 25, 1926, his brief note on electrodynamics of a rotating electron. He remarks in a footnote that this is a summary of a more detailed article to appear in the «Zeitschrift für Physik». The note was published in the issue of May 8, 1926. In the same issue (# 2949), A. Eddington, O. Richardson and F. Lindemann, along with Frenkel, responded to the Goudsmit and Uhlenbeck paper.

As for the large article by Frenkel in the «Zeitschrift für Physik», it contains a criticism of some of Thomas' statements. The corresponding remarks by Frenkel arose from the fact that Thomas had considered in his relativistic theory mechanical and magnetic momenta of an electron as invariant, while in reality they change from one frame to the other according to the Lorentz formulae. Thomas had confined himself to only the rotational motion of an electron and had not taken

54) J.I. means a representation of the rotating sphere as a set of currents running in the planes perpendicular to the rotation axis.

55) ZhRFKhO, phys. ser., v. 50, ## 4–6, p. 143, 1919 (in Russian).

56) On the other hand, one also cannot avoid noticing a characteristic discrepancy here. In 1925 J.I. intensively studied the electrodynamics of point electrons (ZRHFO, phys. ser., v. 7, ## 3–4, p. 393, 1925; Zeitschrift für Physik, Bd. 32, S. 516, 1925). On February 23, 1926, he wrote to Ya.I. Perel'man: «No matter how disgusting the thought of an extended electron is to me, the hypothesis of a rotating electron explains the anomalous Zeeman effect in such an easy and natural way, as well as the similarity between X-ray and optical doublets, nonmagnetism of helium and other noble elements, that I'm starting to believe in it.» (See: Archive of the USSR Ac. of Sci., f. 796, file 3, # 18).

into account its influence on the translation motion, though these drawbacks were redeemed by the positive content of Thomas' paper (a correct explanation of the value of the doublet splitting for alkaline metal spectral lines). Pauli stated that it was this success by Thomas which had led him to the camp of supporters of a rotating electron.

In this paper J.I. first of all suggests, abstaining from any consideration of the inner structure of an electron, and characterizing its properties by definite scalar, vector and tensor quantities. (Indeed, nobody suggests any «classical» model to explain the electric charge of an electron, the former is just a scalar parameter; this electron property has become so accepted that there is no need to «reduce» it to anything.)

J.I reminds us further that electric and magnetic fields are respectively the temporal and spatial parts of the electromagnetic field tensor, which is the result of the Maxwell-Lorentz relativistic electrodynamics. In the frame connected with the electron, the latter does not produce any magnetic field. If one turns to a frame moving relatively to the electron with the velocity v, the electron will create in it the magnetic field

$$\vec{H}' = -\frac{\vec{v}}{c} \times \vec{E},$$

and for small enough v/c, $\vec{E}' = \vec{E}$. Since in the steady frame

$$\vec{E} = Ze\frac{\vec{r}}{r^3}$$

(Z is the charge of a nucleus), then

$$\vec{H}' = -\frac{Ze}{cr^3}\vec{v} \times \vec{r} = -\frac{2Z}{r^3}\vec{M}$$

(M is the orbital magnetic momentum of the electron). This expression for a magnetic field at the point where the electron lies, Uhlenbeck and Goudsmit obtained by passing, on Einstein's recommendation, to the frame in which the electron rests while the nucleus moves in relation to it with the velocity $-\vec{v}$. In this way the corresponding expression for the field $\vec{H}'$ is produced immediately by the Biot-Savart formula. If one transits from the frame where the electric field $\vec{E}$ of the electron equals zero (and the magnetic field equals $\vec{H}$) to the other one, moving against the first with velocity $\vec{v}$, then in this system

$$\vec{E}' = \frac{\vec{v}}{c} \times \vec{H}'$$

and

$$\vec{H}' \approx \vec{H}.$$

The same is also true for electric and magnetic momenta of an electron, which have to be the corresponding components of the momentum tensor. In the frame connected with the nucleus, the electron has to act like a dipole with momentum

$$p = \frac{\vec{v}}{c} \times \vec{\mu},$$

where μ is its magnetic momentum relative to the system where it does not move and has zero electric momentum. This may be stated as follows: similar to how a moving charge (identified with current) creates a magnetic field around it, a moving magnetic momentum has to generate an electric field, i.e., certain electrical momentum may be assigned to it. Frenkel's formulae are in complete agreement with the experimentally obtained magnetic to mechanical electron momentum ratio.

Further in the paper J.I derives rigorously the equation for the rotational electron motion and the translation motion of a rotating electron.

In the letter of August 15, 1926 from Cambridge, Frenkel mentions his talk in «the circle of local physicists». This «circle» is nothing other than the widely known physical seminar organized in Cambridge by Kapitza — the so-called Kapitza-Club. Kapitza accurately kept the seminar diary (journal), where not only the date of session was noted, but also the speakers and participants were able to write anything they wanted to. Frenkel's talk was at the 132nd session of the Kapitza-Club on August 12, 1926. The topic and the speaker's name are in J.I's writing: «The Theory of the Magnetic Electron». And below in German: «Eine echte physikalische Theorie muss mindestens ebenso dunkel und rätselhaft sein wie die Phaenomena die sie behandelt — Nicht wahr, Frenkel und Thomas?» [A true physical theory should be at least as dark and mysterious as the phenomenon it considers. — Shouldn't it, Frenkel and Thomas?]

The signature is that of Paul Ehrenfest.

The next line belongs to some unidentified participant of the seminar. This is a quotation of the well-known Austrian philosopher and logician Wittgenstein (he emigrated to England in 1929): «What may be said at all may be said clearly.»

And, at the end, a conclusive remark made in the characteristic (and terrible!) handwriting of L. Thomas: «But humanity is fallible!»

The nature of these assertions reflects the complexity of the problem discussed. The «mysteriousness» and «darkness» pointed out by Ehrenfest have accompanied the problem of a rotating (magnetic) electron, as we have seen, since the moment of its appearance.

So we know that on 12 August 1926, the physicists who dealt with the origin and further development of the theory of a rotating electron gathered at Kapitza's seminar. One can only guess as to the reasons of the witty exchange of opinions in the Kapitza Club journal. One such guess is: it seemed to Ehrenfest that the arguments of Frenkel and Thomas (who surely were participating in the discussion) are much too simple, and that is why he addressed his remark to both physicists.

To balance this skepticism, let's give the floor to I.E. Tamm and Ya.A. Smorodinskii who contributed to a survey «The studies of J.I. Frenkel on the theory of electrons and atomic nuclei» in 1957. (The survey opened the publication of the corresponding papers by Frenkel in one of the sections of his selected papers, 2nd volume):

«The conjecture about the electron spin put forward by Uhlenbeck and Goudsmit instantly encountered the following problem: to explain the data on the anomalous Zeeman effect, it was necessary to assume the magnetic momentum spin of an electron to be equal to one Bohr magneton, while to clarify the multiplet structure of optical and X-ray spectra one had to take a value two times smaller for it.»

Thomas, in his note of 10 April 1926, suggested an explanation for this seeming contradiction based on the consideration of peculiarities of relativistic kinematics of a charged rotating ball (electron) moving along a circular orbit in the Coulomb atomic nucleus field. However, Thomas' arguments were of a semi-qualitative nature and were not marked by either rigorousness or consistency.

Only two months after this, when J.I. Frenkel had become acquainted with the manuscript of Thomas' note, he put into press his consistent relativistic theory of a rotating electron which implied all the facts observed experimentally ... The paper by J.I. not only reflected an essential progress in the development of our concept of the electron but served also as a basis for the consideration of all the questions related to the rotating electron dynamics, until the moment when Dirac created a new quantum-mechanical relativistic theory of the electron.[57]

We will still have an opportunity to speak about Pauli's skepticism revealed in his assessment of Frenkel's studies and will give several examples of it in connection with the investigations of other physicists, in order to avoid creating the impression that Frenkel was «exceptional» in this respect. Let us finally note that the papers by Thomas and Frenkel feature in Pauli's Nobel Lecture.

3. In his letter of 14 March 1926, Frenkel mentions his paper «On the theory of the elasticity limit and the strength of crystal bodies». It was completed in April, 1926, and the paper was received by the Editors of «Zeitschrift für Physik» on May 7, 1926. O. Stern was quite right: there are lots of formulae in it! However, we confine ourselves here to the consideration of approximate calculations forming the contents of the first section of the paper, which are based on the «model of two saws» simply and clearly introduced by Frenkel. Such a model approach to the problems arising in the process of the development of physics was always very characteristic of Frenkel. The approach is not a unique one: there exists in physics a way of writing general, often very cumbersome and complicated, mathematical expressions which is directly opposite to Frenkel's approach; then by means of this or that simplification, which one usually tries to justify logically, the expressions are reduced to the form which allows the calculations to be carried out. Having

---

57) Ya.A. Smorodinskii and I.E. Tamm in: J.I. Frenkel. «Collection of Selected Papers», v. 2, Leningrad-Moscow: 1958. AN SSSR publ., pp. 455–456 (in Russian).

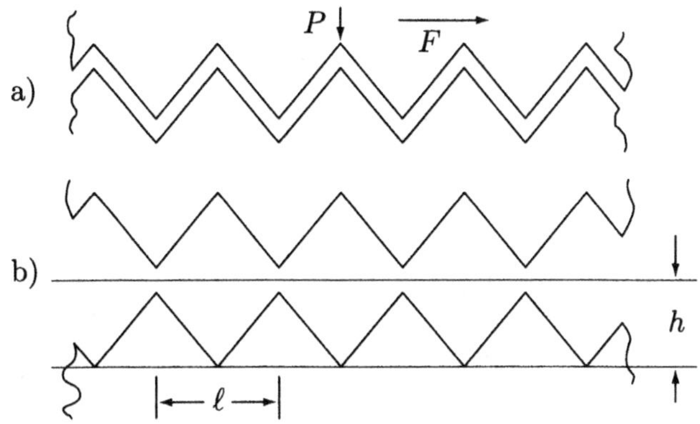

Fig. 1 Frenkel's «saw» model.

thus obtained the final results and then considering the «limit» cases, the adherents of this method usually try to construct the corresponding model.

J.I. Frenkel, with his deep insight and intuition in matters of physics, practically always had a model first. This approach to doing things is especially pronounced in the paper considered here. Note also that in the book by Hirth and Lothe, the section dedicated to theoretical strength under shear begins by noting: «A classical study in this field was carried out by Frenkel» — referring to his article under discussion now.

Frenkel introduces two saws (see Fig. 1, a); the saw tooth is of height h, and the tooth period is l. Both saws are put into contact as shown in the figure, and they are clasped together by force P.

If one assumes the lack of friction between the saws, one can derive from the simplest energy arguments the minimum tangential force F which should be applied to provide the motion of one saw against the other.

How may the onset of such a motion be organized? Obviously, it is enough to lift up the «upper» saw to the height h, simultaneously shifting it for the distance $l/2$ (Fig. 1, b). Equaling the corresponding values of work (energy), we obtain

$$Ph = Fl/2, \tag{1}$$

i.e., the sought-for value of F is

$$F = 2\frac{h}{l}P. \tag{2}$$

Frenkel calls attention to the similarity of (2) with the known relation for a static friction of a solid. As it is known, if some subject is pressed by the force P to a horizontal plane, then the value of the tangential force necessary to start the motion is determined by the formula $F = kP$, where k is the friction coefficient,

$k = 2h/l = \tan\alpha$ in the case of formula (2) (α is the slope of a tooth similar to the so-called friction angle).

A «physical» correspondence between the saw and the real model is the following: atoms of two neighbouring layers play the role of the «teeth» (so that l is the lattice constant a). The value of h proves to be related in a complicated way with the parameters of the atomic interaction. However, Frenkel notes, there is no need at all to calculate this value directly, because to determine F one has to know, as it is seen from (2), not h and P as they separately are, but only their product $Ph = U$ which may be naturally treated as the increase of the layer potential energy under the relative translation of layers by a distance of the order of a. Therefore formula (2) is rewritten as

$$F = \frac{2U}{a}.$$

If one now relates U to a single atom (corresponding surface area $\sim a^2$), then for force F' per unit area one finds

$$F' = F/a^2 = 2U/a^2.$$

Assuming moreover the value of U to be about the latent heat of melting, we obtain the correct order of magnitude for F ($\approx 5 \cdot 10^4$ kg/cm).

As it happened with J.I. Frenkel both earlier and later, he had to stress specifically that such a close coincidence of the theoretical strength limit value with that obtained experimentally (in the experiments where the surface faults had been removed) should not be treated as an absolute one. The adequacy of the model suggested is justified not by the exact numerical coincidence, which is more or less random, but by the equalities of the orders of magnitude.

We will conclude the chapter on J.I.'s life in Germany with one of his last letters of that period.

«Göttingen, 6 October 1926.
... I plan to stay one more week in Göttingen and then move to Berlin. However, perhaps I'll stay here for a longer period. It will depend partly on the development of small studies started the day before yesterday with Frank.[58]

Before parting in Cambridge, Ehrenfest told me that he was very satisfied with me and that I had made, on all the physicists abroad whom I met, an impression which «was beyond all his (Ehrenfest's) expectations». If you ask me whether I am satisfied with myself, I would answer: not quite. I could do more. But a scientific achievement, just as any other one, is partly a matter of luck. Clever things are not «hatched out», and no efforts can help them. I had thought out something clever enough 3 months earlier in Göttingen, but since it happened that a little earlier the

58) J.I. here means Philipp Frank who often visited Göttingen and was there in the autumn of 1926.

same had been done by the other physicist working with Ehrenfest, the matter was left undeveloped and unpublished. Not only the author's priority, but the priority right for the topic is strictly kept here.[59]

For the last few days I have been in a rather sour mood and am using well-tried remedies to break it: the violin and the cinema. The human being is an ungrateful creature (judging by me). What more? Just sit in Göttingen and do your work well, nothing to worry about because nobody hampers your studies. But no such thing, you just want somebody to interfere! While, if someone really does distract you, then you get newly nervous and want to work harder. Doesn't it remind you of the psychology of some quadruped which has to be pulled by the tail to go straight forward?»

It is possible to think that «something clever» worked out by J.I. is related to relativistic quantum mechanics which he pondered over during solitary evening walks around the town, as it is stated in some of his letters from Göttingen (special attention in this respect should be paid to the letters of May 30, June 18, and July 10, 1926). And indeed, in the copy of J.I.'s «Wave mechanics» (part 1) which he used in 1950–1951 when working on a new edition of his book, an additional note made in pencil (on p. 107) has been found. In the remark concerning the relativistic Schrödinger equation J.I. writes: «This equation ... had been found by me and reported at Prof. Hilbert's seminar in Göttingen in May of 1926. I, however, abstained from its publication after I realized that it was repeated exactly by Klein.» The «Non-relativistic Schrödinger equation», about which J.I. writes, is known as the Fock-Klein-Gordon equation.

At the end of October, 1926, J.I. and his family returned to Leningrad.

59) A priority right for a topic was defended not only by Göttingen physicists. This may be illustrated by the example of the famous American chemist M. Gomberg, who discovered free radicals. He concluded one of his papers («Berichte der deutschen chemischen Gesellschaft», Bd. 33, S. 3163, 1900) with the following words: «The investigation will be continued, and I would like to ask that this field be left to me for the time being.» It really is a polite form of something like: «I will continue to work here and do not want anybody to interfere.»

Chapter 4
The Years 1926–1930

In Leningrad J.I. enlarged his efforts in several scientific directions. He started giving the course on wave mechanics not only at the Polytechnic and Physical-Technical Institutes, but at Leningrad University as well. In this way he tried to involve as many people as possible in the studies associated both with the development of the new quantum theory and with its applications to a broad range of macroscopic problems.

J.I. spent a lot of time during the winter and spring of 1927 on the second volume of «Electrodynamics», in which he explained the macroscopic theory of electromagnetic phenomena. For the elaboration of this volume, in contrast to the first one, nothing had been prepared beforehand, and Frenkel had to present an exceptionally large amount of material.

Those sections of this volume which are dedicated to the theory of endovibrators particularly deserve special attention. It is now common knowledge how important the electrodynamics of hollow systems is in modern microwave technology. J.I. was the first to include the elements of this theory in an educational book, which was certainly a service on his part.

And finally, yet another branch of Frenkel's studies was related to the further exploration of the electron theory of metals. In September of 1927, J.I. presented the results at the International Congress in Italy, to where he had been invited in April of that year. Professor Pyotr Petrovich Lazarev was sent to the congress as a representative of the Academy of Sciences of the USSR. The Congress of Physicists was organized on the occasion of the centennial of the death of A. Volta, in the town of Como, on the bank of the very picturesque Lake Como, where Volta had been born. As a congressional delegate, J.I. had got an opportunity to make a trip across Italy, which he took.

Recalling the opening of the Congress, J.I. wrote: «I have to state with satisfaction that among various national banners which have decorated the hall of the Carducci Institute (in Como), a red flag with a sickle and hammer was present. An even greater pleasure for me was the welcome speech delivered in (not very correct, but extremely cordial) Russian, given after similar greetings in the four official languages of the Congress, by the Rector of the University of Pavia, Dr. Rossi.»[1]

The Congress was very well attended. Lorentz, Rutherford, Planck, Bohr, whose names are a legend now, were delegates. Only one member of this glorious galaxy, Einstein, was absent. From Germany came Born, Heisenberg, Sommerfeld, Laue, Pauli, Franck and Stern; from America arrived Millikan, Compton,

---

1) J.I. Frenkel. International Physics Congress in the memory of A. Volta in Como. «Nauchnoe slovo», # 1, p. 64, 1928 (in Russian).

From left to right: W. Pauli, J. Franck, F. Pashen. Como, 1927.

Ya. Frenkel and H. Lorentz. Como, 1927.

From left to right: O. Stern, Ya. Frenkel, A. Smekal. Como, 1927.

Como, 1927. Ya. Frenkel with his colleagues. From left to right: Ya. Frenkel, (unknown person in the background), O. Stern, Wolfgang Pauli, Peter Debye (permission by K. van Meyenn, editor of «Das Gewissen der Physik», dedicated to Pauli, Vieweg, 1988).

Langmuir and Wood; and from England came Aston, Bragg and Eddington. Italian physics was represented by Fermi, Majorana, Levi-Civita. Also among the delegates was P. Debye (arriving from Zürich, Switzerland).

Thirteen out of the 70 congressional delegates had been Nobel Prize winners by 1927, and 7 more of the above mentioned scientists were so honoured later.

Almost all the physicists listed here gave lectures. To explain experimental data on alpha-decay, Rutherford tried to construct «a planetary model of a nucleus» consisting of a very strong core and one or a few helium atoms rotating around it. Lorentz devoted his presentation to a rotating electron. Dutch physicist H. Kramers derived in his communication a dispersion relation for the dielectric permeability. (A relation of this kind appeared again in the physics arena in the 1960s: Kramers's formula served as a preliminary image of the dispersion relation with which elementary particle theory is now armed.)

According to Frenkel's recollections, Robert Wood's lecture on ultrasonic vibrations and, in particular, their biological effect on the most minute (bacteria, red blood particles) and small living organisms (fishes, frogs, mice) was extremely good. But the greatest impression on J.I. himself was made by two presentations. The first of them was the Niels Bohr («Kepler of the presently developing quantum mechanics,» as Frenkel wrote then) lecture about new wave mechanics. There he presented a detailed analysis of the Heisenberg principle and the probabilistic interpretation of the wave mechanics by Born.

The other lecture, given by A. Sommerfeld on the electron theory of metals, the same subject J.I. had spoken about there, was characterized by Frenkel as «one of the most striking».[2]

Let us recall first Frenkel's first work on the theory of metals (1924). The Pauli principle (which had not yet been established then) as applied to one particular atom was its basic assumption. At the point of the total energy minimum, which every system tends to approach, electrons in an atom would have to occupy the orbit closest to the nucleus. Empiric material demonstrated, however, that they populate consequentially orbits more and more remote, up to a comet-like one occupied by the valence electron (in the case of alkaline metals).

In 1926 Fermi and Dirac simultaneously compiled the statistics and obtained the formula for the distribution function of particles with the semiinteger spin obeying the Pauli principle. Electrons in particular are such particles; therefore electron gas must be described by the Fermi-Dirac function. At absolute zero temperature only two electrons with oppositely directed spins are in the state with the smallest energy. The others, in pairs, occupy the states with larger energies.

To describe electron gas, Sommerfeld used the Fermi-Dirac distribution function in a way similar to work by Lorentz who, in his time, had improved Drude's results with the help of the Maxwell distribution function. Sommerfeld's theory agreed closely with the experiment. It implied the Wiedemann-Franz relation.

2) Op. cit., p. 66.

Since the Pauli principle entered it explicity, the theory, just like the old Frenkel one, eliminated catastrophe with the thermal capacity. The mean electron velocities in metal, according to Sommerfeld, exceed the thermal ones by two orders of magnitude, so electron gas contributes practically nothing to the heat capacity of metal. The Hall, Peltier and Volta effects are explained within the framework of Sommerfeld's theory. It also provided an example clearly demonstrating the abilities of new quantum mechanics.

J.I. approached the problem of the electrical conductivity of metals from the other side. As a milestone of his argument he put the de Broglie relation which extended the corpuscular-wave parallelism from photons, the light quanta, to any material subjects. According to the famous de Broglie formula, the wavelength

$$\lambda = \frac{h}{mv}$$

is compared (assigned) to a particle of mass m moving with velocity v. Let us remember that the concept of wave properties inherent in matter that had been put forward by de Broglie as far back as in 1924. It was supported by the experiments of C. Davisson, L. Germer and P.S. Tartakovskii on the diffraction of molecules.

J.I. spread the de Broglie formula over the electrons wandering in a metallic body. At absolute zero temperature the minimum wavelength, for electrons at «the highest level» (i.e., with the greatest velocity), is equal, within the order of magnitude, to the distance between any two neighbouring atoms in the crystal lattice ($\lambda_{\text{min}} \sim n^{-1/3}$, n is the electron concentration).[3]
A temperature rise leads to a distortion of the crystal lattice that occurs not only because of the transition from a perfect lattice to a real one, in accordance with the above-mentioned picture of a real crystal given by Frenkel and Schottky. The other reason is that the thermal motion creates density fluctuations (clusters and voids) in the electron wave path. Thus the situation in metallic bodies is similar to that which takes place under the usual propagation of visible light through the atmosphere: the light is scattered by the air density fluctuations and that is why, as we know, we actually «see» the sky. However, this similarity is not complete. For instance, the coefficient of the visible light scattering by air does not depend on temperature, while in solids it grows with a temperature rise. If, however, the wavelength is smaller than or comparable to the distance between atoms (which is, for instance, the case with X-rays), the scattering coefficient is independent of temperature.

The light intensity variation over the interval dx in the direction, x, of its propagation is proportional to the intensity itself and to the length of the interval,

$$dI = -\mu I dx,$$

3) This result follows immediately from the formula for the maximum momentum of an electron: $p_0 = (2mE_0)^{1/2}$, from which $\lambda_{\text{min}} \sim n^{-1/3}$.

where μ is just the scattering coefficient. Therefore

$$I = I_0 e^{-\mu x}.$$

Comparing this formula with the known Clausius relation for the mean free path of gas particles, J.I. treats the value $1/\mu$ as the mean free path, l, of electrons in metal:

$$l = \frac{1}{\mu}.$$

To determine μ, and hence l, Frenkel uses the Schrödinger equation. He thus obtains the value of the mean free path which agrees with the experimental data (this agreement is revealed if one substitutes the value of l in the formula for the metal's electrical conductivity,

$$\sigma = \frac{4\pi}{3}\frac{e^2 l}{h}\left(\frac{3n}{4\pi}\right)^{2/3},$$

where e and m are the electron charge and mass, and n is the electron gas concentration). Let us remember that in the Drude-Lorentz theories, the first (1924) Frenkel theory and, finally, in the theory of Sommerfeld, quantity l was in no way determined, while its value, derived from the given formula for σ, of an order of hundreds of interatomic distances, seemed to be anomalously large. Only the concept of electron motion in metals as a wave process eliminated these discrepancies. The same notions also explained the role of impurities. Earlier it had seemed strange that, though according to the theory the addition of small quantities of a metal with high σ to one with the «average» value of σ should increase the electrical conductivity, in reality this value for such doped metal went down. From the point of view of the Frenkel theory it was clear that the distortion of the crystal lattice of metal by doping causes an additional electrical (and thermal) resistance to electron waves. The presence of alien atoms makes a crystal even more muddy for these waves, just as atmospheric air becomes turbid because of dust.

One should note that at the Como Congress the schedule was strictly kept. All the contributors were given 15 minutes each; the only exception was made for Bohr, who had a whole hour for his presentation.

J.I. was prepared to keep to this strict regime, but, after he had started, he found out that the majority of the audience was ignorant of his theory of wandering electrons. He reorganized everything in passing. However, at the moment he was finished with the survey of his previous work and was going to explain new results, the bell of the chairman began to tinkle. J.I. continued to speak, paying no attention to this and the warnings which followed. It was almost afternoon; in Como this time was marked by an artillery shot. The shot sounded in the middle of the next attempt to bring the reporter to reason, one may say, «under the bell toll». And J.I., just as if obeying the menacing signal, stopped in midword and got off the podium.

Frenkel went to Italy via Germany, with stops in Berlin and Munich. He wrote in a postcard from Berlin: «Yesterday for two hours talked to Einstein at his place. He rather approved of my theory.»

At the Congress J.I. became convinced of the propitious response to his «Electrodynamics». Skeptical Pauli, though, observed that the book had appeared at an inappropriate time, because the physicists' brains were occupied by new wave mechanics. «Well, Pauli», J.I. retorted, «I won't give you the second volume but will work on «Wave Mechanics» instead.»

In Como J.I. got acquainted with Lorentz, who was already «a classic» then, and talked with I. Langmuir about the problems of molecular physics both scientists were interested in.

But the most interesting experience for Frenkel was his acquaintance with Arnold Sommerfeld, who was his companion during the trip through Italy.

The excerpts from Frenkel's letters to his wife given below represent his «Italian diary».

«Rome, 21 September 1927.
... By 7 p.m. yesterday I was, so to say, undone; an excursion around the city in the morning, with sightseeing at the Forum, Colosseum and other landmarks. Then a prolonged breakfast in the former Caesar's palace; afterwards a trip to the coast (40 km), instead of the visit to Mussolini. I deliberately refused to behold the Duce. It would be unpleasant for me to see how prominent people bend their heads in front of this man. Nor did I mourn the lack of a tuxedo that deprived me of the possibility of attending the Governor's reception in the evening, and spent the time in a cinema with much greater pleasure. I try to regard attentively, as far as possible, the Italian living conditions, discuss political and economics subjects with younger professors and generally try to get to know modern Italy as a whole. Everyone has been trying to convince me that the life of workers is relatively good in terms of both their salaries and their social security benefits as well as their labour protection. Moreover, they assured me that fascism in Italy actually accomplishes the cause of socialism in an evolutionary way. I do not think this is correct, because I know, from the same sources, that the Italian government has tended to pass into private hands a part of the state industry, starting with the railways.

The press in Italy is exclusively fascist. Thus the freedom of the press, if it exists, does so in name only. The Communist press has been strangled (the Communist Party is considered illegal), and the Socialist one has perished due to senility and poverty, since the former «Socialists» remaining in Italy have turned to Fascists. The Italian intelligentsia, in its absolute majority, sympathize with Fascism (in its contemporary «Italian» form). Fascist ideals (i.e., nationalistic ones) are being deeply and firmly rooted in the heads of citizens, starting from their childhood: there are a lot of Fascist children's organizations here. One of the psychological stimuli to the development of nationalism in Italy is the revitalized cult of the great Roman past. «To be worthy of their glorious ancestors» — this is the common formulation of the Italian ideal.

In Italy universal primary education was introduced long ago: however, the secondary schools here provide a rather poor level of education, that, of course, is reflected in the students' performance as well. Very few people deal with pure science in Italy. There are practically no scientific research institutions. University professors have a relatively low salary, and the majority of talented youth go in for technology ...

But enough of politics. A fair amount of written work lies ahead of me. First, I have to revise somewhat the text of my contribution and write a small note about it to a German magazine. Second, I've got a proposal to explain «the success of the electron theory of metals» in nonmathematical form for the international scientific journal «Scientia». Third, I need to proofread the 2nd volume of my «Electrodynamics», write some appendices and send everything to Springer Verlag. I begrudge time spent on such work in Rome! However I'm a little bit tired of the onslaught of impressions and have practically nothing against a short «change of activity».

... My «schedule» is something like this: until Monday (the 26th) I will stay in Rome. Between the 26th and 30th I'll be in Naples. Then I'll go to Messina and Palermo (in Sicily); from there I shall return to Rome for a day on the 4th. The 5th–6th I'll spend in Florence; then the 7th–9th in Venice, the 10th–11th, probably, in Milan, the 12th–13th in Turin. And then from there I'll go to Nice ... I feel a little ashamed that I'm just cooling down (or, rather, warming up) in Italy while you are all plodding along with the daily routine. I'll try instead to compensate you for this upon my return.

Nevertheless, yesterday and the day before I had a rest from sightseeing in Rome (not counting the walks around the city) and started writing. Yesterday I at last finished my contribution (at 2:30 in the morning) and sent it to its destination this morning. One heavy weight has been lifted from my soul. Now I'm going to get rid of the second one, that is, of the accumulated proofs of my «Electrodynamics», which I'll start doing as soon as I finish this letter.

This morning I was in the National Gallery of Modern Italian Art (something like the Luxembourg Museum). Both the paintings and the sculpture are good on average; but it seems to me that there are no outstanding works. I've purchased some reproductions but won't send them by mail.

Then I took a bus and went, along the ancient Appian road, to see the catacombs. They make up a cemetery large enough for approximately a million people. The total length of all the corridors is about 50 kilometers, of which only 20 are more or less restored. In their walls rest the remains of Christians, both martyrs and simple people who had not wished to settle their bones at the same place as heathens. Looking around this cemetery, I thought: what did these Christians struggle and perish for? Was it worthwhile to establish Christianity in the world only to later, in the name of Christ, wipe out the dissidents as ferociously as the one time caesars did away with the Christians? After all, having attained their goal, established firmly their ideal, people always disgraced it where they could. However, I've dived too deeply into philosophy. Let's return to the catacombs. They have existed over 5 centuries, so it's not surprising how gigantic, how enormous they are (in some places corridors and chapels connected by them form five floors). It's interesting to consider what catacombs one would have to construct for the sacrifices of the Inquisition and especially for those of the wars perpetrated by Christian states. However, I'm again falling into philosophy.

On the return journey to Rome I encountered the German physicists Sommerfeld and Gerlach on the bus. We'll definitely go to Naples together on Monday. This meeting was a special pleasure for me because in the two days of my loneliness I'd become completely isolated from my colleagues. And as it happens Sommerfeld is exactly the man with whom I would really like to speak more about something.

Having looked over the catacombs, I, in the most prosaic way, became engaged in satiating myself in the restaurant across the road. I spiced up ham and eggs with a pleasant conversation with a waiter («cameriere») in Italian. From him I found out that the situation for workers in Italy is not so bright. The working day is 10 hours, and sometimes all 12 hours. The maximum salary is 30 lira, which is very little considering the high cost of living in Italy. In general, life is very difficult for the working man. More detailed information on the labour question in Italy I hope to present to you later.

My Italian limps on both feet. However, I somehow manage to explain myself on not very complicated subjects and read newspapers practically without a dictionary. I've bought a small teach-yourself book and am leafing through it when not otherwise occupied. Well, my dear, I have to finish this correspondence and turn to «Electrodynamics»!»

«Rome, 26 September 1927.
... In two hours I'm going (with Sommerfeld) to Naples, from where I will write the details. Yesterday I was in the Tivoli suburb, where I dealt with ... electrodynamics. I got completely covered with dust there (Italians can't boast of cleanliness). In the evening I visited Lazarev, who went today to France, and then enjoyed myself in the company of Sommerfeld and several other physicists. Have just finished proofreading and am in a hurry to get to the post office.»

«Naples, 26 September 1927.
... Traveled to Naples with Sommerfeld. We spoke a lot together about physics, naturally. He is not very young (58 years old); he is very simple and pleasant. It's a pity I wasn't able to become his student 10 years ago. Almost every gifted theoretician (in Germany, and to some extent in other countries) is his pupil ...

Italian cities are as varied in their architecture and planning as their population. On the one hand, there are wide «European» streets, and at their sides are narrow lanes, of course, without what can be called pavements, since their width is less than half the width of the pavement in Berlin. Up above, at the level of the 2nd or 3rd floors, linen is drying, picturesquely hung on rope. There are a lot of such sidestreets in Rome. Trams run along some of them, usually only in one direction, and carts are forbidden to travel in the opposite one because it is impossible for them to miss one another. However, you are acquainted with these images. You've seen them in Nice. Generally speaking, Nice is a typical Italian city. By the way, in Rome and especially in Naples there are probably even more palms than in Nice. Cafes here, like in Nice and generally in France, spread far beyond the limits of their premises, almost to the middle of the streets. This is explained, of course, by the heat: on a stuffy evening nobody is going to sit in a closed room. Generally, the development of street life in Italy is explained in the same way. The abundance of children is especially astonishing and joyful ... And it's of interest that children and adults live, to a certain extent, a common life and even go to bed at the same time ... At 10 in the evening one meets in the street quite a number of mothers and fathers holding the hands of, or carrying in their arms, their progeny, or just sitting together at cafe tables (however, the descendants are usually playing nearby).»

«Naples, 1 October 1927.
... This evening, after a two-day trip to Capri, Sorrento and Amalfi, I returned to Naples.

Two days ago I was in Pompeii and Vesuvius. At Vesuvius we (with Sommerfeld, with whom I did not part company all through the last week) went on horseback (astride): that's why I couldn't walk properly or even sit down until now! Vesuvius is now in its high activity period and makes a grandiose impression (details upon my return).

Yesterday morning we went to Capri. Visited the (rightfully) famous Blue Grotto. On our way back, when getting off the boat, I fell, to my absolute distress, into the water. After that I dried myself for a couple of hours but did not dry completely until the evening. This small inconvenience, however, in no way reduced my pleasure. This morning (at 6 a.m.) we started off by steamboat to Sorrento; from there, with a cabman, we went to Amalfi on the opposite (south) bank of the islet. From Amalfi we took a bus to Vietro and, finally, from Vietro returned by train to Naples at 7 p.m. Before doing anything else I leaped to the mailbox but pitifully, instead of the much awaited letters from Leningrad, found a letter from ... the police, with an invitation to come to see them. However, after negotiations with the hotel manager I declined the invitation and decided to go to Sicily immediately today.»

«Turin, 14 October 1927.
... Yesterday morning I left Venice (a fantastic city!), stayed for several hours in Milan and arrived in Turin in the evening. Today, from morning until 5 p.m. I corrected the proofs for Springer, then looked over the city and spent the evening in the company of two local physicists, speaking Italian all the time. I've astonished

them with my success at it. However, we not only conversed but also saw the splendid film «Casanova». Tomorrow I'll go to Nice; will write in detail from there.»

Frenkel returned from Italy in the first half of November, 1927, and resumed his usual activity: scientific research, postgraduate student supervision, the seminar at the Physical-Technical Institute, lecturing at the Polytechnic Institute.

At the end of the 1920s, the very few theoreticians working at the PTI were joined at the Cabinet of Theoretical Physics headed by V.R. Bursian. He and Yu.A. Krutkov were the oldest PTI members, having joined in 1919. During the following years, the number of theoreticians grew very slowly. In 1921 the PTI was joined by G.A. Grinberg and J.I. himself, and a little later by N.N. Mirolubov, B.N. Finkelstein and G.H. Gorovitz, who became the first of Frenkel's pupils. In 1924 Vladimir Alexandrovich Fock joined the theoretical team, and in the second half of the 1920s it was enlarged by L.D. Landau, D.D. Ivanenko and L.V. Rozenkevich.[4]

In 1930 Matvei Petrovich Bronstein (1906–1938) came to the Theoretical Department of the Institute. In the 7 years spent within its walls, he became a prominent scientist, one of the brightest representatives of the second generation of Soviet theoretical physicists belonging to the galaxy of scientists who started their scientific work at the end of the 1920s and the beginning of the 1930s (V.A. Ambartsumyan, A.A. Andronov, N.N. Bogolyubov, L.D. Landau, M.A. Leontovich, M.A. Markov and others). Matvei Petrovich contributed a lot to the development of physics in this country. He was arrested in 1937 and executed in 1938. A book by G.E. Gorelik and this author[5] is dedicated to him.

V.A. Fock, L.D. Landau, G.A. Grinberg, D.D. Ivanenko and M.P. Bronstein, despite their youth, came straight from the university bench as completely developed scientists and did not need any supervision. We can judge J.I. as a supervisor from the documents available in the personal file of Lev Victorovich Rozenkevich. He started to work at the PTI under Frenkel, and in 1930 went to the Khar'kov PTI, at first heading the Theoretical Department of that institute. Then, after a short period of work at the same department, recently headed by L.D. Landau, Rozenkevich became a head of one of the nuclear laboratories. Simultaneously he was a professor of the Physics and Mechanics Department of the Technological Institute, paying much attention to teaching. In 1935 L.D. Landau, E.M. Lifshits and L.V. Rozenkevich published «Problems in the Course of Theoretical Physics». Later the «Problems» were included in the «Mechanics» by Landau and Pyatigorski, — the first volume of a famous book of theoretical physics.

Lev Victorovich — Levi, as his colleagues called him — was arrested and executed in 1937.

---

4) L.D. Landau and L.V. Rozenkevich were postgraduate students of Frenkel.

5) G.E. Gorelik, V.Ya. Frenkel. «Matvei Petrovich Bronstein and Soviet Theoretical Physics in the Thirties». Birkhäuser, 1994; (Moscow: Nauka, 1990 (in Russian)).

In a report written soon after entering the PTI, L.V. Rozenkevich stated: «The work at the Institute consists of participation in theoretical seminars led by Jacov Il'ich Frenkel for a group of doctoral students, as well as in the general Institute review meetings and the meetings of the Council of the Institute. At the present time, along with the usual study of periodic publications and books, I'm involved in the solution of certain «training problems» suggested by J.I. Frenkel (for example, a clarification of the question about the relativistic interpretation of wave packet motion, and the generalization of the virial theorem).»[6] A little later Rozenkevich wrote: «... I'm finishing my part of the work with Prof. J.I. Frenkel about the light action on metals (it is split into two papers: one on the photoeffect and one on the Compton effect in metals).»[7]

The example of Rozenkevich shows how Jacov Il'ich approached work with his co-workers. A «theoretical minimum» of his requirements of them reduced first of all to their clearly pronounced desire to work in the field of theoretical physics. Just this single desire in those years when physics «had not yet come into fashion», so to say, usually determined their corresponding abilities as well. The presence of such a desire was checked by J.I. in the process of conversation with a new colleague and on the grounds of his solution of the «training problems» mentioned by Rozenkevich. Along with this, J.I. assumed that erudition itself, ascertained by means of the examination procedure, formed insufficient grounds for the acceptance of a person willing to work. He considered all young people capable of absorbing the necessary knowledge in the process of their work, with the greatest profit both for them and the work. J.I. more or less directed and supported this work, tending to provide his pupils with the possibility of gaining faith in their own strength and ability to undertake research without supervision.

We would like to put in several words here about this. The famous Russian physicist P.N. Lebedev (1866–1912) holds the honour of being a founder of the first large school of physics in Russia. His pupils included, among many others, such outstanding physicists as N.N. Andreev, S.I. Vavilov, P.P. Lazarev and P.S. Epstein (who later emigrated to the USA). According to T.P. Kravets's recollections, Lebedev once said something like: «I never had a special school. Talented people need no teachers. I spent my time on people with rather moderate abilities.»

There is much to this semi-jocular remark! The same might be said about Frenkel. In my time alone, in the 1940s, I observed the amount of time he spared for people who, to put it gently, had not been endowed with any bright gifts. Naturally, he also had very talented pupils. But talent predetermines the ability to move on rapidly to independent work after a very short period of time as a pupil. Frenkel's pupils «flew from the nest» quickly. I would add that in the 1920s and

6) Quoted from the paper: V.Ya. Frenkel. Lev Viktorovich Rozenkevich. In: «A.F. Ioffe Memorial Readings», St.Petersburg: «Nauka», 1992 (in Russian).

7) Op. cit. The first of the papers mentioned by Rozenkevich was soon published by him. J.I. had refused to co-author this paper, following in this case his own rule to encourage the most talented of his pupils.

30s, J.I. had extremely limited possibilities for providing people with work at his Department: five to seven people worked simultaneously at the PTI Theoretical Department.

The difference between Frenkel and some other leading Soviet theoreticians was that he, in contrast to them, and similarly to Lebedev, gave time to everyone who exhibited a true love of theoretical physics. I.E. Tamm, a person of the highest human qualities, stuck to other principles. He often repeated: «Love should be mutual; that is, physics also should «love» the man attracted to it; in other words, one should have the necessary abilities.» Tamm categorically refused to support those whose skills gave no hope of successful and fruitful work. He thought that the work in such a case would lead a man rather too soon to hard soul traumas, thus wrecking his life.

To an even greater extent, this viewpoint was characteristic of L.D. Landau. To become his doctoral student, one had to pass, as it is known, several difficult exams in the so-called theoretical minimum. Along with this, Landau also made a special check of the mathematical skills of those who wanted to work with him. The examination usually started with a suggestion to the person being examined to calculate several difficult integrals. From first hand I know of a story of one physicist who tried to overcome the barrier of the Landau theoretical minimum. Lev Davidovich suggested to him a purely mathematical question concerning integral calculation. The candidate asked to be provided with a corresponding mathematical reference book. «Oh, but with a reference book any old caretaker would be able to calculate these integrals», Landau replied. The irascible youngster retorted: «Well then, choose caretakers for your doctoral students!», and refused to take the exam.[8] Of course, Landau never selected caretakers for himself. He took people who demonstrated outstanding capability at the very beginning of their work.

J.I. regarded any person as able, in principle, to deal successfully with a problem carefully chosen to correspond to his own level, and thus that anyone could have quite a happy «romance with physics». I would like to illustrate this with at least one example, though one related to a later period. M.F. Deygen, a member of the Ukrainian Academy of Sciences, recalled: «In 1938, while a student in the Physics Department of Kiev University, I happened to be lucky. I had the luck to have a placement at the then sacred place for the then physicists, the Leningrad PTI. And not even just at the PTI, but at its Theoretical Department headed by Jacov Il'ich Frenkel. There was a mixture of uncertainty and even fear in my enormous joy. There were at least two reasons for this. First, I was sent to J.I. himself and, second, I had a very scanty store of knowledge ... But even the first meeting with Frenkel cheered me a lot. I saw a man with a kind smile who, once acquainted with me, expressed his interest in what I knew. The word «nothing» that came from me fairly enough described the range of my knowledge,

8) Later he carried out first-rate research in the electron theory of solids.

and I waited dejectedly for the fatal end. «Oh», answered J.I., «then I've got an interesting topic for you!»»[9]

Frenkel knew how to find an appropriate problem for every level of skill, for each measure of talent. Often such a «training problem» resulted in the revelation of a great latent talent, and a man went off into his favourite work until the end of his days. Equally, the subject often remained a favourite, even for those whose abilities actually proved moderate.

I would like to supplement Deygen's story with another, characterizing Frenkel not as a head of a school but simply as a person. His pupil, Professor K.S. Shifrin, an outstanding Soviet specialist in optics, narrated: «Sometimes unworthy people abused his kindness, but J.I. regarded it his duty to help all who asked him. I remember how in the late fall of 1951 we walked leisurely along the university embankment[10] (he was already very ill) and I, having collected all my determination, asked him as to why he spent his time and strength on such an unworthy man and bad physicist as N.N. J.I. replied, laughing: «You know, he is so disgusting that nobody wants to help him. But he wouldn't be lost! So I'm bothering with him.»»[11]

Frenkel shared his own scientific plans with A. Sommerfeld in his letter of 18 March 1928. He wrote: «I congratulate you cordially on the beautiful prospect of the round-the-world trip which you wrote to me about. It's a pity, though, if as a result you won't be able to attend the Physics Congress being held in Russia this summer. As for me, I intend to work peacefully at home this year. I have to do a lot to make up for the low productivity of the previous years. I hope that no new books will hinder me (and this for the first time in seven years).»

Though paying due to J.I.'s modesty (he had published about 20 papers during the three previous years; we mentioned the most significant of them earlier), one could not grant him this wish: only a year later, in June 1929, he finished a new book on quantum mechanics published the same year.

So what were the key topics of Frenkel's studies in 1928–1929?

In the first place, these were further applications of quantum mechanics and statistics to the theory of the solid state, especially of metals. We would like to dwell here on two papers published in 1928 which are very characteristic of all of Frenkel's scientific work.

While commenting on the first of them («Elementary theory of magnetic and electrical properties of metals at absolute zero temperature»[12]), Frenkel wrote to Sommerfeld on 8 March 1928: «To me it was always a pleasure to get simple

9) M.F. Deygen in: J.I. Frenkel. Memoirs, Letters, Documents. Leningrad: Nauka, 1986, pp. 200–201 (in Russian).

10) Of the Neva River in St. Petersburg.

11) K.S. Shifrin in op. cit., p. 228 (in Russian).

12) The paper was published in «Zeitschrift für Physik», Bd. 49, S. 31, 1928, and was included in the second volume of «Selected Papers» of J.I. Frenkel.

results in the most simple and intuitive way.» (Let us add that usually only after a simple way to obtain some physical result has been found is the result considered «simple».)

A lot of complicated questions regarding theoretical physics might be, in principle, clarified by means of a comparatively simple argument, though the questions are investigated and the corresponding issues sometimes obtained with the help of rather complicated mathematical tools. A simple solution calls for a physicist with clear vision and intuition.[13] Frenkel not only was capable of that, but the direct and immediate path to physical results grounded on this kind of vision (or analysis) was a typical feature of much of his work. There is no need to translate Frenkel's original papers into a simple language: they are written in it from the very beginning. It was for good reason that he liked to say that for successful work in the field of theoretical physics, the knowledge of differential and integral calculus and the theory of differential equations is enough. It is true, though, that in this (probably deliberately exaggerated) statement about the sufficiency of such a modest mathematical knowledge, there was a silent recognition that a theoretician requires a not insignificant intuition about physics, in the same measure of which Frenkel himself had been so generously endowed.

In those of Frenkel's papers already mentioned, both these sides of his talent are pronounced in a refreshing way. We illustrate this with an example of Frenkel's derivation of the expressions for the zero energy of electron gas and its spin paramagnetic susceptibility based on the Pauli principle.

In Frenkel's imaginative terms, the Pauli principle is a «residence law» governing electron behaviour. The picture of electron distribution over energy levels «recalls the image of a large hotel in which people are trying to find their place on lower floors but are restricted in this by a rule stating that only one married couple may reside in each room (so that the opposite direction of electron axes correspond to a different sex).»[14] In order to make the similarity even more complete, we have to imagine that the whole building consists of similar rooms (the number of which at each floor corresponds to the number of states with a given energy), and that the number of floors is infinitely large. One can imagine that this skyscraper is built up in the phase space [6-dimensional space of momenta p and spatial co-ordinates. — V.F.] of classical statistics, so that all the rooms, i.e., electron phase cells, as they are usually called, have exactly the same volume h^3», where h is the Planck constant. A phase cell may be of any shape. At the temperature of absolute zero, electron distribution over the cells is determined by the condition that the total energy of the system must be minimized. Choosing

13) E.I. Rashba made a good point on this in his article dedicated to the 90th birthday of J.I. Frenkel: «The most prominent feature of J.I. Frenkel's talent seems to be an extraordinary mind when it comes to questions of physics, which sometimes is involuntarily perceived by us now as a clairvoyance.» (UFN, 1984, v. 144, # 2, p. 347 (in Russian)).

14) The Pauli principle does not forbid «divorces» when an electron goes to another free or partly occupied «room», but absolutely rules out a «ménage à trois». (Comment by J.I. Frenkel).

elementary cells in the shape of concentric strata in the momentum space (their volume should be multiplied by a common volume V of electron gas coinciding with the bulk volume of the metal under consideration), according to the Pauli principle, one arrives at

$$4\pi p^2 dp \cdot V = h^3.$$

To N electrons of metal in the momentum space corresponds a sphere with radius p_0 defined by a formula

$$\frac{4\pi}{3}\frac{p_0^3 V}{h^3} = \frac{N}{2},$$

where

$$\frac{N}{2} = Z_0$$

is the number of elementary cells populated by electron pairs. To a cell with radius

$$p < p_0$$

corresponds kinetic energy of an electron pair equal to

$$2\frac{p^2}{2m}.$$

The minimum kinetic energy of the whole electron gas may be written as a sum

$$\sum_{i=1}^{N/2} \frac{p_i^2}{m}$$

over all Z_0 cells. Representing the sum as

$$\frac{1}{m}\sum_{i=1}^{N/2} p_i^2 \frac{4\pi p_i^2 V}{h^3}\Delta p_i$$

and passing from summation to integration (which is admissible for large values of Z_0), for the «zero» energy E_0 one finds:

$$E_0 = \frac{4\pi V}{mh^3}\int_0^{p_0} p^4 dp = \frac{4}{5}\frac{\pi V}{mh^3}p_0^5$$

$$= \frac{4}{5}\frac{\pi V}{mh^3}\left(\frac{3N}{8V}\cdot\frac{h^3}{\pi}\right)^{5/3} = N\frac{3h^2}{10}\left(\frac{3}{8}\frac{n}{\pi}\right)^{2/3},$$

with $n = N/V$ being the electron density. This formula presents one of the simplest and simultaneously most significant issues of quantum mechanics: it demonstrates

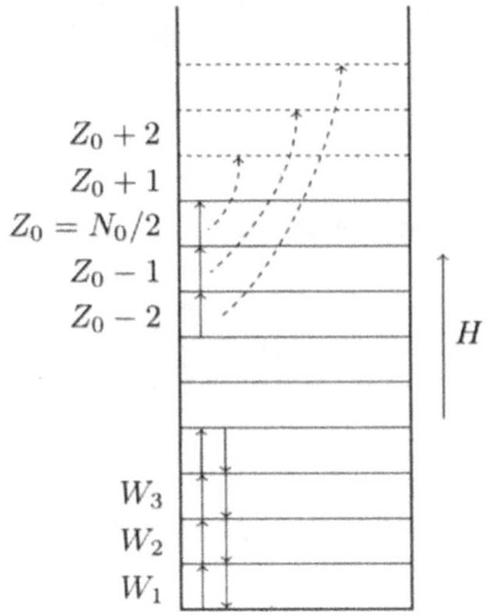

Fig. 2 Diagram showing the derivation of a formula for paramagnetic susceptibility.

that at the temperature of absolute zero electrons in metal possess a non-zero energy determined by their concentration.

Frenkel uses similarly simple arguments when considering the paramagnetism of electron gas.

Let a piece of metal be placed in a weak external magnetic field H. In this field the energy of an electron gas would reach a minimum if all magnetic moments of electrons were oriented along the direction of the magnetic field. But such an orientation cannot happen within each single cell, since it would contradict the Pauli principle. Hence, one of the electrons has to leave its cell and enter the other one, which has energy greater by the value $\Delta\epsilon$. In return, the magnetic energy of such a «divorced» electron pair is reduced by the value $2\mu H$, where μ is the electron magnetic moment. One electron from Z_0 cell (Fig. 2) moves to (Z_0+1) cell; from (Z_0-1) cell to the (Z_0+2) one, et cetera: an electron from (Z_0-k) cell comes to (Z_0+k+1) cell. This will occur until the increase of kinetic energy accompanying this transition does not compensate (more exactly, nullify) the decrease in magnetic energy remaining to be $2\mu H$. If the number of oriented electrons $2k$ is much lower than the total electron number

$$N = 2Z_0 \quad (k \leq Z_0),$$

one may assume the levels within the limits (Z_0-k), (Z_0+k) to be separated by an identical energy interval, and that the kinetic energy growth due to the «divorce» of the last, nth, pair, is $2k\Delta\epsilon_0$. Thus, the last divorced level is given by formula

$$2k\Delta\epsilon_0 = 2\mu H.$$

Introducing the total magnetic moment of all the metal bulk (volume V) and denoting it by M, we have

$$M = 2k\mu,$$

that is,

$$M = \frac{2\mu^2}{\Delta\epsilon_0} H.$$

Since

$$\epsilon_0 = \frac{p_0^2}{2m},$$

$$\Delta\epsilon_0 = \frac{1}{m} p_0 \Delta p_0.$$

From this and the above formulae one easily calculates the value of $\Delta\epsilon_0$:

$$\Delta\epsilon_0 = \frac{1}{m} p_0 \Delta p_0 = \frac{h^3}{4\pi V m p_0} = \frac{h^3 p_0^2}{4\pi V m p_0^3}$$

$$= \frac{p_0^2}{3Z_0 m} = \frac{2}{3} \frac{p_0^2}{Nm} = \frac{4}{3} \frac{\epsilon_0}{N},$$

and, finally for paramagnetic susceptibility defined as the magnetic moment over the volume and the field, one finds

$$\kappa = \frac{M}{VH} = \frac{2\mu^2 H}{\Delta\epsilon_0 VH} = \frac{3N}{2\epsilon_0 V} \mu^2.$$

Replacing ϵ_0 by its average value

$$\bar{\epsilon} = \frac{E_0}{N},$$

corresponding to the absence of magnetic field, and noting that

$$E_0 = \frac{8\pi V}{5mh^3} p_0^3 \frac{p_0^2}{2m} = \frac{3}{5} N\epsilon_0,$$

we arrive at κ an expression identical to the one obtained by Pauli:

$$\kappa = \frac{9}{10} \frac{\mu^2 n}{\bar{\epsilon}}.$$

These methodological findings were used in the well-known monograph of Bethe and Sommerfeld dedicated to the electron theory of metals. Later on they were included in many textbooks, both general and special, though without mentioning the name of their author. This omission is quite natural: otherwise one would have to write a prolix explanatory note to every line of any monograph on physics (and likewise on any other branch of knowledge). In this way lot of Frenkel's methodological ideas once and forever entered educational literature. That is why we found it appropriate to indicate J.I.'s authorship in the case we have just considered.

In the same paper J.I. also obtained absolutely new and important results concerning the nature of ferromagnetism (as he noted in a footnote, these results arose out of a discussion with Ya.G. Dorfman which had encouraged the appearance of the whole «ferromagnetic» section of the paper).

J.I. noticed that an effect similar to the spin-orbital interaction in individual atoms must take place in the electron gas of metals. But free electrons in a metal do not move along unclosed orbits, and their interaction, as J.I. observed, occurs between the proper electron magnetic moment and the total magnetic moment of the electron shell of an ion in a metal crystal lattice. This idea formed the basis for the so-called (*s*-*d*)-exchange model of ferromagnetism developed, originally, by S.V. Vonsovsky.

The second point noted by J.I. Frenkel reduces to the fact that, to explain ferromagnetism, one should exploit not only the Pauli-Fermi statistics, in the same way as had been done by Pauli for the above-mentioned example of paramagnetism of electron gas, but also the notion of the so-called exchange interaction between electrons. This concept had been introduced by Heisenberg (when constructing the helium atom theory). Frenkel was the first to demonstrate that the exchange interaction, and additional energy of electron gas related to it, play a cardinal role in the ferromagnetism phenomenon, which he explained qualitatively. Independently and in more detail this was also explained a little later by Heisenberg himself.[16] It proved that the minimum of the total electron gas energy, for a number of substances, is achieved under the condition that all the electron pairs in the absence of the magnetic field be «divorced», i.e., occupy N steps of the energy staircase and have similarly oriented magnetic moments (even at the temperature absolute zero).

This paper of J.I.'s where the nature of ferromagnetism is explained, is associated with his and Ya.G. Dorfman's study of the problem of spontaneous and induced magnetism of ferromagnetic bodies.[17]

According to the above concept, electrons in a ferromagnetic substance occupy the floors of the phase space sky-scraper in such a way that their spins are oriented in a certain direction. Nevertheless, usually ferromagnetic substances do not generate their own magnetic field without some external one; that is, they are not constant magnets. According to P. Weiss, the reason for this is that they consist of a row of regions, «elementary magnets», or «bunches», as Frenkel and Dorfman called them (the proper modern term is «magnetic domain»), whose total

16) W. Heisenberg. Zeitschrift für Physik, Bd. 49, S. 619, 1928.

17) J.I. Frenkel and Ya.G. Dorfman. Nature, v. 126, p. 274, 1930.

magnetic field is zero. The existence of domains and the shape of their boundaries were revealed when an emulsion of the finest iron powder was placed on the polished surface of a ferromagnetic substance (the method of powder patterns, 1931). The powder particles gather at the places of contact between neighbouring domains, where local magnetic fields are present. They form the so-called Bitter-Akulov strips, which sometimes are of a very quaint shape (for instance, in the case of cobalt). The domain boundaries differ from those of the «microcrystals» of a polycrystal body (the latter are produced by etching of a polished polycrystal surface) by exhibiting a lack of chaos, the presence of a certain structural order.

The zero value of the total magnetic field is similar, to some extent, to the case of bodies composed of dipole molecules. Because of the disorienting action of thermal motion, the latter do not produce a resulting electric field (the essential difference is that there are 10^{12} atoms in each domain).

The reasons why ferromagnetics fragment into domains was first clarified in the paper of Dorfman and Frenkel. This fragmentation proves to result from a competitive action of two types of forces: the usual magnetic ones and specific quantum forces, namely, the exchange forces.
The former are long-range ones; they try to demagnetize a body (two interacting magnetic arrows at rest would position themselves in such a way that the north pole of the first would be opposite the south pole of the second, and vice versa). The exchange forces, on the contrary, tend to arrange electrons in such an order that their spins are parallel. «The action of these factors», wrote J.I. later, «results in the clustering of electrons at small distances in domains, while the latter, under the influence of magnetic forces, take a different orientation which compensates in the mean for their magnetic action.»[18]

The influence of an external magnetic field produces an orientation of the domain magnetic moment along this field. This orientation is reflected macroscopically by the magnetization dependence on the external field (the upper branch of the hysteresis curve). The Earth's magnetic field may play the role of such an external field; its value is large enough for the domain orientation in some materials.

Frenkel and Dorfman determined the average domain size, l; it proves to depend on the size, L, of the ferromagnetic substance, namely,

$$l \approx 0.01 L^{1/2}\ \text{cm}.$$

This kind of dependence of physical properties on the dimension of a body (which does not correspond to its natural origin but is imposed «from the outside») is often met with in physics. For instance, the brittle strength of a body depends on its size, etc. However, the law

$$l \sim L^{1/2}$$

18) J.I. Frenkel. Introduction to the Theory of Metals. Moscow: Fizmatgiz, 1958, p. 114 (in Russian).

does not hold for sizes below

$$L \approx 10^{-5}\ \text{cm};$$

for smaller sizes, fragmentation into domains does not take place. A body of this scale has a one-domain structure predicted in the paper we are describing and supported by experimental evidence. Note that the strongest magnets are manufactured of alloys of such one-domain grains (the alloy of iron, cobalt, copper, alluminium and nickel, called magnico, is produced in Russia).

The theory of Frenkel and Dorfman later allowed people to sort out everything needed for such significant technological processes as ferromagnetic magnetization, hysteresis and so on. In 1935 a more sophisticated theory of the ferromagnetics domain structure was developed by L.D. Landau and E.M. Lifshits.[19]

Just after this paper, J.I. published another one.[20] He gave it the title «Application of the Pauli-Fermi theory to the problem of cohesive forces», which should surely be regarded as ill chosen. Really, one can hardly guess that the paper mainly discusses the structure of atoms and stars, though in accordance with the title, the author actually looks for the cohesive forces which act against the pressure of the degenerate electron gas in the composition of both atomic nuclei and stars.

It is worthwhile to remember at this point that by the end of the 1920s the neutron still had not been found, and therefore atomic nuclei were believed to consist of protons and electrons. J.I. assumed exactly this model, so his attempts to construct the theory of nuclei are of only historical interest now. Nevertheless we would like to reproduce his arguments briefly.

The size of an atomic nucleus (about 10^{-13} cm) had been known since those studies of Rutherford's which led him to the planetary model of an atom. From here the value, by the order of magnitude, for the electron density followed, $10^{39}\,1/\text{cm}^3$. Substituting this value into the expression relating the maximum momentum of particles to their concentration,

$$n = \frac{3}{8}\frac{\pi}{h^3}p_0^3$$

(it is derived above), one finds for the momentum a value of about 10^{-14} CGS. This corresponds to the velocity $\sim 10^{13}$ cm/sec for an electron and $\sim 10^{10}$ cm/sec for a proton. Relying on this reasoning, J.I. used the mechanics of the theory of relativity to calculate the electron kinetic energy. This theory provides $\frac{3}{4}cp_0$ for the mean electron energy, c being the light velocity. Electrostatic forces determined by potential energy U,

$$U \approx -\frac{e^2}{h}p_0,$$

19) L.D. Landau, E.M. Lifshits. Physikalische Zeitschrift der Sowjetunion, Bd. 8, S. 157, 1935.

20) J.I. Frenkel. Zeitschrift für Physik, Bd. 50, S. 234, 1928.

proved to be insufficient to oppose the «blowing up» action of kinetic forces associated with the motion of electrons. Frenkel suggested that the role of cohesion forces in a nucleus might be played by magnetic forces related to the existence of the magnetic moment of a proton and an electron. He observed that from this point of view the presence of the electrical charge of particles in the nucleus plays no role at all.

The further development of physics demonstrated, however, that the particles constituting a nucleus are not held together by magnetic forces. Rather, this is the result of action of extremely short-range nuclear forces.

But the second part of the paper of J.I.'s under consideration, related to the structure of stars, is still regarded valid and has even been properly recognized just recently. So what were J.I.'s thoughts concerning stellar structure?

Back in 1923 Eddington conjectured that stars consist of electrons and «nude» nuclei. The reason for this, in the opinion of Eddington, is a very high temperature which appears under the gradual compression of a star. His hypothesis originated as an attempt to explain the density of a Sirius satellite (600 t/cm), which had been determined not long before.

J.I. noted, however, that a high temperature is not necessary for a gas of nuclei and electrons to appear. Metal, which he had dealt with a lot in those years, exhibits a clear example of how collectivisation of electrons and the accompanying partial atom ionization starts even at the absolute zero temperature and under normal density. Indeed, in the condensation of the metal atoms, when they get closer to one another, a gas of free electrons, which belong to the sample as a whole, is formed. If at the absolute zero of temperature the pressure growth in a star were to continue (that might happen due to the work of the gravitational forces), then a successive separation from an atom of all the electrons, not only the valence ones, would take place. After this remark J.I. repeated the arguments he had used when considering superdense nuclear matter. He performed a quantitative analysis of forces developing in such a superdense star at the temperature absolute zero (in practice, at low enough temperature). This led him to the conclusion that «the existence of superdense stars is a gravitational rather than a temperature effect. Temperature, if it is very high, must cause an opposite effect: its further increase has to yield the drop of density.»[21] Frenkel's ideas about how the pressure of the cold degenerate electron gas withstands the gravitational forces formed a basis of the contemporary theory of white dwarfs.[22] It is a pity that he expressed his ideas in passing, so to say, and did not write a special paper for some astrophysical magazine. As a result, his paper remained unknown to a wide circle of astrophysicists.

---

21) J.I. Frenkel. Collection of Selected Papers, v. 2. Moscow-Leningrad, 1958, p. 121 (in Russian).

22) See, for instance, Ya.B. Zel'dovich and I.D. Novikov, UFN, v. 86, # 3, p. 448 (in Russian). In an article by the same authors published in «Pravda» on 9 March 1975, one reads: «Soviet scientists J. Frenkel and L. Landau made a large contribution to the theory of white dwarfs.» See also: S.A. Kaplan. Fizika zvyozd. Moscow: GIFML, 1961, p. 144 (in Russian).

Finally, one last paper of Frenkel's should be mentioned in this survey of three years of his scientific activity — which is dedicated to the foundations of quantum mechanics. In this paper[23] J.I. transformed the corpuscular equations of the rotating electron motion, which he had derived in 1926, to the corresponding wave equations (recall that the Schrödinger equation does not describe the electron spin). He proceeded to expand his idea of the similarity between electrons and the light quanta, photons, the latter representing a corpuscular image of transverse electromagnetic waves. Therefore J.I. generalized the Maxwell equations in such a way as to make them correspond to the behaviour of a rotating electron. He introduced terms to these equations which accounted for the mass and magnetic moment of an electron.

The equations Frenkel derived in this paper are applicable to the description of vector mesons, particles with the spin of 1 and a non-zero rest mass. Such particles were unknown to physics at the end of the 1920s.

Unfortunately, these equations are usually attributed to the French physicist Proca who obtained them in 1936.[24] In the same year J.I. had shown Proca his paper previously unknown to the latter. The reason for the Proca paper's success was that soon after its publication, mesons were first predicted (by the Japanese physicist Yukawa) and then observed experimentally.

The VI Congress of Physicists, about which J.I. wrote to Sommerfeld, was opened on 5 August 1928 in Moscow, in a very ceremonial atmosphere. The whole first plenary meeting was dedicated to the general problems of wave mechanics. The experimental results on the wave nature of matter were surveyed by A.F. Ioffe, and Frenkel reported the theoretical part. Within the same section young physicists D.D. Ivanenko and L.D. Landau contributed their paper about quantum statistics and the causality principle in modern physics.

The Congress was very large and representative: 400 physicists participated in it (143 from Moscow, 83 from Leningrad and 21 from foreign contries). Among the guests of the Congress from abroad were Born, Debye, Landé, R. Pohl, P. Pringsheim, Schell (Germany), Dirac and Ch. Darwin (England), L. Brillouin (France), and Ph. Frank (Czechoslovakia). The participants worked in various sections; about 200 reports were given.[25] In particular, N.N. Semenov explained his theory of the chain reaction of combustion and explosion (30 years later he and English chemist C. Hinshelwood won the Nobel Prize for their studies of chain reactions). Moscovites G.S. Landsberg and L.I. Mandelstam spoke about their investigations of molecular scattering of light by crystals of quartz and Iceland spar accompanied by the change of the wavelength. This phenomenon was later treated quantitatively and explained in a paper by I.E. Tamm that we shall review below.

---

23) J.I. Frenkel. Zeitschrift für Physik, Bd. 47, S. 786, 1928.

24) A.L. Proca. Journal de Physique et Radium, v. 7, p. 347, 1936.

25) S.I. Vavilov. The VI Congress of Russian Physicists. «Nauchnoe Slovo», # 8, p. 95, 1928 (in Russian).

From left to right: F. Aston, E. Rutherford, A. Sommerfeld, L. Brillouin.

Ya. Frenkel's seminar on theoretical physics. From left to right: L. Gurevich, L. Landau, L. Rosenkevich, A. Arsenieva, Ya. Frenkel, G. Gamow, M. Manchinskii, D. Ivanenko, G. Mandel. Leningrad, 1929 («Anno Quanti XXIX»).

The so-called Volga-Congress, August 1928. From left to right: M. Born, R. Pohl, A. Arsenieva, Yu. Khariton (standing), P.A.M. Dirac.

Max Born wrote about the Mandelstam and Landsberg study: «The phenomenon revealed by Landsberg and Mandelstam in crystals is actually identical to the effect observed by Ch. Raman and his co-worker K. Krishnan in liquids. Russian physics has all the right to be proud of the fact that this important discovery is due to Moscow researchers who worked independently of the Indian ones and at almost the same time (February 20, 1928).»[26]

At the meetings of the theoretical section J.I. presented two of the four of his papers we have discussed.

It is interesting to note that A.F. Ioffe put forward the idea of the «decentralization» of physics at the VI Congress. Soviet physics at that time was completely concentrated in Leningrad and Moscow. This decentralization plan was implemented in Khar'kov, Dnepropetrovsk, Sverdlovsk, Tomsk and other cities, where various research institutions were organized.

The congressional participants were able to learn directly how physics developed in the Volga River region. The point was that the Congress was arranged in a very original manner: on 9 August, after four days in Moscow, the congressmen went to Nyzhny Novgorod and from there by steamship down the Volga River, with visits to Kazan' and Saratov. J.I. wrote from Saratov:

26) Quoted by: M.S. Sominsky. Abram Fyodorovich Ioffe. Moscow-Leningrad, Nauka, 1964, p. 401.

The Volga-Congress. From left to right: (first person not identifiable), P.A.M. Dirac, L. Landau, Ch. Darwin, L. Rozenkevich (standing), O. Richardson, D. Ivanenko, Ya. Frenkel, Ph. Frank, P. Debye, R. Pohl.

«Saratov, 15 August 1928 [to his parents].

... We have had the opportunity of admiring the Volga in all its beauty (especially absorbing at sunrise and sunset), as well as to swim several times a day in its warm waves near the soft sandy shallows. Yesterday, for instance, our steamship came to a deliberate halt in an unpopulated area, across from a magnificient sandy beach, where about 60 or more people swam, dived, etc. for several hours. Among the swimmers were such esteemed persons as Ioffe, Darwin and Debye. Pity the scene was not photographed. Time on board the steamship is flying fast and vigorously. There are a lot of interesting people who are useful or pleasant to speak or chat with. Many happy youngsters, whose company I've been enjoying until late at night or, more precisely, early in the morning. It is not an easy job to concentrate one's thoughts on a letter under circumstances like these.

I started to write this letter in the morning, in an auditorium at Saratov University, before the lectures. I'm continuing it now at 8 p.m. on board the steamship, where I've just returned. The sound of voices, the tinkling of the piano and general distracting noises are all around me in the small lounge where I'm sitting. I cannot write in my cabin because of the poor light. As a result, I feel absolutely perplexed by the abundance of material I would like to tell you about leisurely.

Early today the Congress was declared closed. Summarizing the issues of the 11 days since it opened, one should admit that A.F. Ioffe's idea of a traveling congress was generally a success. If there were any drawbacks, they were mostly caused by an unlucky choice of lectures for «provincial» meetings, especially in

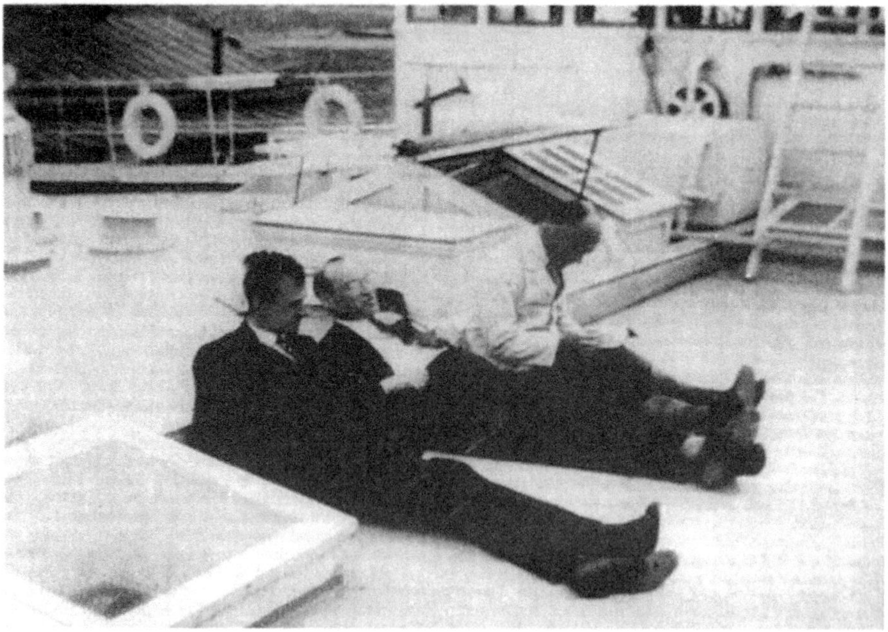

The «Volga-Congress», on the deck of a steamer. From left to right: P.A.M. Dirac, Ya. Frenkel, A. Landé (permission by P. Josephson, author of an article in «Physics Today», September 1988).

Nizhniy and somewhat in Saratov. At an open session for a large audience one should give reports of a popular or semipopular nature.[27] Somewhat complicated presentations remain absolutely incomprehensible for the majority of this type of audience. On the other hand, the closed meetings on board the steamship were extremely interesting and useful for the «active» part of the participants. There were 200 of us on board. The «active» ones were not more than 50, foreign scientists included. In contrast, 400–500 people usually attended the open sessions, the best part of them only remotely related to physics, especially contemporary physics. It's absolutely clear that for these three kinds of people three different types of material were needed. On the steamship I gave 2 lectures on wave mechanics for the «intermediate» audience, which I regard as successful. Several interesting meetings for the «upper» class were held. But the «poor» masses, especially in Nizhny, remained in a position similar to that of the horse served ice-cream instead of oats.»

From Teberda, where J.I. went with a group of young theoreticians, L.D. Landau among them, he wrote to his father:

---

27) J.I.'s lectures in provincial towns in front of large, mostly student, audiences were of exactly this kind.

The Volga-Congress. From left to right: V. Bursian, (standing), P.A.M. Dirac, (unknown person, standing), L. Landau, Ch. Darwin, L. Rozenkevich (standing), O. Richardson (or, perhaps, A. Landé), Mrs A. Shechter, D. Ivanenko, Ya. Frenkel, L. Brillouin, Ph. Frank, P. Debye.

«Teberda, 22 August 1928.
... It is the fourth day now since I've lodged peacefully in this quiet shelter. However, it is not quite peaceful and not completely quiet. Namely, on only the second day we (i.e. my companion and myself) went for a small excursion, about 6 hours long, during which we were constantly racing and generally doing immoderate physical exercises. As a result, I spent yesterday mainly in a sitting position, having occupied my upper limbs with the writing of a book. This morning I resumed this worthy work and continued until dinnertime. At the moment I've just finished with it and am going to undertake a brief excursion on horseback.»

So, half a year after J.I. had eagerly stated in his letter to Sommerfeld that there was no new book on the horizon, we find him writing one. This time he was writing «Wave Mechanics». There was an urgent need for such a book then. J.I. had personally become convinced of it when giving a course of lectures at the Physics and Mechanics Department of Leningrad Polytechnic, as well as when lecturing periodically in front of scientists, primarily PTI experimentalists.

The book was written in Russian, in the hope that it would be published by Gosizdat.[28] However, for the same reasons as with «Electrodynamics», it was initially published in German by Springer Verlag. Frenkel suffered painfully from the abnormal situation with the publication of the book. He wrote on 19 July 1929: «I'm waiting for my German book to appear. With a copy of it I could leave safely.

28) The State Publishing House.

I was informed today that the problem with my Russian book remains as it was. In Moscow they agree to publish it only under the condition of the 100 roubles per page subsidy, while Glavnauka[29] is ready to provide only 50. I hope to get the first positive response to my German book in the near future and on the grounds of this to appeal the decision of the Gosizdat Science Bureau to the Management Board of the Gosizdat or to some higher place.»

Frenkel's «Introduction to Wave Mechanics» appeared in print at the end of the summer of 1929. Along with a monograph by Sommerfeld published a little earlier (also in 1929), these were the first two books in the international literature on physics elucidating this subject (excepting a collection of pioneer papers by Schrödinger concerning new quantum mechanics published in 1928). We shall comment on this work of Frenkel-the-teacher later, when mentioning its other editions.

Along with frequent congresses on physics, conferences on physical chemistry were held regularly in Russia in the 1920 and 1930s (10 such conferences had been organized by 1934). They were mainly aimed at founding of a common language for physicists and chemists, as well as at introducing to chemistry both the theoretical concepts of modern physics — especially quantum mechanics — and also the newest methods of research in physics. Conferences were held in Moscow, Leningrad, Khar'kov and other cities of the Soviet Union. They were organised by leading scientists of the country such as A.I. Rabinovich, P.A. Rebinder, N.N. Semenov, J.I. Frenkel, A.N. Frumkin, and N.A. Shilov. One such conference, the 4th one, took place in 1928 in Moscow and was devoted to the origin of chemical affinity and the structure of molecules. It was attended by many Leningrad scientists who worked at the PTI. J.I. gave a lecture entitled «The Nature of Chemical Affinity» in which he demonstrated the application of the methods of quantum mechanics to the calculation of the energy of interaction between molecules. According to Academician V.N. Kondrat'ev's testimony, Frenkel's paper made a great impression, since quantum mechanics was a novelty or, at the very least, an abstract science not only for chemists, but for many experimental physicists as well. (Viktor Nikolaevich Kondrat'ev recalled that during the conference the well-known Moscow chemist N.A. Shilov had invited the Leningraders to visit his place. When everybody had been seated at the table, general attention was attracted to an enormous cake on which was written in cream: «Impossible to understand, but may be eaten».)

A significant benefit of the physical chemistry conferences lay in the fact that spiritual nourishment became thus not only palatable but necessary for chemists. These congresses played an outstanding role in the process of the consolidation of physics and chemistry and promoted the establishment in the Soviet Union of schools of physical chemistry and chemical physics known the world over.

---

29) The Central Department of Science.

In February of 1929, the general meeting of the Academy of Sciences of the USSR took place. A new group of young scientists, whose scientific activity had been revealed after the Revolution, joined the academic ranks. Among those elected as Corresponding Members of the USSR Academy of Sciences in the Department of Physics and Mathematics were Pyotr Leonidovich Kapitza, Nikolai Nikolaevich Semenov and Jacov Il'ich Frenkel. On February 22, the Indispensable Secretary of the Academy, Academician S. Oldenburg sent this letter to J.I.:

«Dear Jacov Il'ich!
The Academy of Sciences of the USSR, expressing our profound esteem for your scientific services, have elected you as Corresponding Member of the Academy of Sciences of the USSR during its General Meeting this February 13.

Regarding it as my pleasant duty to notify you of this, I hereby inform you that the diploma certifying your rank as stated above will be delivered to you immediately upon processing.

Sincerely yours,
Sergei Oldenburg.»

Approximately at the same time Frenkel obtained an invitation to spend a year in the USA as a Visiting Professor, to give a series of lectures on wave mechanics at Minneapolis University (he had been recommended by Ehrenfest). J.I. accepted the invitation. His departure was planned for September 1930, just after the following Congress of Physicists.

The Congress opened in Odessa at the end of August, 1930. As with the previous ones, it was well attended; this time young physicists from various Republics of the Union participated in it. This fact was reflected in the title of the meeting which, for the first time, read «The First All-Union Congress in Physics». Outstanding foreign physicists were among the guests, in particular, Pauli and Sommerfeld. At Frenkel's invitation, the young German theoretician R. Peierls came to Odessa; K. Ramsauer, F. Simon and F. Houtermans also attended the Congress.

J.I. wrote about the Congress to his parents:

«Odessa, 19 August 1930.
... After breakfast we went to the beach in the company of the German physicists who had arrived yesterday. My old acquaintances were among them — Pauli, and Simon with his wife. Sommerfeld has not arrived yet. Yesterday we met our foreign guests in the hotel they are staying at and spent the evening pleasantly. Pauli has gained some weight but is rocking as energetically as before ...[30]

In the evening of the day before yesterday and yesterday morning we had organisational meetings. This evening the ceremony for the opening of the Congress

30) Pauli, when speaking to someone, always rocked on the tips of his toes.

will be held in the hall of the City Soviet (former stock exchange), and tomorrow I have to give a lecture on wave mechanics at the plenary meeting. It's difficult to satisfy a large audience, and a very mixed one at that, and to have a very broad topic which would require not 1 hour but several sessions to cover everything. Tamm and Pauli will speak after me.»

Along with the problems of wave mechanics, serious attention was also paid at the Congress to electrochemistry. In particular, a joint paper by A.N. Frumkin and J.I. Frenkel was dedicated to this. On the same day Frenkel presented a contribution under the title «Thermal Motion in Solids and Liquids».

Sommerfeld gave a survey of the quantum theory of metals at the Congress. J.I. was his translator. In the process of translating, J.I. sometimes commented on and clarified some points made by the speaker, with the consent of the latter. J.I. judged by the audience's reaction where such comments were necessary and even wrote a number of formulae. After finishing his lecture, heard with interest by the audience, Sommerfeld thanked everybody for their attention and expressed his gratitude to the translator. He added that he himself had found out many new and useful facts during his presentation, nodding with a smile at the formulae written by J.I. on the blackboard. Sommerfeld needed no interpreter to understand them: the language of formulae is an international one.[31]

After the closing of the Congress its delegates made a trip on the steamship «Gruziya» to the Crimea and the Black Sea coast of the Caucasus. Those who participated recalled that J.I. had found a blackboard and chalk somewhere, and on the day after the departure from Odessa there had been a theoretical seminar on board the «Gruziya».

In a postcard sent from Yalta on 25 August 1930, J.I. wrote to his parents: «My dears! So here we are in Yalta. A familiar and pleasant town under only a thin veil of oblivion. On the way back we shall stop here for a longer period, and then I'll walk round all the dear old places ...»

---

31) In 1928, i.e., two years before the events described, Frenkel completed a detailed review of Sommerfeld's paper «Electron theory of metals based on the Fermi statistics». It was published in «Phys. Berichte» (Bd. 9, S. 1051–55, 1928). Sommerfeld knew that this publication went far beyond the scope of a usual review, so his remarks at the Odessa Congress were based on the previous experience as well. The translation of Frenkel's review, because of its importance, was included in the collection of papers: A. Sommerfeld. «The Ways of Knowledge in Physics». Moscow: Nauka, 1973, pp. 103–8 (in Russian).

Chapter 5
In America

Again Frenkel left his home for a long time, and again his frequent and regular letters provide us with the possibility to give him the floor. These letters are the mirror of his soul, and his impetuous, passionate, benevolent personality arises vividly from their pages ...

«New York, 29 September 1930 [to his wife].
I've just visited Columbia University, where I made the acquaintance of some local physicists. It is so pleasant to be met in America in such an honourable and respectful way. Physicists, however, form some kind of a caste, being well known to each other all over the world but sometimes absolutely unknown even to their closest compatriots.

I've also been to the Bell Telephone and Telegraph Company where I met K. Darrow, C. Davisson, R. Bothort and other physicists.»

«Minneapolis, 2 October 1930 [to his parents].
On the morning after a two-day journey in a Pullman carriage I successfully arrived at the place of my temporary job. I was met by Prof. H. Ericsson, Chairman of the Physics Department, and Dr. E. Hill, my deputy. They are both charming, as are all the Americans I have met here. Like most Minneapolis dwellers, Ericsson is of Swedish origin, and perhaps that's why he is lean and tall like a skyscraper. Hill is short and tidy. I don't know how I'll be seen off from America but the welcome, I mean from all the physicists, is fantastic. From the railway terminal I was taken to the university and offered a room at the Campus Club, which is a hostel for unmarried lecturers, and a study at the Physics Building. I liked them both immensely. No sooner had I settled into my apartment and read your letters than I had to start my first lecture. The Americans appreciated my English highly and admired my eloquence. The audience consists of about 25 young people, 20–25 years old, men exclusively. How would you like it? As for me, I've gotten so used to the indifference of women to my subject that I took their absence to be quite natural. Besides students, some professors were present too, among them Ericsson, Tate, Editor of the «Physical Review», and some chemists.

I won't venture to describe Minneapolis because I haven't had the chance to leave the university site yet. The university consists of a large number of beautiful buildings where various faculties are situated. There are more than 12,000 students and over 1,000 lecturers at the university.

I'm enjoying my stay here immensely, both the people and the place itself. As for the conditions, well, what can I say? Imagine, my duties reduced to three lectures a week of an hour's duration each. If I don't succeed here I'll have to admit I'm not cut out for science, but I hope I will, the more so as Hill is a very gifted and pleasant co-worker.»

Towards the end of the 1920s. Physico-Technical Institute, Leningrad. Meeting of the Scientific Counsel. From left to right: Ya. Frenkel, A. Ioffe, V. Bursian (seen from behind), N. Dobronravov.

«Minneapolis, 3 October 1930 [to his wife].
I'm getting plenty of new impressions of the people whom I have to see and the places I have to visit. Last night there was a reception for the teaching staff and their wives given by the Dean of the School of Science, Literature and Fine Arts, to which the physics department belongs. I met a lot of people, and it turns out that I'm the only Soviet citizen among this elite. I was frequently asked about Russia and people expressed sympathy for «the Russian Experiment». Today I was invited for dinner at the Ericsson's. What a charming and natural couple they are! Ericsson gave me his daughter's violin, so I'm going to hold concerts at the Campus Club soon. After the dinner, at 8 p.m., we went to an exhibition of local artists, which had just opened. Most of them imitate the Western European charlatans, but still there are two brilliant masters. Perhaps I'll ask one of them for guidance.

Today my second lecture took place. It was as successful as the first one. My audience consisted of 14 students and 11 professors and lecturers. They say it's a record for my subject; J. Van Vleck's lectures on wave mechanics gathered only 6 listeners. We shall see what happens next. My English causes no trouble at all. My American friends have already gotten used to it and have stopped their compliments on my linguistic talents.

I have not started studying yet. I'm greatly distracted by various meeting and talks which have little to do with science. Yesterday and the day before I spent a lot of time examining the laboratories and experimental work being done. Some of it is very interesting.»

Edward Hill. Minneapolis, 1931 (photograph taken by Ya. Frenkel).

«Minneapolis, 12 October 1930 [to his parents].
I've already told you that so far there haven't been any representatives of the female sex in my audience. At last the first women appeared, apparently attracted by my good reputation as a lecturer. Who do you think it was? A nun! Meanwhile there is a real flower garden of pretty American girls outside the lecture hall! Hell knows where she came from. I hope she will escape soon and stop irritating me with her gloomy look, as if stressing the unpopularity of my subject among the representatives of the fair sex.

I enjoy reading increasingly sympathetic and well-informed articles about the USSR in the American press. Americans have begun to believe in the success of the Russian experiment and the fulfillment of the five year plan in particular. I'm going to write to «Izvestiya» about that and America in general.»

«Minneapolis, 15 October 1930 [to his wife].
Today I started writing my first paper on the electrical resistance of contacts. I want to finish it soon. Next week I'll write another one, which I more or less worked out in Leningrad, so that I can present both to the Program Committee of the Congress of the American Physical Society, which is to be held in Chicago at the end of November. I hope I'll get used to working bit by bit. What I need is at least one more good report for the Congress of the American Association which is to take place in Cleveland, 29–31 December.

The past few days I have felt fresher. Yesterday we went horseback riding with one of my neighbours! It's next to impossible to find a horse in the streets of American cities; that's why riding looks so strange here. However, there are a few stables in the city specially for hiring out horses. We spent about an hour riding along the special road for horsemen around one of the picturesque lakes which make up the landscape of Minneapolis. Riding through the city streets is dangerous because of cars and slippery asphalt. Because of riding I've got aching legs, but nevertheless it was such a pleasant occupation that I will try it once more. Today I made the acquaintance of Burton, the locally known artist who teaches painting and drawing at the Architectural Department. Next week I'll start my lessons, 3 or 4 times a week, together with other students and outsiders. He's such a pleasant man that I'm sure we'll become friends soon.

That's all the news. Oh, I've forgotten to say that today I was interviewed for an article in a university newspaper about me and Russia. I'll send it to you as soon as it comes out. Please tell my parents everything I've just written to you. I'm not in the mood for another long letter now, and there is no need to repeat the same thing again.

Having gotten my first salary, I paid all my debts and really don't know what to do with the rest of it. Minneapolis is a modest city and I'm a bad debaucher.»

«Minneapolis, 19 October 1930 [to his parents].
I've just read in «The New York Times» a telegram from Moscow about a sudden and radical improvement in the supply of the food and consumer goods which up to now we were short of. Now they are beginning to appear in the state and cooperative stores, causing thereby a sharp drop in prices in the private sector of the economy. My heart filled with joy for Russia in general and for you especially. I hope it's the very beginning of a long awaited turning point and that the great hardships we suffered last year will never be back. I'll impatiently keep up with the course of events to come.

I'm gradually getting used to my work. Begun with «ppp», it's going on «crescendo», but has reached «m.f.» only, or maybe even «p». The first paper is being finished at the moment, the second one waiting its turn. These are old projects of mine, begun but not completed in Leningrad, where I was far too disturbed to work.»

«Minneapolis, 21 October 1930 [to his wife].
This morning I got up at 8 a.m and at 8.30 had already started painting an old man who posed for Burton's pupils at the engineering building studio. Nobody uses oils here. At 10.20 a.m. I stopped painting to give my lecture. After that I painted for about 40 minutes more and then went to my office at the physics building. On lecture days I have 2–4 office hours there for my audience and other public. That's where I'll be writing my article. I find it satisfactory. My only wish is to finish it as soon as possible so that I can turn to more interesting subjects.»

«Minneapolis, 23 October 1930 [to his wife].
Recently the student newspaper published my interview, entitled «Frenkel Sees Prosperity for Russia under Soviets». I'm sending you the press-cutting. Naturally, some of my statements were distorted. I never spoke of the reconstruction of the «former power» of Russia but rather of its great future. I characterized my scientific services very modestly and that's why I'm «known as a technical author». As compensation I'm sending you the notification about a lecture I'm to give on 27 October, at the local department of the chemical society. You see that the chemists spare no efforts to advertise me.

Today the Editorial Board of the «Technologist», a journal edited by a group from the technical faculties, asked me for an article about the latest technological and industrial achievements in Russia. I need some materials for it, preferably illustrated. Please address this matter to the Röntgen Institute, to Fakidov or Braude, and ask them to send what's necessary. It would be good to supply me with illustrated journals on this subject in the future.»

«Minneapolis, 29 October 1930 [to his wife].
Yesterday I gave a report about Russia which was a success. Moreover, the audience was very enthusiastic and interrupted me many times with unplanned applause. After the lecture, which lasted an hour, I was asked different questions for half an hour. I got several invitations for private visits and perhaps I'll accept some of them. I have enough time and sometimes I'm bored in the evenings. I've not really gotten accustomed to work because I sleep badly and seldom feel cheerful through a whole day. I need more exercise, so I went horseback riding again, this time feeling much more confident than the first time. I even quit trotting and rushed into a gallop! Can you imagine your husband galloping? How peculiar: I'm learning horseback riding in America, not automobile sports! However, I'm going to take up auto sports soon.»

«Minneapolis, 30 October 1930 [to his wife].
I've just returned home from a visit. It was my first «non-physics» visit. I went to see one of the local professors of forest science, Prof. Zonn. I met him during my

report about Russia. He and his wife left Russia as long ago as 1897. Before that they were both involved in revolutionary activity. By the way, he graduated from the same school as Lenin, though they were in different years. Now the family seems much more bourgeois, but the old training can still be seen. I enjoyed being at the Zonn's and I will go there again.»

«On the train from Minneapolis to Chicago, 3 November 1930 [to his parents].
Today I finished my third picture — a portrait of an old man. I «cooked» it in three sittings and with success. This is not just my own opinion. I don't know what my scientific achievements will be like, but my artistic ones seem guaranteed. Well, there's no trouble with physics either. The only thing I really need is some external pressure to ensure the necessary pace. I get such pressure from the lectures I'm to give in Urbana and the forthcoming congress in Chicago. As a result the pace has started to quicken.»

«Minneapolis, 7 November 1930 [to his wife].
I had dinner at Prof. Loomis'. Though comparably young, a bit older than me, he heads the Physics Department at Illinois University. A few young physicists were present. Curiously, in decent American families the hosts switch off the electricity during dinner and light up candles to make it cozy. Yesterday I had lunch at the University Club together with a crowd of professors in various subjects. I had a long talk with them about Russia. Within this public-gastronomic framework some place was found for my activities as a physicist: lectures, the inspection of laboratories, introductions to specific works, consulting and scientific talks.

At my request I've been given a number of illustrated publications, about Russia — partly in English — from the American-Soviet Trading Council. I use them for written and oral statements. So far I haven't rejected any proposal of this kind, feeling sure it's my duty to use any opportunity for «Soviet propaganda». Soon I'm to talk about Russia at the Faculty Club where I have been invited as a permanent member, which is considered very flattering here, and then at the International Students' Club and some city club. Usually speeches like these are followed by lunch or dinner and thus take lots of time.

I'm going to work hard for the next few days to come so that I can finish my article on the absorption of light in solids. I have managed to solve all the problems, so I hope this work won't take long.

I was greatly impressed by the talking picture «All Quiet on the Western Front». I'm sure you remember this book. Sometimes it was frightening, even terrifying. You see, although my nerves have become dulled it was so hard to watch and listen. It's a brilliant work which keeps strictly to the original text of the book. I couldn't imagine more convincing anti-military propaganda. I'm filled with emotion even now. Surely the impression I've gotten will never be swept from my soul.»

«Minneapolis, 9 November 1930 [to his parents].
Yesterday I put on my dinner jacket and all the accessories, including a starched

shirtfront, a special fashionable collar and a bow tie, and thus dressed arrived at Prof. L. Coffman's, the President of the University. He visited Moscow and Leningrad a couple of years ago to see what the system of public education in Soviet Russia was like. He got the most favourable impressions then.

Yesterday there was a traditional reception for the teaching staff and their wives, or husbands if this was the case, in order for people to meet the new members of the teaching family. Headed by the host and hostess, the newcomers stand in line while other guests pass by shaking hands, introducing themselves and saying «How do you do» or «I'm very glad to meet you». Having found my place in that honourable rank I hoped this ceremony would be over in half an hour maximum. In fact it lasted for $2\frac{1}{2}$ hours, during which time I exchanged greetings with at least 600 persons of both sexes. In the end my head began to spin and if, my best beloved parents, you had emerged in front of me dressed in a dinner jacket and a long dress with almost a train but with a very low-cut back, I would probably have extended my hand and said: «How do you do? I'm very glad to meet you.»

To celebrate Sunday I got up late today, had an hour's stroll, worked a little and then turned to a portrait (in oil) of one of my neighbours.

Then I went to Prof. Zonn's for a Russian-style dinner. At 9 p.m. I was at home again, worked until 11.45 p.m. and then started this letter which I'm finishing now. The only thing to add is that I'm about to nod off, just like M. Zostchenko's chief accountant did, who put down all his thoughts and actions in the form of an official report.»

«Minneapolis, 11 November 1930 [to his wife].
The last few days I have spent working more or less continuously, which has brought me deep internal satisfaction due first to the process itself and second to successful advancement. I'm also pleased with my success in art. Today I turned to Marget's portrait once again and admired the quickness, easiness and success of the work. Together with this very Marget I just returned from a chamber music party. The programme included Schumann's quintet performed by Prof. Frenkel (1st violin), Dr. Marget (2nd violin), Dr. Hill (viola) and two others playing the piano and violoncello. It lasted for 3 hours; we were all very pleased with our performance and have decided to do it regularly.»

«Minneapolis, 14 November 1930 [to his wife].
After a long break I had a headache this morning which, however, didn't prevent me from giving a successful lecture. (The previous one was awful.) Every time, after such a failure, I promise to prepare properly next time, but I have only endurance enough for a few cases. By the way, after the lecture Ericsson showed me a letter from Prof. Loomis in Urbana thanking him for «bringing us in touch with Prof. Frenkel whose lectures here our graduate students appreciated more than any other they have heard in some time» and adds: «You are very fortunate to have Prof. Frenkel with you for a whole year.»

Yesterday when writing my paper on light absorption in solids I ran into an unexpected obstacle. This time I really don't know whether I'll manage to overcome it. Anyhow I was greatly upset, the more so as I was eager to finish the paper as soon as possible.

In a month, on December the 15th, the Christmas holidays begin. I'd like to spend them in New York, or at least the first half until December 29. I promised to give a report at Columbia University before the Christmas holidays begin. They have them from December 21 on. The Congress of the American Association for the Advancement of Science and the American Physical Society opens on December 30 in Cleveland. I suppose I'll give a report there if I have time enough to finish my current work. I must do it. From Cleveland I'm going back to Minneapolis, perhaps visiting Detroit and Niagara Falls on my way.»

«Minneapolis, 17 November 1930 [to his wife].
Yesterday I finished the picture of Marget. It took me four sittings. The whole population of the Campus Club and crowds of guests came on a pilgrimage to my «studio» to see this work of art. Ericsson was so excited that I won't dare to reproduce his compliments. I decided to reproduce him instead, thus thanking him at least a little for his permanent care of me.

Scientific research goes on quite satisfactorily. The obstacle I wrote to you about last time has now been successfully overcome. At last the way is clear, and I hope that next Monday this delayed work will be completed.

I read very little. Reading has been mostly reduced to mine and Hill's joint surveys of current publications. Hill is a nice fellow. His outlook and mannerisms are more like those of a German assistant than of an American one.»

«Minneapolis, 18 November 1930 [to his wife].
I've just returned home from the club where I gave my $n+1$-th report about Russia. The n-th one happened in the morning and the $n-1$-th one on Saturday, three days ago. They all were successful, but I'm getting tired of them. I constructed my evening report very differently from the previous ones. First I tried to describe the impressions the audience would have gotten in Russia as tourists and then supplemented them with more general considerations. As for my Saturday report, I'm sending you its abridged and distorted version in the form of an article from a newspaper.

I feel so tired now. It was a long speech after which I was asked lots of questions. The air in the hall was very stuffy and smokey.

The growing American interest in Russia first appeared this year. This is mostly due to the crisis that America is suffering from. It has shaken their minds rather than their economic institutions. In newspapers or conversations one can see doubts in the sacredness of private property. It's such a new thing for them! Still, more and more Americans are becoming interested in the results of denying or limiting private property that can be, and actually has been, achieved in Russia. Perhaps that's the reason for the popularity of my lectures about Russia.

Today between the two reports I went to the annual tea party held by chemists (and their wives) for better contacts with physicists (and their wives). Electricity was replaced with candles, which is an American custom.

All in all, I'm very glad to have no obligation to go anywhere or make reports for the whole week. I only have to write the article about Russia I've promised a local technical journal.»

«Minneapolis, 21 November 1930 [to his wife].
The day before yesterday I received a letter from Granat Publishers. For a long time I didn't dare open it, being aware the matter concerns an article on Einstein I've promised them. I have to write it in an extraordinary hurry after my scientific paper is finished.

Today I got a letter from Prof. J. Lennard-Jones in Bristol. Following my request to send me the English translation of my «Wellenmechanik» he informed me that the manuscript is already with the University Press publishers, who «are favourably disposed to consider its publication». He finished his letter with the following words «It would be most satisfactory to have an English translation of your book. I have spoken about it to many English physicists and they speak in the highest terms of its excellence.»

Now I really don't know myself where to publish the book: with the local publishers, Oxford University Press or in the American Chemical Society series. In any case I'll have to rewrite it, essentially, by 50% percent, in the near future. As for the Russian edition, I'll agree on it with Romm (Head of Gostechizdat) as soon as I get the letter he was going to write me. I'm very pleased with the success of my book. I guess the revised edition will be even more of a success.

The audience has not become bored yet with my lectures. I think I was right in having constructed the course in such a way that the maximum of physics is combined with the minimum of mathematics. I have put off the mathematics until the end of the course.

Today for the second time Hill and I went to play handball in one of the two rooms on the ground floor of the physics building specially assigned for sports. Having taken a shower after the game I felt an extraordinary freshness in my body and a cheerfulness in my mind. I agreed with Hill to go on with our sports regularly twice a week. I'm so lucky with assistants. It was Levi and Abbat in Leningrad [this is what Y.I. and his co-workers called L.V. Rozenkevich and M.P. Bronstein — the gifted theoreticians — V.F.]. They are all pretty nice. I consider Abbat the most talented but Hill is a brilliant young man, too. I'm going to initiate talks on the Rockefeller scholarship for M.P. Bronstein. Most likely it will happen in New York. I've had a letter from Levi recently. I think his discovery concerning superconductivity is false. Nevertheless he's a fine fellow; he's so persistent!»

«Minneapolis, 24 November 1930 [to his wife].
I've been working hard lately and today I'm finally finishing my paper. Although it has cost me a great effort I'm very content with it now and consider it my best work

on quantum theory. I'm sending it to the editor tomorrow. It's so convenient to have the «Physical Review» editorial board nearby in the same physics building. It's one more incentive to study intensively. Another incentive is to justify my trip and show myself in a most favourable way. My minimum task consists in the presentation of a report at every local congress of the American Physical Society. Before I leave for Russia five congresses are to be held. These external motivations are now being reinforced by a strong inner desire — I mean curiosity and satisfaction brought about by creative activity. I'm sure just these feelings will be of prime significance. Obviously, I had to settle just like a glass of muddy water so that I could once more feel unselfish interest in science. I have a number of ideas now for further studies partly in the same direction — that is, light absorption in solids — and partly in others. Because of my studies I have almost stopped reading newspapers and journals, the more so as their contents aren't very striking.»

«Minneapolis, 26 November 1930 [to his wife].
Today I had to give my seventh brief report about Russia before the local Optimist club, where various people with optimistic minds gather. To prolong high spirits a few optimistic songs were sung in chorus after the lunch. Then my optimistic report was listened to. Although I try to vary my speeches to maintain my own interest I'm getting bored of them. Next week two more reports are expected.

Yesterday I finally sent my paper to the editor, but instead of having a rest I have to quickly write an article for «Granat» about Einstein. Tomorrow I'm leaving for the Chicago congress with a group of local physicists. I expect to be back on the 30th or 1st and then go on with lectures until December 10th, and afterwards go to Iowa and finally to New York.

I've just been interrupted by a phone call from some organization in St. Paul, most likely the Cosmopolitan Club. They asked for a report about Russia. What shall I do? God help me! I answered that I'll do it willingly but only after the Christmas holidays. They seem content with that. By the way, one more report I promised the day before yesterday is scheduled for March. Just think how dreadfully troubled I am with these reports! Nevertheless I'm still glad, and I regret that my voice as a Soviet citizen protecting, moreover glorifying, the new Russia can be heard only in Minneapolis and not all over America.»

«Minneapolis, 27 November 1930 [to his parents].
Today America celebrates the so-called Thanksgiving Day, which is the day of gratitude for the first fruits of the labour which the first pilgrims offered to Heaven. On this occasion I was invited to the Ericsson's for dinner, after which we spent two hours with Mrs. Ericsson playing music from our traditional repertoire: Vernyavskii's «Legend», Vietan's ballad, «Scene de Ballet» by Berio, «Devil's Trills» by Tartini and «Gypsy Melodies» by Sarasate. The audience of Prof. Ericsson personally appeared greatly pleased with the concert. I suggested regarding it as interest on the capital lent to me in the form of the violin.»

«Chicago, 30 November 1930 [to his wife].
Today is the third and last day of my stay in Chicago. I spent the two previous days at the conference and therefore had almost no chance to see the city, but today I undertook a long stroll through the whole eastern part to the centre. I was greatly impressed having unexpectedly found myself in a black environment. White people disappeared; everyone was black instead. Among them there were fat bourgeois, young dandies, children, proletariats; in a word, all social groups were represented. I had thought Chicago's black district was small, but in fact it's a real city which I walked through for more than an hour without meeting a single white man.

The central part of Chicago gives an impression as majestic as that of the centre of New York. I didn't wander there for long but dropped into the Art Museum. It took me more than two hours to see all its remarkably rich collections, especially the French impressionists. The newest works of American artists were exhibited too.

The congress itself, as a matter of fact, was rather dull, but it was interesting to get in touch with a number of American physicists. It was nice to confirm my popularity here, especially with the youth. In spite of its brevity, my report was a success. Everyone was limited to ten minutes, timed by an alarm clock. We didn't have to resort to a gun this time. I was invited to Purdue and Ann-Arbor. In brief, the Americans showed respect for me.

By the way, from time to time in Chicago I came across respectably dressed persons who in Russia might be taken for bourgeois, but in fact they asked me for help in a trembling voice: «Sir, have you some money to give me? I'm unemployed and tired and hungry!» Meanwhile the stores are overflowing with goods. A real abundance of goods! A real joy for Russia was to overcome the capitalist system!»

«Minneapolis, 9 December 1930 [to his wife].
You'll probably get this letter, together with my portrait as a gift, just on the day or on the eve of our anniversary. Watching myself in a mirror I thought about you to keep myself smiling, which I hope I managed to depict. Burton, my artistic supervisor, approved this portrait.

What a pity I can't roll up into a small tube just like this portrait and go back home to Leningrad for a stay as a post parcel. One of the justifications, however, for living apart, if, of course, it's not too long, consists of the refreshing of feelings. Closeness needs to be altered by separation so that it becomes a real celebration. Celebrations cannot be eternal; only day-to-day routine may last forever.»

«On a train, 13 December 1930 [to his wife].
I'm still feeling impressed by my visit to Iowa City. I spent two days there talking almost non-stop about scientific and half-scientific subjects, but always with nice and highly cultured people. Prof. G. Stewart, head of the Department of Physics, cordially invited me to stay at his place. He's a real American and of the best kind — joyful, energetic and ingenuous. Every occupation is a game for him.

Speaking about his X-ray studies of liquids he used the following wording: «I like to play that.» He told me about his work in detail, and I became vividly interested (I hadn't heard about it before) because his results that prove the close likeness between liquids and solids confirm my recent ideas. I formulated them for the first time in 1925 when I was with the Roentgen Institute. I'm going to work them out in more detail soon.

There is a small but capable group of young experimental physicists in Iowa. Among the most interesting research workers are A. Ellet, E. Tyndall and J. Eldridge. They all are well known for their work, especially Ellet.

In the Stewarts' waiting-room I saw with excitement a number of pictures of my acquaintances and friends who had visited Iowa as lecturers before: Ehrenfest, Brillouin, Dirac, G. Wentzel, Schrödinger. From now on your husband has been accepted into this brilliant society. Soon I'll send you my pictures taken in the morning in front of the institute building. Yesterday night, after my second lecture on applications of wave mechanics to the theory of viscosity of gases (it's my original work), I was rewarded by dinner at the University Club. Afterwards I made my regular report about Russia (number 12!!!). It seemed more successful than the previous ones. About 30–40 people were present, mostly physicists and their wives. Everybody was pleased.

On the previous day I had my dinner at the Tyndalls'. Just like ours, their family consists of a charming wife and two kids of 5 and 6 years of age. I played a lot with the younger; we both enjoyed it. I «took» pennies and other coins out of his nose from where they «had grown». My passion for our own children has even improved my tenderness towards children in general; therefore I'm on best terms with the younger child. However, the same refers to the elder one, too.

Enclosed you'll find Mrs. Stewart's reprints on asthma. Her work seems so vital to me that it needs to be translated and published in one of our medical journals. Would you undertake this? I've written to Volodya [his brother — V.F.] about the same thing. I've always been interested in medicine; that's why my talk with Mrs. Stewart still remains one of my most interesting impressions of Iowa. I'm looking forward to seeing Stewart at the Cleveland congress. If you and Volodya don't handle the popularization of these works in Leningrad I'll do it myself by writing an article for the «Vechernyaya Krasnaya Gazeta». I'm going to start familiarizing the Russian public with America with the help of printed words, and, when I am back, with spoken ones as well.»

«New York, 16 December 1930 [to his wife].
Now I have to start an article about Russia I promised long ago so that it can appear in the January issue of the Minnesota journal «Technologist». Of course I'll send you the reprints.

My further plans for New York studies imply rewriting or, to be precise, writing anew the first chapter of «Wave Mechanics». I confess that I have no wish to do that, but I'll have to force myself to.

I wrote to Bronstein long ago; hasn't he received my letter yet? By the way, today I got a pretty good letter from Dirac with a brief description of his impressions of Brussels.

I'm awfully glad that the situation with the food supply in Russia is beginning to improve. No doubt in a couple of years our country will be the happiest one.»

«New York, 17 December 1930 [to his wife].
Yesterday I went to Columbia University. The reception was warm. I was shown the laboratories and acquainted with the work being done. The Physics Institute of Columbia University is situated in an impressive 13-story building. Its internal content, however, doesn't fit completely with its outward appearance, at least in respect to the research. There is not much difference between Columbia University and the Illinois, Minneapolis or Iowa universities in this respect. Theoretical physics is represented by Mr. I. Rabi, a nice young man who has just returned from Europe where he had been working with O. Stern and M. Born for a year. Another New York university has got G. Breit, a gifted theoretician of Russian origin who has been living in America for a long time. Tomorrow night the New York theoreticians are gathering for a seminar where I'll be the guest and speaker. As for my «address», i.e., my report before a general physics seminar at Columbia University, Prof. G. Pegram, Dean of the Physics Department, has asked me to put it off until the end of February because this week is scheduled for the lecture by Prof. P. Scherrer from Zürich who is going back to Europe by the end of December. The next seminar, which corresponds to the Roentgen Institute Council session, will take place after my return to Minneapolis. Thus the day after tomorrow I'll be at Columbia University again, first at dinner with Scherrer, Pegram and others, then at Scherrer's lecture.

I guess Americans' interest in Russia is primarily due to the sharply increasing economic crisis, especially unemployment. In New York alone there are about 700,000 unemployed (not including members of their families). Just think how, in such terrible need, they are getting nothing but charity from various philanthropic societies. In America, the most advanced capitalist country, there is no insurance against unemployment such as exists even in less advanced European countries. Moreover, the trade union movement involves only a thin layer of working aristocracy with very reactionary views. Great need and crisis, I hope, will stir up the proletariat masses, which are much less developed politically than those in Europe. They tend to ascribe their sufferings to «bad luck», not to the capitalist system. I'll touch on this point in more detail in my letter tomorrow to my parents because of its particular interest to my dad.»

«New York, 20 December 1930 [to his wife].
Yesterday I made a detailed report about my two latest works for the local theoreticians and other persons who have something to do with theory at the theoretical seminar of Columbia University. Rabi, Breit, O. Halpern from Vienna, Prof. Breit's assistant, and others were present, all in all about 25 listeners. Only yesterday I met

Breit. He's an extremely talented theoretician and a good-looking young man of about 30. My report attracted vivid interest, thus confirming my own impressions of the success of my latest American work [on electrical properties of contacts and excitons — V.F.].

In the very latest paper a number of new and thrilling questions are formulated. Although their solution is already outlined I'm going to devote to it the next three papers. It would be excellent if I had time to finish them by the next Physical Society Congress, to be held in New York on 29 February. Either before or just after it I'll make a survey at Columbia University.

I presented Breit with some of my ideas concerning quantum electrodynamics and suggested jointly developing them. He accepted my suggestion with noticeable enthusiasm and we decided to start immediately. After this letter is completed I'll go to see him and have a three-hour talk. Then we'll go to the university to have dinner with Scherrer and to listen to his lecture.

It's not worth getting upset because I have no chance to communicate with physicists better than I in America. One hardly needs constant communication with them. I saw Pauli and Dirac this summer and don't think there is much use in further talks with them or Bohr, etc. I've written you already that I have plenty of fresh ideas, partly brought from Russia, which I hope to master without any help from outside.

Yesterday I went to the American-Soviet Trading Council. I got a warm welcome there, and they expressed their contentedness with my lecturing activity. My proposals, in particular the one concerning an exhibition of Russian editions in America, were approved enthusiastically. I'll soon be there again to see the ASTC Chairman, P. Bogdanov.

Having begun, my studies are going on with a crescendo. Today I began rewriting my book. You know yourself how difficult is to start a new work.»

«New York, 29 December 1930 [to his wife].
Yesterday I spent the whole day in the city. I went to the ceremonial meeting of the Industrial Democracy League dedicated to the 25th anniversary of its foundation, then to a dinner for the same occasion. I've seen and heard a number of outstanding activists of American radicalism and socialism, namely Thomas, a candidate for the presidency, Hillquist, Lapler, Brown and others. J. Russell was present at the dinner as an honored guest. He is an Irish poet who devoted many years to establishing cooperation between Irish farmers. He had the chance to boast of his brilliant results in this field. For me, a Soviet citizen, Irish collective farms seem a child's play in comparison with ours, both in respect to scale and rate of growth.

Enclosed you'll find the two issues of a new magazine «The Unemployed». It will help you form an opinion about the real size of this disaster in contemporary America. Note that unemployment has touched not only workers in an intrinsic sense but office employees and intellectuals too. When you have looked the magazines through give them to Fakidov and Braude so that our public organizations can familiarize themselves with the political situation in America.

The Industrial Democracy League is a purely educational organization founded as a student circle on the initiative of Jack London, Upton Sinclair and other famous American citizens. Nowadays there are lots of departments in a number of American universities which try to involve wide masses of students and intellectuals in their movement. The results, however, have been poor so far. The total number of members is as small as several thousand. Besides, it has no influence on the government or on the masses at all. Events, however, especially unemployment, are stimulating revolutionary activity and inciting the League and the Socialist Party to more active actions. Publishing the magazine is one such action.»

«Cleveland, 30 December 1930 [to his wife].
After having slept in a hotel and had my lunch I rushed to the conference which was to open at 10 o'clock in the morning. Fortunately, I was the third to speak and therefore wasn't late for my own report. As a matter of fact, it was interesting but I guess it was less appreciated by the audience than the previous one, partly because it appeared dashed off because of a shortage of time. Ten minutes seems a long time when you are listening to somebody, but if you are speaking yourself it's next to nothing. No sooner had I said a couple of words than the clock rang. Two minutes later another bell followed and the chairman, Prof. W. Swann, attacked me so furiously that I had to break off.

Along with a great number of recent friends I met a few old ones, e.g., Profs. R. Millikan, P. Bridgman and A. Landé. I also familiarized myself with a great number of physicists whom I had heard of.

Thus a couple of new invitations are likely to arrive — to Madison, Schenectady and Harvard. I note with satisfaction that Americans treat me very carefully and in a friendly manner. Quite frankly, I think my popularity has not been really earned.

Today I spent the day at the sessions and afterwards went to the dinner for the members of the American Physical Society. Then until 10.30 p.m. I talked to various physicists, mostly about my latest work and its probable development, which I'm going to take up in Minneapolis.»

«On a train, 6 January 1931 [to his wife].
Yesterday I had a talk with Bogdanov, the ASTC chairman. I liked his outlook and remarks. The purpose of my visit was to point to: 1) the need for more intensive circulation among the American intellectual elite of the English language magazines devoted to Russia which are published here; 2) the desirability of circulating Russian language editions published in Russia, both fiction and partly scientific literature 3) the possibility of attracting masses of American workers and engineers to Russia for payment in rubles. My first remark was accepted as a guidance for actions. As for the others, it was suggested that I write an article

to «Izvestya» since ASTC, although quite in agreement with me on the matter, has no chance of doing anything. I'll write this article soon; perhaps others will follow. So far I have dealt primarily with telling Americans about Russia. From now on I'd like to study American life more thoroughly so that I can tell more about it to Russians. My dominant impressions of America are of sharp contrasts: rich and poor, knowledge and ignorance, morality and amorality, and so on.»

«Lafayette, 7 January 1931 [to his parents].
I arrived here at 1 p.m. and was met by Prof. K. Lark-Horovitz, head of the Physics Department. First of all he lodged me in an excellent room in the University Club, a very impressive building where students and professors have meals and rest. Although Purdue University is comparably small, about 5000 students, it's widely known for its technical and agricultural faculties. Physics began to develop here only two years ago thanks to intensive support from Prof. Horovitz and the state authorities. At present, the Physics Department is well organized and numbers more students than that in Minneapolis. Curiously, there are only 3 women per 2000 students at the technical faculties. Women engineers aren't popular in America. Generally, women make up 10% of all students.

A little later, I and Prof. Horovitz had an auto tour through the city. He showed me the university buildings scattered over a vast area. Then I was taken on an excursion to the Physics Institute. I was shown the library where I saw my «Electrodynamics» and «Wave Mechanics» being shown off, just like in all the other university libraries I've been to. Then I came back «home». I have had too many homes lately! A short nap revived me and I gave a long lecture. I didn't like it myself, but the comments were favourable; I don't know, however, if they were quite frank. The audience was very heterogeneous and I was not sure the manner I had chosen was the proper one. I hope tomorrow's lecture will be better. What I need first of all is a good sleep, so I am about to finish this letter. It's already past 11.30 p.m. After the lecture I was rewarded by dinner at the local club and then I gave ... what do you think? Of course, as usual, a report about Russia. The task was facilitated by a copy of my article, so I read it aloud and answered numerous questions. This time there was no reason to doubt my success. Discussions lasted until 9 p.m., and then I played a couple of chess games. These were my first games of chess the whole time I have been in America. I won both, thus proving that Russia outstrips America in some respects.

Tomorrow I'm to see the most interesting technical faculties more carefully and to finish my inspection of the Physics Institute. My lecture is scheduled for 4 p.m. At 6 p.m. I'll be in a railroad car again.»

«On a train, 8 January 1931 [to his wife].
I've had time to visit a couple of institutes of the Agricultural Department, namely the Institute for Diseases of Plants and the Institute for Dairy Products. It appears that diseases of plants are very similar to those of animals. Beside parasites they might be caused by viruses, i.e., microscopic organisms. In this case implantation

of a tiny amount of juice extracted from any part of an ill plant into a healthy one makes the latter fall ill. In natural conditions this infection is carried by insects.

At the Dairy Product Institute there is a lecture hall where lectures and practice in preparing ice-cream are conducted. Isn't it funny! Imagine students overeating ice-cream!

During my inspection of Purdue University I thought with astonishment that I don't actually know much about my own university, I mean the one in Minneapolis. We have a habit of looking for something that's far away without noticing what's nearby. I decided to look at the University of Minneapolis after my return as if I had arrived for a brief visit.»

«Minneapolis, 9 January 1931 [to his wife].
I was cordially met by Ericsson, Hill and other listeners. After the lecture I became engrossed in looking through my correspondence: 2 letters from you, 2 from my parents, 2 from Bronstein, one each from Dorfman, Simon, and Bothe, plus a number of parcels and heaps of newspapers.

I was deeply moved by Bronstein's letter in which he expressed doubts about his own talents and advised me to get the Rockefeller scholarship for somebody else. I think his doubts are absolutely unreasonable, being quite sure he will be a good researcher and scientific worker. In Odessa I agreed with Sommerfeld about Bronstein's training with him.

Information from Bronstein about the expulsion of 6 students of the 4th year, F. Vol'kenshtein among them, made me so furious that I jumped up with anger and walked back and forth about my room to calm my nerves a little. I'll write A. Bubnov (Minister of Education) an indignant letter on the matter; I hope it will have an effect.

My letter is coming to an end. I have to write 20 letters more! How would you like it? It's enough work for three days.»

«Minneapolis, 14 January 1931 [to his wife].
I'm eager to know how you are doing? Do you go out? I have lots of other questions to ask you, but there is no point in doing this because I'll get the answers even before this letter reaches you. Sometimes it seems to me that our letters form some kind of a double necklace decorating the part of the globe between Leningrad and Minneapolis. While I'm writing this letter, 5 or 6 previous letters to you and 4 to my parents are on the way. Some are sailing across the ocean, others are traveling by train. About the same flow comes in the opposite direction. These counter-moving letters know nothing of each other. Having met they don't stop for a while to exchange contents but rush by, just like unacquainted people. Sometimes they are outpaced by fast telegrams; then it's likely that the distance between us is not long at all. On Monday morning when I got your telegram sent on Saturday evening I had a feeling that I, too, could be in Leningrad in a few hours. Alas, I'll be there only in 5 1/2 or 6 months. Maybe we'll meet somewhere in Europe. Time will tell.

Today I began rewriting my book. I haven't really finished my answers to Dorfman, Bronstein and B. Finkelstein. Their letters are saturated with various scientific items which have to be thought about before answering them. So much time is needed. I've written a complaint to Bubnov instead, concerning the expelled students, in which I asked him to inquire into the matter and reinstate them for the sake of justice and the state.»

«Minneapolis, 15 January 1931 [to his wife].
Yesterday I sent a complaint to Bubnov concerning the expulsion of the fourth year students. Glad to know they have been reinstated. I'll write to Bubnov again today; sorry for having anticipated this event.

Luckily I have made an essential step forward in the development of my new work. Hope it will go on more easily. The «Physical Review» issue with my previous (second) article has just appeared.»

«Minneapolis, 16 January 1931 [to his parents].
Today it has been a week since my return to Minneapolis. The past few days I have been busy with solving a new problem. The solution seems closer now at the expense of putting off all other activities, music in particular. I haven't even touched the violin for the whole week. However, tonight I listened to a incommensurably more qualified violinist at a symphony concert. Unfortunately, the programme was ultra-modern.

I have to reluctantly admit that I have no skills in the proper distribution and efficient expenditure of time; it simply slips away through my fingers. Alas, nothing can be done about that. Generally speaking I have nothing to blame myself for; anyhow I've had time to do something in science and art.

The first part of my article about Russia is to appear soon with Lenin's portrait and the declaration that «Communism = Soviet power + electrification». I'll send you the reprints. Another part is to appear in February. On Monday I'm leaving for Rochester, Minnesota, which is 80 kilometers from Minneapolis, specifically to give the report about Russia to the personnel of the Mayo Brothers' Clinic, the best and largest hospital in the United States. I'll also be introduced to the most interesting research there. What a pity I'm quite healthy, as otherwise I'd have a chance of being treated. I wish I could share my health with Sarra!»

«Minneapolis, 20 January 1931 [to his wife].
A couple of hours have passed since I came back home from Rochester where I had given a 3 1/2 hour report about the USSR for an audience of 50 people, mostly Mayo hospital personnel and local intellectuals. I've already stopped counting speeches of this kind. The one yesterday was the 16th or the 18th. A tremendous success. I persuaded them completely; moreover I persuaded myself, or to be precise, excited myself so greatly that I felt an irresistible attraction for my socialist fatherland not because of you, the children and relatives. I read the Soviet New Year newspapers with an excitement I hadn't felt for a long time. The only thing I'm sorry about is

that although my audiences are usually selected ones they are small, which lessens the useful effect of my propaganda.

In the morning I was met by Abraham Shird, Director of the Biophysics Laboratory, a charming elderly but lively American who showed me around his laboratories and introduced me to the work being done. Then he took me to the Institute for Experimental Medicine located 3–4 miles from the city. We arrived there just at the end of an operation on two miserable dogs and the beginning of an operation on a just as miserable a rabbit. The sight was not a pleasant one — I even felt sick by the end — but all in all I enjoyed my visit because of talks on the effect of venom on an animal organism and the regeneration of an animal's liver from a small piece of it (in contrast to the human one). Physiology is a fascinating science; what a pity there is no theoretical way of studying it analogous to theoretical physics, or at least in a non-bloody experimental way. Of course one can get used to anything, bloody operations included, but still I'm glad I don't have to.»

«Minneapolis, 23 January 1931 [to his wife].
Today, after a week's silence, I finally got your letter of January 7 where you mentioned a moral teaching letter sent a couple of days before. Unfortunately or fortunately, I haven't gotten it yet.

Yesterday I read a report for chemists on chemical statics and applications of wave mechanics. Next Thursday I'll give a lecture on chemical dynamics. Local chemists are as ignorant in theoretical physics as their colleagues in Leningrad. By the way, tell Kolya Semenov that Lind sends him his best regards and asks for reprints of his works. Kolya has gained an immense popularity among American chemists.»

«Minneapolis, 27 January 1931 [to his wife].
Here I was interrupted by a pianist whom I recently met, and agreed to play together with. He fetched me in his own car, which is usual here. After a pleasant evening I came back home at 2 a.m. We played Beethoven's and Grieg's sonatas. At first I was perturbed by the apparent superiority of my partner but then I warmed up to our mutual enjoyment. We agreed to repeat this chamber music party in the near future.

Perhaps you aren't inclined to look sympathetically upon my musical and other amusements because they draw me away from my studies, but you should realize that I simply could not live just being busy with physics all the time. I often feel such a terrible melancholy that I can't always suppress it by working intensively. In cases like these music, painting and reading are the best distractions. Great doses of loneliness are hard to bear.

As for my scientific studies I took trouble over them for a couple of weeks and thus have cleared up some essential points and then I put them off for some time to take up reading serious books on physics, which I have neglected for a long time. In fact, I don't even remember when I last read a book like this from the very beginning up to the end. It must have been eight years ago. Now I'm reading Dirac's book with delight. Some of its fragments I have already looked through in Kharkov where I saw them as proofs. I'll finish it in a few days and then I'll have a couple more and then turn to my own work of rewriting the «Wave mechanics». You know, I have come to the conclusion that great creative contractions are no more than a waste of energy. Good ideas are born and grow in our subconsciousness if, of course, there is proper mental feeding. Then they easily come to the surface of our consciousness with no great throes.

You shouldn't think I have become conceited here. It's true that I didn't expect such warm greetings. I lack self-confidence and don't highly appreciate my articles, except for two or three at most. I don't try to achieve the impossible or to exhaust myself by hard work in a hope for worldwide fame. What is it to me? My nature is such that I'm not inclined to concentrate on one single subject. Attempts to narrow the sphere of my activity always make me uneasy. Now I feel the flowing of time more distinctly than ever, the transient nature of the world, the fleetingness of life. I feel it painfully and think the only thing that can compensate for it is the richness of living. I love science and know what creative joy is, but there are lots of other things I love too. Why should I give them all up? Excessive self-restriction to science empties my soul and turns the happiness of creation into torture.

I couldn't fall asleep last night. I was thinking about home, distinctly saw myself in Lesnoe park, in our apartment. I was moving through the rooms, their images bright but not those of my beloved. As for me, the frequent reproducing of an image wipes away its freshness. What about you? I remember Frau Simon telling me in Odessa that the images of people whom she was absolutely indifferent to were very clear in her mind whereas those of her own children were dim.»

«Minneapolis, 28 January 1931 [to his parents].
I was just present at the demonstration of several scientific talking pictures staged by the General Electric Company. They reproduce lectures of outstanding scientists such as G. Bragg from London, Rutherford from Cambridge and Langmuir from the USA. The sound could not be very clearly heard because of the noise from the equipment placed in the same lecture hall and the bad acoustics. The picture and the soundtrack were precisely synchronized, which brought me to the sad conclusion that unemployment will soon touch professors as well, at least in the capitalist world. Only one good lecturer is now needed to serve the whole country every year without really repeating his course of lectures. Jokes aside, I think talking pictures will force out fair lecturers soon; only the best will stay. Others will have to requalify as commentators. I think that the education of young specialists in the USSR can be greatly facilitated by talking pictures. I'm going to write an article for «Izvestiya» on the matter. I'm sending you a fragment of the talking picture of Langmuir's lecture. The sound is reproduced with the aid of a narrow strip on the edge of the film. A teeth-like shadow is converted into an electric current using a photodetector. The current acts upon a telephone membrane, thus

generating sound. See and admire the simplicity of this invention. Give it to Sarra to show to the interested dwellers of Lesnoe.

Yesterday I took a long walk along the high bank of the Mississippi, admiring local sights and cars rushing by, and thinking about quantum mechanics. After all, good ideas can enter my brain even outside the study. My profession as a theoretician provides me with the opportunity of being idle under the pretense of deeply reflecting.

After the lecture, while I was reading my fresh correspondence, a stranger came to discuss with me, as an expert, his original theory about the effect of radio on the weather. He seemed deeply convinced that the good weather that has now come to stay in America is due to the radio waves emitted by numerous American broadcasting stations dispersing clouds. I must say that when speaking to this gentleman — by the way he's a lawyer with nothing to do with physics, I was neither tolerant nor patient and showed him the door in 15 minutes. Letters and newspapers are much more interesting than his theories about weather. Most Americans are inclined to explain it by the Bolshevik effect upon the sky making it hot. I wonder, why hasn't it happened in the USSR?»

«Minneapolis, 2 February 1931 [to his wife].
Last week I busied myself with self-education, reading Dirac's book with great delight. Now I'm going to broaden my sphere of interests by reading some literature on electrical engineering and biophysics. There is not enough space for me in one branch of science! So far I have supplemented physics with the «History of American Civilization» but it moves slowly. I hope I'll finish the picture of Ericsson tomorrow; then I'll take away my brushes and turn to music. This week I'm going to play with my pianist, and also I'll take part in a trio and play with Hill Bach's concert for two violins.»

«Minneapolis, 4 February 1931 [to his wife].
I've just returned from the home of Mr. Gell, an old music fan and a violoncellist, where we gave a concert with Hill and another violinist, trying to play properly Beethoven's and Mozart's string quartets. These exercises made me remember clearly the similar ones at Andreev's with Nicolai Nicolaevich and Lev Sergeevich [Vagin, — V.F.] taking part. Thus my whole American life has grown covered with a veil of unreality as if I had never left Leningrad at all. I am eager to be back with my small family.»

«Minneapolis, 8 February 1931 [to his parents].
Throughout the next $2\frac{1}{2}$ weeks I'll be preparing my reports and studying some scientific issues I had put off before. I'll try to reduce the frequency of my visits to a minimum. The common notion of Americans as very businesslike people for whom time is money doesn't fit in with the reality completely. They waste plenty of time on talks or reading newspapers, especially after lunch and dinner. Dinner, however, usually occurs at the end of the working time. In general, they enjoy having a good time. Germans are much more systematic and stingy in their attitude

toward time than Americans. All in all, Americans are nice people, open-hearted and ingenuous. They are easy to deal with, but intellectual talks with them are sometimes boring.»

«Minneapolis, 10 February 1931 [to his wife].
Today I was woken up by a phone call from Western Union and read your telegram. Looking into the mirror yesterday I saw grey hair on my temples. That gave me no pleasure. Thirty-seven years is something! Once I thought I would never be grey haired because my hair would fall out before turning grey. Anyhow, better grey hair than no hair at all; that's some consolation!

This morning I got an official invitation from Yale University. Since it is located just half-way between New York and Boston I'll call in on the way there and give a lecture. I began writing it yesterday because I expect to devote it to the general issue of the method of analogies in physics. It seems a fascinating subject which needs to be worked out. My nature is such that I can do it in the form of a special paper only.»

«Minneapolis, 19 February 1931 [to his wife].
My «Wellenmechanik» has been accepted for publication at Oxford University Press. That is decided once and for all. Now I have to push hard in order to compensate for the lag and be in time by the end of March or middle of April. During my stay in New York I must rewrite the first part at least.

I wrote you once that I had managed to move from dead centre where I had stood still for almost $1\,^1/_2$ months. Inspiration came to me and helped me to advance the point I was interested in so far that I began writing a new paper for the «Physical Review». So I'll have to write with both hands to carry out my plans.

I forgot to tell you that McGraw-Hill, one of the biggest scientific publishers in America, whom Ericsson addressed with a proposal to publish my «Electrodynamics», replied that they are greatly interested in it and that they are sending their representative for negotiations. I still think that since my books are available for Americans and they find them useful, they should be available for Russians too, in spite of the opinions of the authorities at Gostechizdat.

Enclosed you'll find a press-clipping from yesterday's newspaper where a crazy pastor criticized me severely. Don't worry my darling. I won't be sent out of America, and if I am you might be glad, mightn't you? There is a photo of Prof. Ericsson, too. In my opinion, my painting bears more likeness to him than this photo. Based on his original works, Ericsson gave a brilliant talk yesterday about the mobility of ions in the air. We seldom talk about scientific subjects so it was a pleasant surprise for me to listen to the distinct and persuasive arguments in his report.»

«Boston, 2 March 1931 [to his wife].
I began my exploration of Harvard University with the physics laboratory, which

Professor H. Ericsson. Minneapolis, 1931 (photograph taken by Ya. Frenkel).

Ya. Frenkel and E. Condon. Minneapolis, 1931.

appeared a rather archaic institution. Prof. P. Bridgman, however, is working there, successfully studying the effect of high pressure upon various properties of solids. Nearby a new building is being constructed. It seems worthy of Harvard University's glory. Apparently my lecture was a success; at least Prof. Bridgman enthusiastically approved it. My audience was large as usual. I tried to be comprehensible, and I guess I succeeded.

After the lecture I had dinner where, besides Bridgman, some other physicists were present, among them J. Slater, one of the best American theoreticians, and G. Kistyakovskii, that same young physical chemist with whom Semenov intensively exchanged letters once. He's a nice young man who stands out among Americans because of his typically Russian face.

After the dinner we talked about Russia for about two hours. I can't get rid of this!»

«New York, 4 March 1931 [to his wife].
My speech yesterday at Yale about the analogy method in physics should be considered among the most successful ones as witnessed by the comments and applause. I arrived in New Haven at 2 p.m. Prof. J. Zeleny, Head of the Department of Physics, greeted me in a friendly manner. It appeared he had held the same position at Minneapolis for 20 years, and Prof. Ericsson was one of his pupils, though they are almost contemporaries. New Haven is a nice town located between the rocks and the sea shore. As for the beauty of its buildings, Yale exceeds all the other universities I've seen before, especially its new library, a giant Gothic hall. Curiously, women are not permitted to study at Yale and Harvard. What an anachronism! Really, American women have every reason to envy Soviet women who have vast opportunities open to them. Here a woman can be a teacher at most. Of course, there are women doctors, lawyers or even engineers here, but it's so hard for them to occupy a decent position; hence there are very few such women. Thus an American woman who wants to feed herself on her own salary must be a secretary, a salesgirl or a teacher.

In New Haven I was hospitably lodged at Prof. Zeleny's. In the evening after dinner all the other Yale physicists gathered there. We had an interesting talk about the quantum theory and, of course, Russia.

Yesterday evening I listened to Einstein on the radio, who was honoured on the eve of his departure to Europe. I didn't even try to meet him; it would have been hopeless.»

«New York, 5 March 1931 [to his parents].
My dears, I went to Princeton University, made my acquaintance with local physicists, had a talk and meal with them and gave a brief report about my latest work. All in all, it was a nice and useful day. Princeton is a pretty town with splendid university buildings resembling Oxford and Cambridge. As for imitation of their British prototypes, American universities resemble a millionaire trying to lay

a foundation of noble origin under his wealth, but English traditions don't suit simple-hearted Americans.

By the way, W. Wilson was the president of Princeton University until 1912. They say he was a very imperious man who got along only with those who unquestioningly obeyed him. Because of this non-complaisance Americans disliked him during his presidency and now they don't commemorate him at all, while utter dullards like Harding and Coolidge are widely celebrated.»

«New York, 7 March 1931 [to his wife].
I spent last night with my former compatriot, at present professor at Columbia University, Department of Mechanics. On the eve he attended my lecture. He fully confirmed my notion that American «businesslikeness» is a myth. I wrote you that, as for wasting time, Americans are hardly second to Russians. Then my interlocutor said that American technology is a product of the unification of American capital and initiative with foreign brains. It's so curious but, undeniably, a number of remarkable technical inventions in America — e.g., television, which is to be implemented in practice soon — were made by Russian engineers who generally are valued higher than American ones, while we invite Americans to manage our industry. After the talk we went to the roof of one of the highest New York skyscrapers. From the 68th floor we admired the panorama of New York at night. It was a fantastic sight. When I'm back I'll try to describe it in more detail; now I have no wish to compete with Gogol.

If I described every new impression in detail I would have to quit all other activity, even getting new impressions itself.

It's funny that on March 3 there was a socialist card game at Fridman's. About 25 local socialists played bridge until 1 a.m. in order to get money for the party's needs. Such socialists won't build socialism.»

«On a train, 10 March 1931 [to his wife].
Yesterday before my departure I went to the board of directors of the Rockefeller Foundation with the matter concerning Bronstein. They greeted me cordially and recommended I apply directly to Paris where all European affairs are determined. I'll take the matter up with them soon.

I've been to a number of other institutions, Columbia University among them where I spent a few hours with Breit and Rabi. In the evening I went to the Guild Theater which corresponds to our Art Theater. Unfortunately, I understood only half; besides, I was tired and sleepy. As a result, my sensitivity was decreased and I got less pleasure than expected.»

«Minneapolis, 12 March 1931 [to his wife].
It has been half a year since my departure. Looking back I'd say it has been a long time. So many things have been seen, heard and felt. Four months left till my return. They will pass more quickly not only because four is actually less than six but because I've got accustomed to the new surroundings which causes a virtual reduction of time. Einstein didn't forsee reduction of this kind!

I have plenty of work to do in connection with rewriting the English translation of «Wellenmechanik». Although it draws me away from my so-called research work, it generally is a pleasant occupation. My third paper for the «Physical Review» and the report for the physical congress to be held in Washington on 1–2 of May need to be completed immediately. Another distracting factor is the compulsory visits.»

«Minneapolis, 17 March 1931 [to his wife].
I've been working diligently for a few days on my book without doing anything else. I just finished two chapters, which makes up about 20 percent of the whole amount of work to be done, and have given it to Hill to look through. I have to work at a record speed in order to finish it in about a month. I am taking a short break now to finish a paper I began long ago. I feel a bit tired from the latest activity though glad to have moved from dead centre. Perhaps it is my fortune to be addicted to work.

Yesterday I read in «Evening Leningrad» about our institute becoming «patron» to several factories and other undertakings in order to strengthen its ties with workers and industry. Semenov must be the number one man in this affair. What a nice fellow he is!»

«New York, 23 March 1931 [to his wife].
On my way to New York, i.e., on a train, I managed to write 30 pages of an additional chapter of my book. Here I'm working less productively, but I still wrote 15 pages yesterday.

Lately I've been rewriting my book, which is progressing rather successfully both in respect to quality and quantity. I hope I'll finish it in 3–4 weeks, and then I'll send it to Oxford. Students are employed in typing up 5 copies, one for me. Shall I send a copy to Leningrad for translation into Russian? Who will do that? Would you do part of it yourself? However, the question has to be settled with the State Publishers first.

I'm going back to Minneapolis tomorrow, making a one-day stopover at Schenectady for a visit to the General Electric Company where I have been invited. You must already have become used to my mobility, which even astonishes Americans. Americans are hardly more mobile than Russians so I can boast I know America better than most Americans. After my visit to the Pacific Coast my education in this field will be completed.»

«Schenectady, 30 March 1931 [to his wife].
Today I arrived in Schenectady at 5 a.m. Although I slept for two hours in a hotel I can hardly stand after a long and hard day. In the morning I saw the General Electric Company's labs. I made my acquaintance with the new works of I. Langmuir and his co-workers. I'll write Dorfman in more detail about one of them concerning magnetism. I saw the famous Coolidge electron gun which induces the strikingly beautiful fluorescence of various minerals. I was especially

interested in those of Langmuir's works that provide the opportunity to find out the nature of ball lightning.

I was met enthusiastically, and then a breakfast was given in my honour. After seeing the labs I gave a report based on one of my papers on metallic contacts. The audience, which consisted of such outstanding employees such as Drs. J. Whitney, Langmuir, E. Hall, W. Coolidge and others, was very pleased with my work.

I spent the evening with a young physician, V. Rozhanskii, Dmitri Appolinarievich's distant relative, and his colleagues from Union College and their wives. Of course, we talked about Russia. By the end I felt terribly tired and could hardly keep up the conversation.»

«Minneapolis, 1 April 1931 [to his wife].
I was surprised by the news that Peierls has married Eugenia Kannegisser. Now I see why he came to Russia so willingly and studied Russian so diligently. Really, physical congresses are bridal fairs. My best regards to the newlyweds. Is Peierls going to stay in Russia any longer?

... Now I'm completely engrossed in my book. During my stay in New York, the round trip included, I wrote about 100 pages. I hope it'll be finished in 3–3 1/2 weeks. Unfortunately I guess I'll be drawn away by numerous public duties one doesn't have in the Soviet Union, i.e., visits, meals with visitors (by the way, it's a splendid custom worth adopting) and so on and so forth. Today I went to a dinner in honour of an English scientist, political economist and socialist named H. Lasky. After the dinner, as usual, he shared his impressions of the situation in Western Europe. They are rather gloomy. He thinks a new war is quite likely to break out in Europe soon. I took the floor in order to prove that Russia represents the only hope for peace in Europe. If only the war doesn't break out this or next year, God save us from that, then not a single state will dare start it for no other reason than that it will immediately turn into a Communist revolution all over the West.»

«Minneapolis, 3 April 1931 [to his parents].
Last night I was invited to a dinner in honour of Prof. Lasky, a well-known English specialist and a politician not in a practical but in a theoretical sense; I mean he heads the sub-faculty of political economics at London University. Ten professors from the local university were present. After dinner we talked about various subjects, mostly political. Lasky is a very moderate socialist, but he nevertheless considers the Russian Revolution the greatest historical event since the birth of Christ. He fears that a possible war in Europe would be aimed mostly against Russia.

By the way, he told us about his meeting with Dreyfus in a Paris cafe, the very Dreyfus the whole world was speaking about once and who is completely forgotten now. Dreyfus is about 70 now. He participated in the last war as a captain.

I just met a reporter from one of the Minneapolis newspapers. We talked for about an hour, and I finally seemed to persuade him completely. We'll see what he writes. You will see, too; let Sarra read it for you.

Now it's 6.30 p.m. I have to leave for a dinner organized by Ericsson in honour of A. Haas. He has given lectures here for a small audience, the first about relativity theory and wave mechanics, the second about Gibbs' works on statistical mechanics. The former was fair, the latter much better. In general, he's not a very original or interesting person. He's extremely sensitive to praise and naively immodest. For instance, he expressed regret at not being informed about the translation of his book into Russian because this fact could have been noted in his biography.»

«Minneapolis, 9 April 1931 [to his wife].
Having read the excited reviews of the ballet news I thought with shame that I haven't taken advantage of my stay in America; in this country where dancing is the favourite occupation of the young and middle aged, I haven't learned to dance. Although I can still make up for lost time — you remember there are 2 $1/2$ months left until my departure — I will have to dance on the chair I'm riveted to with my book. I'm stuck rewriting it completely. In fact it's not rewriting but writing anew. Poor translator! The editor won't be any better off either because the length of the book is growing. All this, however, and my work too, is justified by the improvement in quality. I want to make it a masterpiece. What a pity I delayed beginning this work! Now I have to spur myself on but I feel I'll be worn out soon. In two weeks, however, I'm leaving for a new journey. It will be an amusement and a pleasure for me. First to Madison (April 23) where A. Frumkin worked. I'll give a lecture there about my latest work. Then to New York for a lecture about Russia for local socialists, and then a visit to Princeton. From April 30 until May 2, I'll be in Washington. Then I'll call in on my way back in Philadelphia, at Cornell University and Michigan University.

My impressions of America are gradually crystallizing into more or less accurate formulations. The dominant one is of America as a country of great contradictions, not record figures. I do not mean only the generally known contradictions, such as that between greatly advanced capitalism and the almost complete absence of the proletariat movement, or between plenty of wealth on the one hand and plenty of poverty on the other. Americans consider themselves individualists and America a country of freedom, yet they allow their government to regulate petty details of their private life, whereas the slightest intrusion of the government into commercial affairs is taken as a scandalous encroachment on personal freedom.»

«Minneapolis, 11 April 1931 [to his wife].
It's about 11 p.m. now and before going to bed I'd like to write you a few words, to have some kind of a talk with you. I'm dead tired. I was busy with my book all day long except for an 1 $1/2$ hour break for painting — the model was Ericsson's

secretary who types my papers — and a 2-hour extra lecture. I didn't even have time to eat from 9 a.m. until 8 p.m., except for nuts and chocolate one of the students treated me to. Well, I worked hard but I'm very much pleased with the results. In comparison with the new edition of my book, the old one seems very unsatisfactory. Because of my forthcoming trip to the East, I'm in a terrible hurry, trying to write as much as possible so that when I return I can finish it before my next trip to the West.

You see, I write you about this work so frequently, so you may judge for yourself how thrilled I am with it. This book is to some extent an examination for me of my scientific development in the $2\,^1/_2$ years since writing the German edition. I can confirm with satisfaction that there has been great progress. By the way, I'm going to include my latest work on wave mechanics in it.

Yesterday I went to a concert of a symphony orchestra, the last this season, and had a chance to compare American music («Chanticleer» by Mason) with Russian music («Pathetic» Symphony by Tchaikovsky). Poor Americans!

It has become colder in Minneapolis which, however, doesn't prevent plants from budding. Lawns at the university campus and along the pavements are shining with fresh grass, and the tops of the trees, just like a new-born chick, are covered with the tender transparent green and yellow down of opening buds. The same picture will be seen in Leningrad in our Polytechnicheskii Park by the time you get this letter. You know, intensive writing is linked in my mind with my life in Lesnoe so tightly, that even now working on my book, I feel as if I were in my study at the Röntgen Institute a couple of steps from home. It's so strange to think during a break that I'm 10,000 kilometers from home in distance and three months in time.»

«Minneapolis, 15 April 1931 [to his wife].
The day after tomorrow I'll give a lecture entitled «What does Einstein mean?» i.e., about relativity theory, for a wide audience at the scientific society «Sigma-Xi». If it's a success I'll convert it into a small pamphlet.»

«Minneapolis, 20 April 1931 [to his parents].
There is no special news. On Friday in spite of a terrible headache that tortured me the whole day and the previous day as well, dressed in my dinner jacket I gave a speech for an audience of five hundred on relativity theory. Yesterday I gave my last speech on Russia in Saint Paul, a city grown out of Minneapolis but not united with it administratively. The audience was quite democratic. A representative of the local department of the Communist Party took part in the discussion.

What do you think of the overthrow of the monarchy in Spain? I am pleased.

Today in the evening I'm leaving with the Ericssons for a chamber music party. I and a teacher from a music school with whom I once played sonatas by Grieg and Beethoven are taking part. Yesterday with Ericsson and Hill I saw the Ford automobile plant, similar to ours in Nizhnii Novgorod. The plant puts out 600 cars per day assembled from parts sent from Detroit. Of course, assembling is not all they do. I was particularly impressed by the glass department. Imagine a glass ribbon 12 meters wide flowing permanently for 18 months. From the front huge sheets are cut off and then polished.»

«On a train, 24 April 1931 [to his parents.]
In Madison I gave an hour-long lecture on my latest work. In the evening I learned about the work being done at the physics laboratory, and then I had a talk on various theoretical subjects with Madison theoreticians J. Van Vleck and R. Fowler from Cambridge, who is conducting a seminar in America. By the way, Fowler and Kapitza are the editors of the English edition of that very series of science books in which my «Wave Mechanics» is to appear. Since I've gained the right to move freely in space and time, I think I'll take the third part of it to London myself on my way home to Leningrad. As for the second part, I mean to finish it in May before my trip to California. Yesterday after the lecture I was invited for dinner at the University Club. After dinner, to my great delight, I didn't give a lecture, but on the contrary listened to a report about India.

I left Madison this morning. Tomorrow I'll be in Philadelphia at Prof. W. Swann's (he is the head of a scientific institution corresponding to something like our Röntgen Institute), then 3–4 days in New York and then Washington.»

«Philadelphia, 25 April 1931 [to his wife].
I'll finish the day begun with a postcard to you with another letter to you. I spent most of the day at Swarthmore, a picturesque suburb of Philadelphia where the Bartol Research Foundation that Swann heads is located. The Institute is a tiny one, 25 employees only, but their work is excellent. Swann resembles an artist in looks; in fact he is an artist, a brilliant violoncellist, though I wouldn't say the same about him as a physicist. At Swarthmore I met A. Compton — you know, that same Compton who discovered an effect later named after him, and a pair of other physicists. Having seen the labs I had lunch with the other guests at Swann's. Then I had a talk with one of the research workers and wrote a small paper in my mind. I'll prepare it for publication soon. On my way from Madison I «did» another one. Unfortunately, I had a headache due to lack of sleep.

I spent the evening with the Swanns and the Comptons at the hotel where I am staying. We had dinner together and had a good time speaking about our Minneapolis friends. Swann and Compton once both were professors at Minneapolis for long periods. I had met Mrs. Compton in Italy; she's charming. Among other things, she told me about such a funny incident. Once in Stockholm her husband asked Prof. M. Siegbahn, known for his strict puritanism: «Are you married?» and the response was: «Sometimes». It appeared that Siegbahn had heard «merry» instead of «married».»

«Washington, 1 May 1931 [to his wife].
Only after having written the date did I remember that it's May Day today. This date is not observed in Washington in any way except for the session of the Physical Society. I had the pleasure of meeting Ch. Darwin and a number of

other physicists here. In general America has now become a gathering point for European theoreticians. I wonder how long this situation will last — perhaps until America provides itself with a sufficient number of its own theoreticians. Yesterday after a brief yet adventurous flight, no sooner had I time to calm down than I had to present a report at a small conference organized by Breit. Yesterday evening Fowler made a speech on the same rostrum, and today Darwin's turn came. One of the local physicists drove us around the city for two hours, showing us its remarkable places.

I've just returned to the hotel from the dinner held for the members of the Physical Society. There was a crowd of 600 people there. The dinner itself was mediocre in spite of the rather high price, 3 dollars per «plate», as we say. The speakers who followed were worse still except for Darwin, who said that physics, the mother of other sciences that gave birth to physical chemistry 40 years ago, has done the same to «mathematical chemistry» and is temporarily exhausted by childbirth at the moment.

I'm exhausted too though haven't given birth to my book yet. However, I have given birth to two small things I'm going to publish in the «Physical Review».»

«New York, 4 May 1931 [to his wife].
Send Abraham Fedorovich greetings from L. Loeb whom I met in Washington yesterday. He's divine! He's been insistently inviting me to stay in California for a week after the congress is over to visit Berkeley University where the summer term will begin on June 21. I think, however, that I'll have to refrain from this to avoid over-delaying my return. Besides, Loeb has promised to arrange an official invitation to the congress for me, which implies giving a lecture on my original theory of the absorption spectra of crystals at a plenary session devoted to the discussion about crystals.

I'm very pleased with my stay in Washington. I met and talked to a large number of physicists, among them a few Englishmen (Darwin) and Germans (W. Heitler), listened to several interesting lectures and gave a lecture myself.

Today I saw New York from the top of the newly-constructed Empire State Building. It's 20 meters higher than the Eiffel Tower but the view from it of the adjacent giants whom it proudly rises above is much more effective than the sight of Paris from the top of the Eiffel Tower.

Having finished my investigation of New York and eaten a simple dinner, I was in time for part of a Mass by Bach performed by a chorus of 200 or even 300 singers. When it is performed in Leningrad we'll have to listen to it from the very beginning up to the end. It's enchanting!»

«Ann Arbor, 7 May 1931 [to his wife].
Everything is O.K. I arrived here about midday and had success with the lecture on the «Abuse of Analogies in Physics and Quantum Theory Especially». It's the improved version of a talk I gave at Yale 2 months ago. There is a nice theoretical group here consisting of two Dutchmen (S. Goudsmit and G. Uhlenbeck, a former

pupil of Ehrenfest's), and a German. I knew them all before I came here; I met Goudsmit and Uhlenbeck in Germany in 1926. I was very glad to have a talk with them and share my ideas. In half an hour I'm to speak before a rather closed audience, especially interested in theory, and tell them about about my last and next to last papers. The latest achievements have been obtained during this trip. Sommerfeld and Pauli are coming to Ann Arbor for the summer term. Perhaps I'll see them in Pasadena.»

«Minneapolis, 10 May 1931 [to his parents].
Among other letters whom do you think I got one from? You'd never guess: from H. Stimpson, the State Secretary, that is Minister of Foreign Affairs, signed by the head of the Eastern European Department. What's the matter, you ask? It's simply as follows: a month ago I sent Stimpson two issues of the «Minnesota Technologist» with my article about Russia and a letter in which among other things I mentioned that I would be in Washington soon and would be glad to have a talk with him if he'd like. My suggestion was received and approved; on behalf of Stimpson I was asked to call in at the State Department during my stay in Washington. Unfortunately, the invitation came too late, but I might visit Washington once again at the end of July just before my departure in order to lobby for my fatherland.»

«Minneapolis, 13 May 1931 [to his parents].
There is no news. I write my book diligently and am very pleased with the results. In general, I feel intellectual enthusiasm due to which unclear subjects easily become clear and solutions to various problems come into my mind without any effort. What a pity it has come so late! I wonder if it will last for long.

The old book has been destroyed completely. A new skyscraper is being constructed on the ruins. I'm disturbed by the question of whether I will have time to finish it before I leave for California. I wouldn't like to mess about with it in Leningrad, but there is still a heap of work to do. If the weather is favourable and my sea journey to Europe less «Homeric», I'll have time to do a lot in that week, so the book will arrive in England in an absolutely complete form.

Fowler was here today and drew me away from my book for the whole day. He's an awfully nice man and a very interesting physicist. In the evening a small group of professors and their wives gathered at the Ericsson's so I had to entertain them by playing music.»

«Minneapolis, 17 May 1931 [to his parents].
I'm busy hatching my book, stopping only when my brain refuses to work any longer. Today is Sunday; it's 6 o'clock in the evening. I started working at 10.30 in the morning after a hearty breakfast and stopped at 2 p.m. to have a rest and play the violin. How wonderful that at the proper time you didn't spare money on my music and art lessons. The relaxation and pleasure I feel are sometimes mixed with gratitude that in my childhood I was offered the opportunity to lay the hard foundations for these occupations. I've just finished a new chapter. I've

written 85 pages this week which is less than planned (120–130 pages) but various circumstances prevented me from fulfilling my plans 100 percent.»

«Minneapolis, 17 May 1931 [to his wife].
What a curious country America is! I feel as if I discovered it after Columbus. When I'm back I'll be able to talk about lots of interesting things, but writing about everything is impossible. I have neither enough time nor energy. Even without that, I have to work on my book 6–7 hours a day. After some delay I have picked up a considerable pace but nevertheless I won't be ready in time for my departure from Minneapolis, so I'll have to take it to California with me. I don't know for certain when I'll move there because I have decided to stay here a bit longer to avoid dashing off from my lecture course.»

«Minneapolis, 26 May 1931 [to his wife].
On Sunday, the day before yesterday, I and some local physicists went for a picnic organized at the summer residence of one of them at a picturesque spot on a river bank about 30 miles from the city. I thought that if my head couldn't have a rest then I'd let my hand have one! It's so tired that I can hardly write this letter. I spent the whole day there enjoying splendid air without the smell of gasoline inherent to modern cities. The central point was, of course, the snack, which included the customary ice-cream. I've gotten so used to it that you, my darling, will have to treat me to it more frequently. Then we had some fun each in his own way. I myself had a boat ride; because of this my muscles are still aching.

Today I got hold of and read with delight «The Evening Leningrad» where Ioffe's picture and speech had been published. I greatly sympathize with his plans and passions but in general I read very little now, restricting myself to looking through articles about Russia in newspapers and magazines. When I'm back I might be more interested in articles about America than now. It's typical of human psychology — I mean, to be more interested in what's far away than what's nearby.

It is 6 p.m. already. I have to leave for the session at the Faculty Club where I had the honour of being accepted soon after my arrival. I've been there only three times this year but today is the farewell session which cannot be avoided. I'm leaving Minneapolis in 1 1/2 weeks!»

«Minneapolis, 5 June 1931 [to his wife].
In fact it is no longer the 5th; it is already 3 a.m. Why did I choose such a strange time for writing this letter? It's simply as follows: half an hour ago I woke up after a short sleep and having despaired of falling asleep again I started writing. Let this letter make you or me sleepy.

Today I had a farewell visit at the Ericsson's which ended with a compulsory performance of Wieniawski's «Legend». A few people who had been invited were present too. With one of them I had an exciting discussion about American politics and economics.

I'm leaving the day after tomorrow, on June 7, by the route I described to you once. In three weeks I'll be in New York unless I'm delayed in Berkeley for a few days. Unfortunately, I'm still under heavy pressure from my yet uncompleted book. The last few days I have been so frequently interrupted that I have fallen behind schedule drastically. I'll try to make up for this on my way. Quite frankly, I'm very pleased with this book. I think it's my masterpiece and probably I'll quit writing books after it's finished except maybe for the third volume of «Electrodynamics».

It's a bit sad to part from people whom I've had time to get used to but it's so delightful to think about returning home.

6 June. I fell asleep at last and I'm going on with this letter in the morning. I had a pleasant surprise in my office. There was a bunch of flowers decorating my table and filling the room with a splendid perfume. One of my students just brought me a leather suitcase which was bought at my request (what wastefulness) and some gramophone records. I feel a slight «Reisefieber». My colleagues and students are likely to give me a ceremonial farewell.

It's time to finish. I have got plenty of work to do, especially writing the book, which I cannot neglect even for today or tomorrow. Otherwise I would have to finish it in Leningrad. Don't worry if my next letters are delayed; remember I'll be moving west next week, which is moving away from you. «Pour mieux sauter, il faut reculer», as Frenchmen say.»

«Los Angeles, 15 June 1931 [to his wife].
I'm in the aphelion of my orbit at last. Yesterday I admired the waves of the Pacific Ocean in a seaside suburb of Los Angeles named Venice where bathing and various entertainments are both available.

The conference began today. I now feel quite comfortable at such congresses; they say I bring excitement with my critical speeches. Having listened to a talk, Americans either ask questions or keep silent. They don't criticize for some reason; maybe it is politeness or doubts about their own correctness. As for me, I could hardly be singled out for polite manners or excessive hesitation in expressing myself, which apparently impresses Americans.

... I'm writing these words in a tram on my way home from Hill's. I called in at Hill's being quite sure he had brought me my correspondence but in vain. I got a letter from L. Brillouin instead in which he informed me that I have received my visa for France.»

«Los Angeles, 18 June 1931 [to his wife].
I'm on my way from Santa Catalina, some kind of a Californian Capri. In an hour I'm leaving for Mount Wilson to see the famous observatory where I expect to meet those of my colleagues staying here on the occasion of the Congress.

I'll go to Berkeley tomorrow with a stopover for a couple of days in the famous Yosemite Valley to see the giant trees growing around. In the days since my arrival in Los Angeles I've completely gotten out of hand, I mean I've quit working on the book. I'll take it up as soon as I sit in the carriage. Really I'm a bit ashamed of having such a pleasant, or more correctly, an interesting time here while you are so busy working but, agree, I've been working so hard lately that

I deserved to have some rest and, besides, there is plenty of work still ahead. I'll have to write on the train, in New York, on board the ship and most probably even in France. Even so I'm not quite sure I'll be through with it before I'm home.»

«Los Angeles, 20 June 1931 [to his parents].
The day before yesterday I went to Mount Wilson where the famous observatory is located. The Earth is even more beautiful from Mount Wilson than are the stars. Los Angeles and its suburbs Pasadena, Hollywood, etc., look like a shining sea of lights.

The observatory with its telescopes and towers looks very impressive. Nearby there is a hotel with crowds of tourists. In the morning a sea of light clouds floats over Los Angeles, spread beneath the feet of a spectator on Mount Wilson.

Hollywood, a direct continuation of Los Angeles, is a nice town resembling Pasadena. Unexpectedly its cinematographic nature is not outwardly obvious. I haven't seen any movie stars, but never mind, because it's beautiful enough even without them.

In the morning I took up my writing and hardly left the hotel. I wrote a little but flew into a rage again, so I'll have to keep on working on the train. Fowler consoled me by saying I shouldn't worry about the volume but added that it would be highly desirable to send it off before the end of July, i.e., before I'm back in Russia. I have to go faster.»

«Southampton, 14 July 1931 [to his wife].
My journey reminds me of an anecdote about Russian merchants traveling abroad. In every town they came to they first of all got drunk, then caused a riot and eventually woke up in prison. Thus Berlin, Paris and London, where, among other things, they cut fog with sticks, differed in their reminiscences from the Sanduny Russian baths only in that it was warm, wet and light in Sanduny; dark, wet and cold in Berlin (in a local prison!); dark, dry and cold in Paris; and dark, wet and warm in London. My book has successfully replaced a prison for me. Mostly because of it I made no acquaintances on board the ship. I spent most of the time on deck and therefore got terribly sunburnt and weather-beaten.»

«On a plane, 20 July 1931 [to his wife].
I'm writing this letter under rather curious circumstances, namely in an airplane carrying me from London to Paris. The very fact that I'm able to write under such conditions is proof that I'm fine. This giant biplane with 15 passengers and 3 crew members is flying absolutely smoothly without the slightest tossing. Complete comfort is hindered only by the roar of the two motors. The plane is still above England. England resembles an endless green carpet with dark embroidery formed by trees and occasional constructions.

A second son has been born to Kapitza. Because of this I stayed at a hotel and not at his place in Cambridge. On July 17 I gave a talk. There in England the matter concerning the terms of publishing and finishing my book was finally settled. It will be published in three volumes. The first one has been handed over long ago, the second one is almost completed. As for the third one, I'd like to put it off for a while so as not to feel tied down with the urgency of completing it at home.

I returned from Cambridge yesterday and spent the evening at Prof. Lasky's in charming company.

We are flying through the clouds, which from time to time appear either beneath or above or alongside us and which sometimes cover the whole plane. It might rain soon. The plane has started shaking a bit. We are approaching the sea; I can already see the shoreline. We are flying over Dover or some other port. We are already above the Straits of Dover. Nothing can be seen except water beneath and clouds above and sometimes steamboats. The plane is flying smoothly again because there are no air pockets above the water.

I'll continue my story. I met Dirac in Cambridge and we had a nice evening together at Kapitza's. Both they and Anna send their best regards to you. In $1\,^1/_2$ weeks Dirac is going back to America until the end of the year.

The French shore is already within sight ahead. It's very close already. The plane is so fast that it even outpaces me though I'm famous for quick writing. The day before yesterday Dirac noted that he is equal to one tenth of a Frenkel in this respect.

I'll be at Brillouin's today. Tomorrow I'll visit the Colonial Exhibition and perhaps Metal'nikov. Tomorrow or the day after I'll leave for Nice.

While writing these words the plane has reached the shore and is flying over land again.

It's high time for me to reach my shore in Lesnoe.»

On his way to Leningrad Yakov Ilych wrote a letter to his assistant E. Hill. We have reproduced below a fragment of this letter.

«Riga, 5 August 1931.
I spent 5 days in England and then flew to Paris; it was the luckiest flight in my life. There I saw the Colonial Exhibition and then went to Nice for a few days. I then returned to Paris where I met Ioffe, Brillouin and other physicists. On 31 July, I arrived in Berlin to take part in Schrödinger's seminar where Peierls gave a talk about the work he has done with Landau on the relativistic form of the uncertainty principle. I kept silent for some time and then attacked what I called «Theoretische Experimentalphysik», which results from an overestimation of the significance of the Heisenberg relationships. I suggested my own, or to be precise Newton's, alternative principle, a [probabilistic, but nevertheless actual, — V.F.] «passing through, or reflection» which was accepted by Schrödinger with interest and approval. Besides this, I had a discussion with Peierls about my theory of light absorption in solids. Peierls considers it false and suggests a more strict dynamic theory which seems equivalent to mine in terms of the results.

I saw Dirac in Cambridge. He has abandoned his proton theory and introduced me to a new though unsuccessful attempt to explain the atomic nature of electricity. The paper is to appear soon in «Proceedings of the Royal Society».»

The above letters of J.I.'s do not need any special comment. However, we would like to make just one exception associated with the work on the physics of excitons which he completed in the USA. But first, two short remarks:

The first of them concerns the lecture of J.I.'s at Yale University, mentioned in the letter of 4 March 1931, on the analogy method and, more broadly, on cognition theory. The text of this presentation is partly extant and was later published by me in the collection of popular science papers by Frenkel (J.I. Frenkel. «Na zare novoy fiziki». Leningrad-Moscow, Nauka, 1970 (in Russian)). This lecture is especially interesting since J.I. used the analogy method in a number of his own works, such as those on nuclear physics.

J.I. many times mentioned his work on «Wave Mechanics», the work he had given himself to as a «drunkard», in his own words. The book was published in England in 1932–1936 (two printings of the first volume), in the USSR (1933, 1934), and then republished in the USA in 1950. It produced a large number of reviewer responses and was highly regarded; two examples of its high evaluation are demonstrated in the following chapters.

Let us now proceed with the story of excitons.

Frenkel was accused more than once of anthropomorphism in a popular exposition of some peculiarities in the behaviour of particles, quasi-particles and even physical phenomena. This rather harmless and worthy anthropomorphism manifests itself also in the fact that in the physical literature many remarkable dates celebrate the «life» of particles. For example, many books and papers have been dedicated to the anniversaries of the discoveries of the neutron, positron, meson, the phenomenon of heavy nuclei splitting, the origin of new quantum physics (i.e., the birth of light quanta, photons), the theory of relativity and so on. However, the tradition may be more accurately called «terramorphism» in the case of quasi-particles. The 500th anniversary of the discovery of America was celebrated in 1992, though the continent in question had been formed on the Earth surface God knows how long ago. Similarly, neutrons, positrons, mesons, etc., existed in nature for many billions of years before their discovery by physicists.

A fundamental collection of papers entitled «Excitons»[1] came to readers in 1982, on the 50th anniversary of the publication of Frenkel's paper about light absorption in solids, that is, the discovery of the exciton. A little later, E.I. Rashba wrote a paper on the occasion of another jubilee, 90 years after Frenkel was born[2]. There he called the excitons of Frenkel «the pearl of his creative legacy». It is

---

1) Ed. E.I. Rashba, M.D. Struge. — Amsterdam: North Holland, 1982.

2) Prediction of Excitons (on the 90th anniversary of the birth of J.I. Frenkel). UFN, 1984, v. 144, # 2, p. 347 (in Russian).

always interesting to learn about the author's own attitude to his scientific works[3]. Frenkel stated in his biography (1946) that he considered his treatises on the kinetic theory of the condensed state as the most important of his scientific results. We will never know whether he would have changed his opinion had he lived 3 or 4 years longer, when papers were published on the discovery of the exciton in direct experiments (1951). It was after their appearance that the triumph of excitons in physics and in the optics of solids started.

Frenkel had turned to the problem of the transformation of light into heat in crystals back in 1929 in his little-known paper «The quantum theory of absorption of light» (Nature, 1924, v. 124, # 3133, pp. 758–9), where an attempt was made to investigate the phenomena of the dissipation of optical energy in crystals. As one may see from the exciton papers by Frenkel (1931), he knew well the experimental studies by J. Becquerel (1907), as well as those by I.V. Obreimov and V. de Haas from the late 1920s with the rare earth and iodine spectra at low temperatures which had been carried out in the famous Leiden laboratory. In this experiment the scientist observed light absorption in the narrow lines which are characteristic of excitons, as we know now. J. Slater, whom Frenkel met in the USA, strongly believes that the optical studies of crystals carried out by R. Pohl and his school in Göttingen in the same period provided one more impetus for the appearance of Frenkel's papers.

Aside from these «experimental» foundations, Frenkel undoubtedly had his own theoretical ideas which impelled him to turn to crystal optics. I have written about them in the Russian edition of this book, and they are rather convincingly expounded in the paper by Rashba (cited above). The main idea distills to the fact that the mechanism of the absorption of light by crystals has to be similar to that occurring in gases, where the light energy stored in the excited atoms turns into the thermal energy of gas (in the process of their collisions with other atoms). How does this process develop in crystals? The answer to this question came in Frenkel's papers on the exciton.

The unfamiliarity and complexity of the very formulation of such a question was stressed in the memoirs of Frenkel written by S.I. Pekar. He wrote that in the 1930s at the seminar held at the PTI it was revealed that, to the majority of the participants, the experimentally observed light absorption in a dielectric which did not lead to photoconductivity «... seemed to be absolutely paradoxical, since, according to the band theory recognized by everybody in those years, any excitation of electrons in the undoped dielectric leads to the appearance of a free electron and a hole in the conductivity band ... J.I. asked the bewildered audience: «And light absorption in gas which does not lead to photoconductivity, is it clear?» The voices were heard: «Sure it is clear! It is the photoexcitation of molecules [or atoms, we might add — V.F.] without ionization.» «So then consider a dielectric

3) I can refer, on this matter, to my conversation with I.E. Tamm who told me that he valued his paper on the exchange nature of nuclear forces (1934) much more highly than the papers on the theory of the Vavilov-Cherenkov radiation which earned him (and I.M. Frank) the Nobel Prize.

as a compressed gas,» said J.I. ... The whole idea of the exciton light absorption was in this phrase, clarifying the paradox, as well as the demonstration of the drawbacks of band theory»[4]. At that time there was indeed no place for excitons within the framework of the Bloch band theory!

Passing on to the consideration of the Frenkel paper itself, we would like to make some preliminary remarks. Let us imagine a perfect crystal consisting of atoms of just one sort. They are similar to the strings, tuned to one certain frequency, of a many-stringed but one-toned instrument. If a ringing tuning fork of the same frequency is put close to this instrument, then all the strings are in a resonance: you excite one of them, and all the others begin oscillating. This simple analogy (first used by I.M. Lifshits) allows one to understand roughly that any excitation imposed on one of the lattice atoms is not localized in it but affects the lattice as a whole, transitting from one atom to the other in the form of an excitation wave. But quantum mechanics demonstrate that any motion has both a wave and particle nature: a particle is assigned to any wave. Einstein first made such an assignment, and it led to the origin of the concept of light quanta, photons. Various excitation waves arising in a crystal are also quantified; they are transferred in separate portions called quasi-particles in solid state physics. The usual particles, such as atoms, molecules, and ions, are structural units of matter, its construction material, elementary bricks of a condensed system. But quasi-particles are elementary carriers of motion in condensed matter whose atoms are involved in mutual interaction. Practically all progress in the study of the quantum physics of solids (and, generally, condensed bodies) is related to the fundamental concept of quasi-particles, «the method of quasi-particles is now the most powerful way of treating quantum mechanical problems for solids.»[5].

So what kind of waves can exist in a crystal? First of all they are elastic waves associated with the oscillations of atoms in the crystal lattice. Debye introduced the notion of such waves in the process of developing the theory of the heat capacity of solids. They described well the features of heat capacity, leading to the Dulong-Petit law at high temperatures, and at low ones to the Debye law (proportional to the cube of temperature).

To describe the particle aspect of the elastic wave propagation phenomenon, Tamm in 1930 introduced the notion of the «quanta of sound», thus quantifying the acoustic field. The reason for this consideration and its crucial test was the Raman-Landsberg-Mandelstam effect mentioned earlier. Tamm's «sound quanta» were typical quasi-particles. We would like to note that Frenkel was their godfather: he called them phonons in the first volume of «Wave Mechanics».[6] A philological

---

4) S.I. Pekar in: «J.I. Frenkel. Memoirs. Letters. Documents.» Leningrad: Nauka, 1986, p. 198 (in Russian).

5) I.M. Lifshits. Quasi-particles in contemporary physics. In: «To the Depth of the Atom». Moscow: Nauka, 1964, p. 219 (in Russian).

6) On this subject see: C.T. Walkes, G.A. Slack. «Who named the -on's?» Am. J. Phys., 1970, v. 38, # 12, p. 1380.

analogy[7] between photons and phonons depends directly on the physical one. For instance, Tamm's explanation of the Raman scattering was similar to the explanation of the Compton effect. In the latter case a photon provides an electron with some part of its energy (and momentum), and in the former case some of the energy goes to a photon. The resistance of metals is described in terms of electron scattering by phonons in the modern theory of electrical conductivity.

Frenkel's excitons came chronologically after the phonons quasi-particles. The idea for his exciton paper was formulated very briefly by its author in the summary of the first publication: «Starting from the analogy between a crystal and molecule, it is shown that the electronic excitation, which forms the first step in the process of light absorption, is not confined to a particular atom, but is diluted among all of them in the form of «excitation waves» similar to the sound waves which are used to describe the heat motion in the same crystal.»[8]

During his visit to the United States J.I. concentrated his attention exactly on such currentless (i.e., without photoconductivity) absorption of light in dielectric crystals. «In this case,» he wrote later, «light absorption also proves to be possible in general; however, it results not in the appearance of mutually a free electron and hole, but in the origin of an excited neutral atom. This originating state somewhere does not stay in the place where it was formed but moves uniformly along this or that row of atoms as some particle which I call «an exciton» («an excitation quantum»).[9] The motion of an exciton is characterized by a wave function of the same type as that for the motion of an electron or positive hole, and for not large values of quasi-momentum p it is characterized by the additional energy $p^2/2m$, where m is the effective mass of the exciton (this energy is added to the atom excitation energy corresponding to the formation of an unmoving exciton).»[10]

By the time Frenkel's paper on the exciton appeared, the theories of both exterior and interior photoeffects had already been developed. The first of them (1905) explained the laws of the electron emission from the surface of metals exposed to light. It was in this paper that Einstein introduced the concept of photons, thus connecting his work with that of Planck. Einstein's notions were later used to explain the photoconductivity phenomenon occurring in semiconductor crystals exposed to light — the photons knock out electrons from the atoms inside the crystals. What would, however, happen if the light frequency, ν, were smaller than the atom ionization energy, E_i: $\nu < E_i/h$ (h is the Planck constant)? Experiments demonstrated that both the exterior and interior photo effects are absent in such a case; there was no electron emission then and the photoconductivity was not observed, as a rule. The light absorption is «controlled» by the energy

7) «photos» means light in Latin; «phone» means sound.

8) J. Frenkel. On the transformation of light into heat in solids. I. Phys. Rev., 1931, v. 37, # 1, p. 17.

9) From the Latin word «excito» meaning «I excite».

10) J.I. Frenkel. Introduction to the Theory of Metals. Fizmatgiz, Moscow, 1959, p. 156 (in Russian).

and momentum conservation laws. The latter leads to the selection rule for the «exciton» transitions by allowing for a single state of all the spectrum of values corresponding to the broad enough exciton absorption band.

In his second paper on light absorption by dielectrics Frenkel wrote: «This result provides an explanation of the remarkable phenomenon referred to in paper 1, that the absorption spectra of solid bodies which at ordinary temperatures consist of continuous bands, at very low (liquid hydrogen or liquid helium) temperatures become more or less linear, like those of gases.»[11]

The other significant circumstance associated with excitons and stressed in the Rashba paper reduces to the fact that the features of the exciton absorption of light are determined by the structure of a perfect crystal and not by the impurities or the lattice distortions, to which various effects had been often related. In other words, the results of Frenkel and those later involved in the development of the exciton theory may be obtained «from the primary principles», that is, the features of the structure of the lattice atoms (or molecules) and their interaction in the crystal.

Because of the translational symmetry of the crystal, an excitation like an electron or hole confined initially to some particular atom (or molecule), starts to propagate along the crystal in the form of the excitation wave. Frenkel explains the effect of the translational symmetry by means of an analogy between the crystal and a simpler subject, a molecule. The optical excitation of one of the molecule's atoms is not localized but belongs to the molecule as a whole, «wandering» from one atom to the next. The same analogy allows one to understand why the transfer of optical energy from the excited molecule to the lattice is possible, though the magnitude of the lattice oscillation quantum is approximately 100 times smaller than that of the excitation quantum. Such a great difference seems to eliminate the possibility of this energy being accepted by either one of the oscillators or a hundred of them simultaneously (according to the selection rules). The point here, as well as in the then-known case of the isolated molecule, is that the electron states of the excited and unexcited molecules are different. Therefore the usual (prohibitive) selection rules cannot be applied in this situation. The same goes for the case of the excited (i.e., containing an exciton) and normal (unexcited) crystal.

Returning to the fate of an excitation wave traveling along the crystal, we would like to remind the reader that, according to the «wave-particle duality», some particle image corresponds to such wave motion, and this image is precisely the exciton.

Let us now dwell on the initial responses to Frenkel's first paper on excitons. We learn from his letters that the first reaction to it from Zürich, where Pauli's assistants H. Bethe, F. Bloch, and R. Peierls were then working on the problems of quantum physics of solids, was reserved, if not completely negative. Peierls, who

---

11) J.Frenkel. «On the transformation of light into heat in solids». II. Phys. Rev., 1931, v. 37, # 10. p. 1276.

went to Leningrad from Switzerland, took the news there. J.I. wrote about this to his wife in the letter of 1 April 1931 cited above: «You mentioned that my large paper in the «Physical Review» had been criticized in Zürich and pronounced false. Not only because of my talks with American theoreticians but also because of an inner belief as well, I'm sure they are mistaken. The fact that Pauli considered my work «falsch» [«false»] proves nothing but that it's not «trivial».»

This letter is frequently quoted in papers on the physics of excitons.[12] Recall that W. Pauli was the author of a categorical statement which he often used when evaluating this or that paper of his colleagues, namely: «Entweder falsch oder trivial», i.e., «it is either false or trivial». It is worthwhile to note that the case with the excitons is not the only one where he proved to be wrong. For instance, Pauli rejected the idea of the rotating electron (1925), objected to the conception of the charge independence of nuclear forces (1937), expressed strong doubts regarding the effect of the parity non-conservation in weak interactions (1957). As for solid state physics, Pauli stated many times: «Ich mag diese Physik des festen Körpers nicht» («I really do not like this solid state physics»), and, introducing one of his staff members to a colleague (and again in the same period!), he said rather scornfully about his works in this field: «He is doing some rubbish.» Such remarks are even more surprising since it was Pauli actually who discovered the grounds of quantum solid state physics by inventing his famous prohibition principle, and he also wrote his well-known paper on the paramagnetism of electron gas using this principle. In 1956, in a conversation with L. Pauling, he remarked, not without some surprise, that in all his life he wrote just one single non-mathematical paper (more correctly, one not so mathematically tailored as all the rest of his work) paper. And precisely this paper won him the Nobel Prize, — they were speaking, naturally, about the Pauli principle.

In June 1931, W. Pauli was in Pasadena, California, at one of the congresses of the American Physical Society. Frenkel also attended the meeting. On 15 June 1931 he wrote to his relatives : «I've met [here] a number of my old friends, among them Fowler and Pauli. The latter is as fat this year as he was the previous one. In an hour he, Heitler and another German physicist are coming to meet me at the hotel to have some fun.» We do not know what kind of fun these then-young theoreticians were going to organize, but undoubtedly Pauli and Frenkel discussed the problems of light absorption further. One can infer this from letters written by Pauli from the United States to Peierls in Zürich. In a letter written on 1 July 1931, along with a sharp (and unjust, as we now know) criticism of Peierls' work on the theory of electrical conductivity («In Dreck soll man nicht wühlen» — «One must not rummage in dirt», Pauli categorically advised his assistant), there also appear the following lines: «If you are able to treat the questions of light absorption more

---

12) See the cited paper by Rashba and also: B.P. Zakharchenya «The discovery of the exciton-theory and experiment». Fizika i Tekhnika Poluprovodnikov, 1984, v. 18, # 11, pp. 1940–3; A.A. Kaplyansky. Excitons in semiconductors. In: «Physics: Problems, History, People». Leningrad, Nauka, 1986, pp. 5–19 (all in Russian).

W. Pauli (drawing by Ya. Frenkel, 1937).

precisely than within the order of magnitude estimation, that would be very good. But I am very skeptical of this. Frenkel will probably soon come to Zürich or Leipzig, and perhaps you will be able to convince him of this.» Later Pauli wrote: «... We should pay more attention to our disagreements with Frenkel. Would you be able to meet and convince him? He will be in Europe.»[13]

13) Wolfgang Pauli. «Wissenschaftliche Briefwechsel mit Bohr, Einstein, Heisenberg u.a. Band II: 1929–1958». Springer Verlag, 1982, S. 45 und 89 (in German).

R. Peierls was the first to suggest in 1932, and later to develop, the study of excitons, having criticized Frenkel's assertions that the crystal lattice expansion should accompany the appearance of excitons in it, and that such a lattice was in a steady state.

Interesting material regarding Frenkel's papers on excitons may be found in a book by the well-known American theoretical physicist J. Slater who entered the arena of theoretical physics back in the 1920s (and who probably has no equal in this respect among his colleagues in the United States). We preface these quotations with a general appraisal of Frenkel by Slater. He wrote: «I was quite different from most scientists and in fact from most university people, among whom there were a very great number of Communist sympathizers. But in spite of any political feelings, Frenkel was a great favourite with me as well as with Americans in general. He was a physicist of the type I preferred, writing clearly and comprehensively. His books on wave mechanics written in English, were among the most usable of all those that were published in the early days. He made many distinguished contributions to physics. It was a great pity that he died in the 1950s; he would have been a great help to his country in the difficult days that followed.»[14] We would like to note here that Slater published in Review of Scientific Instruments a very complementary review of the first volume of Frenkel's «Wave Mechanics».

Slater recalled: A. Hippel (a student of R. Pohl's), who emigrated to the United States and organized an experimental laboratory to study dielectrics at the Massachusetts Institute of Technology. The first work he told Slater about was related to the problem of excitons and developed Frenkel's ideas as applied to the mechanism of light excitation propagation in NaCl. These studies were started at MIT by W. Shokley, whose first report at an MIT colloquium in March 1933 was dedicated to the survey of Frenkel's papers on excitons. Slater, Hippel and Shokley provided a foundation for the exciton studies at MIT.

In 1936 Frenkel published one more paper on excitons[15]. Without going into its details, we note just one importantfeature about it, which Pekar stressed (and which is, generally speaking, of a broader character, illustrating Frenkel's creative methods). Pekar wrote: «Sometimes not even a model, but a precise, clear, imaginative expression by J.I. actually provided the formulation of some problem later solved by other theoreticians. For instance, ... in 1936 J.I., when discussing the question of local lattice distortion under the motion of a conductivity electron, wrote that «a light particle behaves itself as if it pulls a heavy weight of atomic displacements». In this imaginative expression J.I. anticipated the theory of polarons started in 1946,» a theory formulated by Pekar himself, we should note.

---

14) J.C. Slater. «Solid State and Molecular Theory. A Scientific Biography». New York, 1957, p. 158.

15) On the absorption of light and trapping of electrons and positive holes in crystalline dielectrics. Phys. Z. Sowietun., 1936, Bd. 9, H. 2/3, S. 158–86.

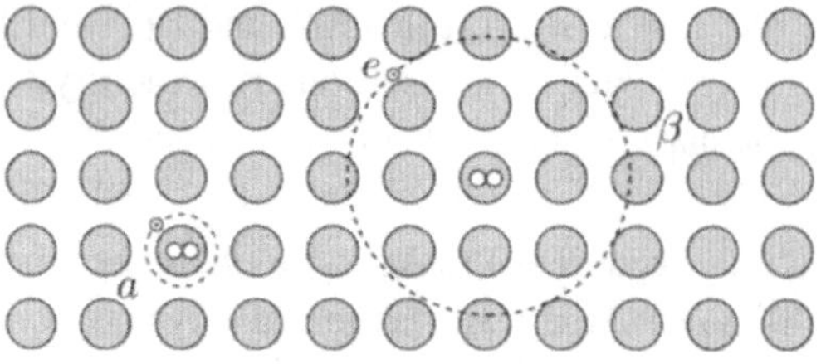

Fig. 3 Schematic representation of an exciton in a crystal lattice. a) The Frenkel exciton. b) The Wannier-Mott exciton.

In a 1936 paper Frenkel generalized critically his previous results in order to account for the arguments of Peierls. Previously he had dealt mostly with the excitons to which a weak excitation corresponds. He called them «small» excitons. They exist in molecular crystals and so are now called «molecular excitons».

The field of action of «small» excitons is molecular (organic) crystals. Several molecules enter the elementary cell of such crystals, and any of them may be optically excited. And to this corresponds the splitting of the optical (exciton) spectrum into several branches, several narrow bands, in accordance with the Frenkel prediction.

Yet in the first papers Frenkel also considered the other possibility, namely, the existence of «megaexcitons». These form a system of an electron and a hole separated by a distance of several lattice constants. Such an «excitation atom», resembling either hydrogen or a positronium atom, propagates as a whole along the crystal. It has a rather large radius, ϵ times greater than the Bohr atom radius, where ϵ is the dielectric constant of the crystal (see Fig. 3).

Megaexcitons occur in semiconductors, for instance, in copper peroxide, a «classical» material of practically all the semiconductor experiments of the 1930s. Megaexcitons are also formed under light absorption in the other «favourites» of semiconductor materials — germanium, gallium arsenide and others. Fig. 4 presents the exciton levels in Cu_2 (see well-known Ch. Kittel' book[16]).

Significant papers by the American physicist G. Wannier (1937) and the English physicist N. Mott (1938) were also dedicated to megaexcitons. Now such excitons are called the Wannier-Mott excitons. The exciton mechanism of light absorption is observed also in polymers. There are grounds for believing that excitons participate in the phenomena of photosynthesis in biopolymers.[17]

16) Ch. Kittel. «Introduction to Solid State Physics» (Russian version: M.: Nauka, 1963, p. 559).

17) For more details we refer the reader for more details to: Polarons and Excitons. Ed. by C.G. Cuper and C.D. Whitefield. Edinburgh and London, 1961; R. Knox. Theory of Excitons. New York, Acad. Press, 1963; D.L. Dexter, R. Knox. Excitons. New York, Interscience Publ., 1965.

Ya. Frenkel working on his paper while having his shoes shined. The street scene was photographed by Dr. O. Heil with his «hidden camera». Borzkomi, Georgia, Summer 1932.

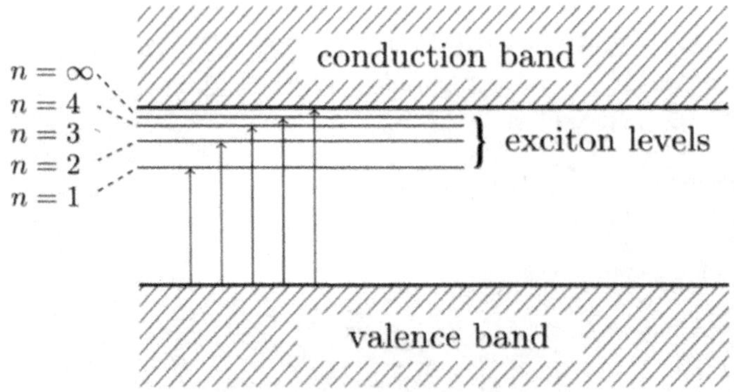

Fig. 4 Exciton energy levels*. The frequencies of the exciton's absorption lines of Cu_2O are described by an empirical formula

$$\nu(\mathrm{cm}^{-1}) = 17.508 - 800 \cdot \frac{1}{n^2}.$$

The contribution of Soviet physicists (A.S. Davydov, V.P. Zhuze, E.F. Gross, S.M. Ryvkin and others) to the new exciton chapter of the solid state physics is known the world over. To the studies which are now being carried out in this direction, the name of their pioneer, Jacov Il'ich Frenkel, is indissolubly tied.

*) Ch. Kittel. Introduction to Solid State Physics. N.Y., Wiley. 1957.

Chapter 6
Prewar Years

During the prewar years Frenkel published more than 100 papers. Their distribution through time is not uniform: the peak of the «article» activity coincides with the low point of that of the «book». This correlation could not escape J.I.'s own notice. However, his work on the books, and textbooks in particular, he regarded as his civic duty, so he had chosen to make it a priority, harming thus to a certain extent his research studies.

Perhaps it was exactly in the 1930s when the full variety of Frenkel's scientific interests manifested itself with a particular clarity. J.I. was then working in practically all branches of physics, in tight collaboration with experimental laboratories of the LPTI and the leading scientists of the Institute, such as A.F. Ioffe, A.P. Aleksandrov, S.E. Bresler, P.P. Kobeko, A.V. Stepanov and many others. I.E. Tamm wrote about Frenkel: «Frenkel combined his passion for the fundamental problems of physics with an interest in concrete physical experiments. The door to his office in the Physical-Technical Institute opened to the assembly hall, and on the days when no meetings were held there, people formed kind of a queue to speak to J.I. (like in a medical doctor's waiting-room). Both theoreticians and experimenters from the LPTI and other Leningrad and non-Leningrad institutes all wanted to tell Frenkel about their research and/or to consult with him.»[1]

It is impossible to recount in any detail the principal results obtained by Frenkel during that decade. We thus confine ourselves to only a very brief summary of these results, making an exception for J.I.'s works on nuclear physics, in view of their value and the role they played in the corresponding branch of physics in the coming years.

1. In the field of the mechanical properties of solids one should first of all note J.I.'s papers from 1938–39 (written with T.A. Kontorova) on the theory of plastic deformation. It was the first microscopic theory of plasticity ever to appear. It demonstrated that a special type of collective atom displacement is possible in a crystal lattice, where the atoms move from one state of equilibrium to the other, leading to the plastic shear. The collective nature of motion is caused by particle interaction. For such collective particle motion to show, some minimal energy, W_0, is necessary, just as it follows from the experimental data. The magnitude of W_0 can be estimated from theoretical arguments. If one arranges the values obtained from these estimates for various substances in a series (for instance, in the order of W_0 growth), the substances turn out to be arranged in the order of the increasing values of their experimentally measured flow limits. Frenkel and Kontorova obtained the now well-known relation of this energy, W_0, to the energy, W, of the chain of particles involved in shear, the shear propagation velocity, w,

---

1) I.E. Tamm. UFN, v. 76, n. 3, p. 427, 1962 (in Russian).

and the sound velocity, w_0, of a crystal:

$$W = \frac{W_0}{\sqrt{1 - \frac{w^2}{w_0^2}}}.$$

As we see, it is similar, in its form, to the famous relations of the theory of relativity and superimposes a «relativistic» limit on the plastic deformation velocity.

By the end of the 1940s the Frenkel and Kontorova results were obtained by F. Franck and D. Eshelby using macroscopic elasticity theory. The beginning of the 1950s was marked by the explosive development of the study of plastic deformation and dislocations, still going on today as well. Quite a number of monographs are dedicated to the subject now. One can state that the Frenkel-Kontorova theory still is the unique microscopic theory of crystal plasticity. Despite the fact that it was developed with the help of a very simple model, it has been successfully used to solve various concrete and practical problems associated with plasticity phenomenon. Several papers by Russian and foreign physicists have been devoted to the further refinement of this «classical model»[2].

This work by Frenkel and Kontorova obtained new life after the onset of the «soliton boom», which is still in progress. A singular solution found by the authors and describing the motion of a chain of atoms (dislocation) in a crystal lattice presents a beautiful example of a propagating soliton. This is exactly «the new in the old» of which Frenkel wrote in the above-cited preface to his «Wave Mechanics». The Frenkel-Kontorova soliton is now being discussed in hundreds of papers. We refer only to a wonderful book by A.N. Filippov, «Multifaced Soliton» («Mnogolikiy soliton», 2nd ed., Moscow, Nauka, 1990, in Russian), a whole chapter of which («The Frenkel soliton») is dedicated to this work by J.I., as well as to the collection of papers edited by N. Mott («The Beginning of the Solid State Physics», L., The Royal Soc., 1980), where it is discussed in minute detail by A.K. Seeger.

2. In the field of electrical properties of dielectrics and semiconductors, J.I.'s was exceptionally prolific in the period we are describing; about 15 of his articles deal with these problems.

Among these papers special attention should be paid to an investigation, carried out with A.F. Ioffe in 1932, concerning the nature of rectification at the contact between metal and a semiconductor. To explain the mechanism of this phenomenon, the authors used the idea of a tunnel of electrons through the restraining layer, the tunnel being very narrow, though exceeding interatomic scales.[3]

---

2) V.L. Indenbom, I. Kratochvil. Checkoslov. phys. journ., v. 13, p. 814, 1963 («Mobility of dislocation in the Frenkel-Kontorova model»); W. Atkinson & N. Canberra, Phys. Rev., v. 138, p. 763A, 1965 («Motion of the Frenkel-Kontorova dislocation in one-dimensional crystal»).

3) A little earlier Frenkel had considered the same rectification mechanism but at the contact of two metals (Phys. Rev., 1931, v. 37, 31, p. 102).

2nd All-union Conference on Semiconductor Physics. Kiev, Spring 1934. From left to right: A. Goldman, (?), A. Ioffe, P. Sharavskii, Ya. Frenkel, Ig. Kurchatov, (?), V. Zhuse, L. Nemenov, D. Nasledov, Yu. Maslakovec, A. Aleksandrov.

Examination of the Frenkel-Ioffe theory in the following years has demonstrated that another mechanism actually provides current rectification at the contact (for example, between Cu and Cu_2O). «It required 25 years of rapid development in semiconductor physics and technology to implement the idea of J.I. Frenkel and A.F. Ioffe in the tunnel diode invented by Japanese scientist L. Esaki in 1959,» reads one review devoted to the tunnel diodes[4]. Thus the study of Frenkel and Ioffe which had seemed incorrect was made use of again.

A little later, in 1934, J.I. presented, for the first time, a theoretical explanation of the photomagnetic effect of I. Kikoin and M. Noskov (in semiconductors) which was developed further in the later works of other scientists, in particular, Landau and Lifshits[5].

In 1935–36, Frenkel published a new important paper on the exciton theory and the colour centres in alkaline-galoidal crystals. The notions suggested there were later exploited and diversified by S.I. Pekar and his school.

---

4) V.I. Fistul' & P.Z. Shvarz. Tunnel diodes. UFN, v. 77, # 1, p. 109, 1962. In Russia the quantum-mechanics theory of the tunnel effect was further promoted in well-known papers by L.V. Keldysh.

5) L.D. Landau & E.M. Lifshits. Physikalische Zeitschrift der Sowjetunion, Bd. 9, H. 5, S. 475, 1936 (in German).

In 1936, Frenkel put forward the concept of the heat capacity anomaly in crystals of galoidal-hydrogen compounds, stating that it is related not to the transition from rotational swinging to free rotational motion, as L. Pauling imagined, but to the liquidation of the long-range order in the orientation distribution under the conservation of the short-range order and the associated with it type of thermal motion, i.e., rotational swinging (Frenkel called it orientation melting).

Professor S.E. Bresler wrote about this paper: «Some new concepts introduced by J.I. Frenkel in physics proved to be profoundly fruitful only after a number of years had passed, when experimentalists had found direct proofs of the effects predicted by theory. Such has been the destiny of his theory of orientation lattice melting. Liquid crystals provide a brilliant example of orientation melting occurring at a higher than usual melting temperature; they were treated from this point of view by V.N. Tsvetkov. The solid polar crystals HCl, HBr and NH_4Cl serve, according to contemporary data, as examples of substances where first orientation melting, and then the usual melting, takes place.»[6]

3. A whole string of significant results were obtained by Frenkel in polymer physics. He developed the theory of the elasticity, viscosity and swelling of rubber-like substances (1938). In the process of the study of thermal motion of long molecular chains, carried out with S.E. Bresler (1939), it was shown that linear molecules, for instance carbohydrogen chains, are similar not to chains with freely rotating links, but to bands possessing some torsional elasticity. The presence of this elasticity is related to the non-free, hindered rotation of the chain which is responsible for the mechanical properties of rubber (elasticity). The classical study by Frenkel and Bresler is included in textbooks[7] and still retains its significance.

4. A paper by Frenkel on biophysics dedicated to the mechanism of muscular activity (1938), relates to the paper just described. J.I. takes as a basis here the concept that muscular shortening is caused by a process similar to the volcanisation of rubber, which occurs because of the ion extraction in chemical reactions accompanying the muscle irritation.

In 1939 at a conference on the influence of radioactivity on living organisms Frenkel reported on the action of ultrasound oscillations and pointed out the possibility of using USO in the struggle with cancer diseases (the necessary technique developed in the USSR by S.Ya. Sokolov allowed one to concentrate an ultrasound wave beam at the place where a cancer tumour was suspected).

5. As always, Frenkel worked hard in the field of molecular physics. His works (on melting theory, the theory of heterophase fluctuations, the kinetic and electrical properties of liquids, etc.) formed a kind of a bridge which led him from his early studies of thermal motion in liquids and solids to the construction of the

---

6) S.E. Bresler. J.I. Frenkel's studies on molecular physics. In: J.I. Frenkel, Collection of Selected Papers, v. 2, 1958, p. 231 (in Russian).

7) See cf. L.D. Landau and E.M. Lifshits. Statistical physics. Moscow: Nauka, 1964, p. 560 (in Russian).

kinetic theory of liquids completed by the middle of the 1940s with the publication of a monograph on the subject. In some ways leading up to this monograph was a short book by J.I. entitled «The Theory of Solids and Liquids» (1934).

Concluding the brief list of J.I. Frenkel's main achievements, we would like to note that quite a number of his treatises in the period of time considered were dedicated to general questions of theoretical physics, in particular, the foundations of the statistical equlibrium and irreversible process theory (1932), the superconductivity theory (1933–34), the theory of helium II superfluidity (1939) and its capillary properties (1940), magnetism, the theory of metals and so on.

Let us now proceed with a more detailed survey of J.I.'s results in nuclear physics, first describing the state of this science at the beginning of the 1930s.

The year 1932 was made remarkable for nuclear physics by the discovery of the neutron and positron, which joined (and doubled) the family of elementary particles composed, before this, of the electron and proton.

The discovery of the neutron is associated in particular with the name of the English physicist Chadwick, whose investigations concluded a long series of studies on the identification of radiation carried out by W. Bothe and R. Bekker in 1930. The radiation appeared as a result of the bombardment of beryllium by α-particles (from a natural source) and, in the opinion of German scientists, resembled γ-rays. Later F. Joliot-Curie proved radiation's non-electromagnetic nature.

The experimental founding of the positron was accepted by the physics community with enthusiasm. Congratulatory letters to Dirac, who had predicted the positron, were, in particular, from the Soviet physicists I.E. Tamm (5 June 1933)[8] and J.I. Frenkel. On 5 May 1933, Frenkel wrote to Dirac: «I still am under the spell of the unexpected discovery of the positive electron as an experimental fact. It indeed brilliantly confirms your whole theory! What do you think about protons? If one ought to treat them as elementary particles, then we also have to assume the existence of negative protons. If otherwise the former are a combination of neutrons and positive electrons, a question arises: Why does a neutron associate with positive, but not negative, electrons? And if, moreover, neutrons are to be regarded as elementary particles satisfying the Pauli principle, then there must also be «anti-neutrons» which correspond to the few open positions remaining in the distribution patterns of neutrons at the states with negative energy.»[9]

Dirac spoke about the anti-proton in the fall of 1933 at the Solvay Congress.[10] The anti-proton was found experimentally in 1955, and the anti-neutron the next year after.

---

8) Tamm's letter is published in: «Paul Dirac and Twentieth Century Physics» Moscow: Nauka, 1990, pp. 163–165 (in Russian).

9) Frenkel's letter to Dirac was published in several places, in particular in the book: «Frenkel, Recollections», and in the collection of papers mentioned in the previous footnote.

10) The analysis of this statement by Dirac at the Congress is given in: J. Mehra. «The golden age of theoretical physics». In: Aspects of Quantum Theory / Ed. by A. Salam and E.P. Wigner. Cambr. Univ. Press, 1972.

P.A.M. Dirac. Oil portrait by Ya. Frenkel. Leningrad, 1936.

The experimental discoveries described resulted in a great increase in nuclear research all over the world. In the same year, 1932, studies of nuclei were started within the walls of the PTI as well. A very strong group headed by A.F. Ioffe, I.V. Kurchatov and D.V. Skobel'tsyn was organized there. The theoreticians M.P. Bronstein, D.D. Ivanenko, and S.A. Bobkovsky (a postgraduate student of J.I.) were also drawn to this work; a nuclear seminar was started. There were several laboratories in the PTI nuclear physics group after only 2 years of its existence;

Ig. Tamm, towards the end of the 1950s. Tamm's seminar on theoretical physics in the Lebedev Physical Institute.

the number of institute staff members involved in the sphere of these researches was growing.

In 1932, D.D. Ivanenko came forward with the idea that atomic nuclei consist not of protons and electrons, as it had been assumed earlier, but of protons and neutrons. This idea[11] removed the difficulties physicists studying magnetic and statistical properties of atomic nuclei had faced.

In September of 1933 the First All-Union Conference on the Atomic Nucleus, timed to fall on the 15th anniversary of the Leningrad PTI gathered. Leading physicists from the USSR and foreign guests (P. Dirac, F. Perrin, the Joliot-Curies, F. Rosetti, V. Weisskopf) participated in it, and until the beginning of the war such conferences were held in the USSR practically every year.

From 14 to 20 March 1936, the session of the Physics and Mathematics division of the USSR Academy of Sciences was held in Moscow. I.E. Tamm gave there a special lecture on the problem of the atomic nucleus. He presented a survey of the problem as it had been approached during the previous five years. He paid special attention to a paper by N. Bohr published in the 29 February 1936 issue of «Nature». In this paper Bohr put forward the concept, which proved to be extremely fruitful, of the compound-nucleus. According to Bohr, a nuclear

11) The idea was developed further, most notably in a fundamental paper by Heisenberg who referred directly to Ivanenko.

reaction in which a nucleus-target and neutron-projectile are involved has two stages. In the first one a neutron is absorbed by the nucleus, and this compound splits, forming a new nucleus and emitting this or that particle, only after some period of time (which is greater than the «flight» time for the distance equal to the size of the nucleus by a factor of approximately one million).

Tamm's report was discussed by Skobel'tsyn, Kurchatov, Mysovsky and also Frenkel. We reproduce here a fragment of Frenkel's speech: «I would first like to say a few words concerning the Bohr theory Tamm has told us about. It applies to more than the most complicated nuclei consisting of a large number of particles. It is always that the essential degree of the structure complication enters together with some simplification of the theory. The intermediate cases are the most difficult to treat theoretically.

«It seems to me that Bohr's ideas may be tailored quantitatively to describe heavy nuclei. Tamm's arguments on the Bohr theory make it clear that a complex nucleus is, to a certain extent, similar to either a solid or a liquid consisting of a large number of particles strongly bonded to each other. This leads to the quite natural question of whether it's possible to regard the energy transferred from the neutron stuck to the nucleus to the whole set of particles constituting this nucleus, as a thermal energy of a kind pertinent to the nucleus. Isn't it then possible to characterize its state by some temperature corresponding to this energy and the number of particles? Roughly speaking, if this energy is large enough that one is able to use classical statistics here, then the temperature of a nucleus proves to be inversely proportional to the number of particles in it. Hence, we can imagine the matter ... in such a way that, due to the energy release associated with the capture of a new particle (a kind of absorption), the nucleus is heated. The heated nucleus has some chance of being partly vaporized, and this vaporization is just the disintegration which we recognize as a neutron, or proton, or α-particle flying out.»[12]

In this short communication Frenkel formulates a new approach to the description of the droplet model of a nucleus.[13] His first article on the subject was published in 1936, having been submitted to the editors of the «Physikalische Zeitschrift der Sowjetunion»[14] a week after his speech at the session. The programme outlined above was implemented there. For neutrons with high energy, W, the temperature, T, of the excited compound-nucleus consisting of n protons and neutrons is related to T by the equality $W = 3(n-2)kT$. It was also shown that, in accordance with the experimental data, the lifetime of the compound-nucleus

12) The notion of a nucleus similar to a liquid drop was first formulated by G.A. Gamov in 1929 and later developed by W. von Weizsäcker. A thermodynamic aspect of this model had not, however, been at all covered by them.

13) J.I. Frenkel. Speech on the I.E. Tamm report. Izvestiya AN SSSR, ser. fiz., v. 13, 1936, p. 343. Included in the collection of papers: «Neutron», ed. by I.M. Frank. Moscow: Nauka, 1983, pp. 283–85.

14) A journal published in Khar'kov in German and English since 1932.

is much greater than the period of γ-oscillations (that is, than $\tau = 1/\nu$, where ν is the frequency of electromagnetic radiation emitted by the nucleus) and that the lifetime increases with the atomic number of the nucleus.

On returning from the March session to Leningrad, Frenkel found a letter from Bohr mailed, by a pure coincidence, 2 days before the beginning of the session where his paper on the compound-nucleus was so heatedly discussed. Bohr wrote:[15]

«Dear Frenkel, I often return in my thoughts to our stay in Russia, a sojourn which became an unforgettable event for my wife and me. I have been thinking for a long time about writing to you and asking whether you would like to come to Copenhagen to participate in our little annual conference on atomic physics. This year it starts around June 10, and directly after it is finished, the 2nd International Congress on the Unity of Science will open here, at which the problems of causality in physics and biology will be discussed. A number of scientists from different countries, both philosophy specialists and researchers, have consented to come; I was invited to be part of the Organizing Committee and to deliver the introductory lecture at the Congress[16]. One of these days a special invitation will be forwarded to you on behalf of the Organizing Committee, and we hope you will find it possible to accept. At our less official conference, besides the questions of atomic physics itself, their relationship with biology will also be pondered. We expect a number of the former staff members of our institute to come, such as Kramers, Heisenberg, Pauli, Dirac and Klein, as well as some guests. We would be very glad if you and other Russian friends — Landau, Tamm and Fock, to whom I am writing simultaneously, will be able to participate. As for the funds necessary for your participation in both congresses, we can offer to pay for your trip from Russia and back, as well as for your expenses when staying in Copenhagen.

With cordial regards to you and your family, and also to all our common Russian friends from my wife and me,

Yours, Bohr.»

In reply to Bohr, Frenkel wrote:

«Leningrad, 22 March 1936
Dear Professor Bohr,
I wish to express to you my heartiest thanks for your kind letter and the invitation to participate at the Conference of an Atomic Physics and the Congress on the Unity of Science. I should be delighted to accept your invitation. I am afraid however that I shall not be able to obtain the requisite permission. I shall know more about it in a couple of weeks and shall write you again then.

---

15) We are happy to take the opportunity to thank Professor A. Bohr who kindly provided us with copies of the letters by N. Bohr and J.I. Frenkel cited below. Bohr's letters were written in German; Frenkel answered him in English.

16) The report is published in «Philosophy of Science», v. 4, p. 289, 1937.

We have read and discussed with the keenest interest your paper in «Nature» on the theory of complex nuclei. I must confess to you that I immediately attempted to apply to it the general principles of the theory of solid bodies, and, introducing the idea of temperature to describe the state of the nucleus excited by the «adsorption» of an additional neutron, obtained the probability of its disintegration using the well-known expressions for the probability of the evaporation process.

The validity of this expression rests on the assumption that the fundamental vibration frequencies of the nucleus correspond to energy quanta, which are small enough compared to the «absorption» energy, the former amounting to 0.5 MeV, while the latter is of the order 10 MeV. I am not sure whether this assumption is correct or not. The result is at any rate encouraging for it corresponds to a mean life of the excited nucleus short enough not to be observed directly and long enough to obliterate the sharpness of the γ-ray lines.

Do you think the idea is a sound one? It is rather amusing to apply to a heavy nucleus a number of mechanical and thermal notions specifying the behaviour of a small solid or liquid body [...]

How are you, dear Professor Bohr? Besides the scientific proceedings in Copenhagen, I feel very much attracted to it because of your presence, and I shall be extremely sorry if my trip to Denmark is rendered impossible.

With kindest regards to Mrs. Bohr and to yourself from my wife and myself, Yours, J. Frenkel»

Bohr responded as follows:

«2 May 1936
Dear Frenkel!
Thank you very much for your friendly letter. I greatly hope that you will be able to come to our conference, which will be held, and it is final now, during the week from 14 June to 20 June. If I am able, in one or other way, to help with the organization of your and your colleagues' visit, please write, or better, send me a telegram as to what I can do and to whom I should appeal.

The analogy between the emitting of a neutron from an excited nucleus and the well-known vaporization process, which you mention in your letter, though tempting, would be sound only in the case of very «high» excitations, when not only one, but a number of neutrons may leave the nucleus, as is indicated in the end of my paper[17]. For the nuclear reactions known, the excitation of an intermediate system often is, probably, large enough for the levels to form a continuous spectrum, and the nucleus formed in the result of a neutron loss transits to a normal state or to one of the deeper discrete spectrum levels. In this case the situation is essentially different from that which takes place under the usual liquid vaporization, where even for the lowest temperatures one may consider the total

17) The corresponding fragment of Bohr's paper is cited below.

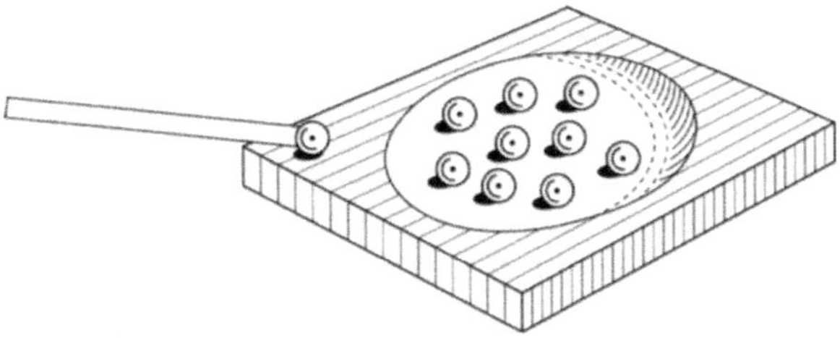

Fig. 5 The model presented at Bohr's lecture which demonstrated the interaction between a neutron and a nucleus' nucleons.

energy of the system to be sufficient for the freeing (flying out) of a practically infinite number of vapour molecules.

All these questions will be touched upon in a paper by me and Kalckar which, I hope, will be ready by the beginning of the conference. It would be very good if we could discuss them with you ...

May I ... conclude with expressing my best wishes to you, your wife and all our common Leningrad friends.
Until the upcoming meeting! Yours, N. Bohr»

During the three years after the appearance of papers by Bohr and Frenkel, the study of nuclear physics (and of the statistical theory of nuclei in particular) underwent evolutionary development.

On 2 September 1937 in a letter to E. Hill (his former assistant at the University of Minneapolis) J.I. wrote: «I was glad to learn of the big advances in the development of nuclear research work at Minneapolis. We are also doing some work in this line at the Physico-Technical Institute, as you surely know from the publications of A. Alikhanov, Kurchatov and others. Unfortunately, we are very poorly equipped, having at our disposal neither our own radium nor any high voltage apparatus. We hope to get both in the near future.

«I am gradually exploring the nuclear field and hope to assimilate soon the huge body of knowledge collected in Bethe's review[18]. Some research in this field is carried out under my guidance by a few co-workers (the radioactive scattering of fast neutrons by heavy nuclei, the statistical theory of α-decay, and the nature of the nuclear forces).»

N. Bohr visited the Soviet Union for the second time in 1937. In June he gave a lecture «On the transformation of atomic nuclei» in Moscow. In this lecture[19] Bohr dwelt on Frenkel's results. He said: «When discussing the question of the decay of a complex system [compound-nucleus — V.F.] with the freeing of some material particles, the use of thermodynamic analogies is fruitful. The case of the

18) H. Bethe. Review of Modern Physics, v. 8, # 2; v. 9, # 2, 1937.

19) N. Bohr. UFN, v. 18, p. 337, 1938 (in Russian).

neutron emission in which no forces act outside the limits of the typical nuclear distances, presents a successful analogy of either liquid or solid body vaporization at low temperatures.» Bohr puts in a footnote here: «Frenkel was the first to suggest applying the usual vaporization formula to the probability of the neutron flight out of a compound-nucleus. A more detailed study on the general statistical mechanics is given in a paper by Weisskopf (in press).»

The further development of the investigation of the capillary nucleus model was provided in an article by Bohr and Kalckar[20] (a special section of it is devoted to Frenkel's results), in the paper by Weisskopf mentioned by Bohr, in a simultaneously published paper by J.I. himself («On the statistical theory of the decay of the heavy nuclei»[21], which was the exposition of his report at the II All-Union Conference on Atomic Nucleus (Moscow, September 20–26, 1937), and in papers by L.D. Landau and by V. Weisskopf.[22]

At the beginning of 1939 the evolutionary development of nuclear physics took a revolutionary leap: the phenomenon of the fission of heavy nuclei, uranium in the first turn, under the action of neutrons, was discovered. We are speaking about the epoch-making paper by O. Hahn and F. Strassmann (published in «Naturwissenschaften» on 6 January 1939). Thanks to their fine analysis the German scientists were able to prove that under the neutron irradiation of uranium, isotopes of barium and molybdenum are formed, i.e., that uranium splits into two approximately equal parts instead of forming transuraniums, as Fermi thought.

The first qualitative explanation of the uranium decay phenomenon was given by O. Frisch and L. Meitner («Nature», 11 February 1939; the note is dated 16 January) who used the capillary nucleus model. They treated splitting from the classical viewpoint, assuming that it is implied by vibrations of the nuclear droplet as a whole. No calculations were given in the paper; its essence was just this idea. One should say that the possibility of such vibrations of a nucleus-droplet had been discussed by Ehrenfest in his letters to Bohr back in 1929, and it was mentioned by Bohr in a lecture in Moscow (1937). In the illustrative drawing used in this lecture Bohr even depicted such «oscillations of the shape» of a nuclear sphere, the «ripples» on the spherical surface of a nucleus. But he did not then make any conjectures that such vibrations might lead to the splitting of the droplet.

A week after the Frisch and Meitner note, Frisch produced an experimental confirmation of the hypothesis of the splitting of a uranium nucleus (also in «Nature»).

Bohr, staying in the United States at the time, was informed by phone about the Frisch and Meitner interpretation of the Hahn and Strassmann results. Due to

---

20) Its Russian translation is published in UFN, v. 20, p. 1, 1938; Bohr refers to this paper in his letter to Frenkel of May 2, 1936.

21) Izv. AN SSSR, OMEN, # 1–2, p. 233, 1938.

22) See Physikalische Zeitschrift der Sowjetunion, Bd. 11, S. 556, 1937 and Phys. Rev., 1937, v. 52, p. 295.

this, the results of the European physicists' research were soon confirmed (and developed further) by American scientists.

Simultaneously, the processes of uranium splitting were successfully studied in the Soviet Union. At the first stages the most intensive research was carried out within the walls of the Radium Institute in Leningrad. There N.A. Perfilov investigated the splitting of uranium in the Wilson chamber, and A.P. Zhdanov, L.V. Mysovsky and M.Yu. Mysovskaya — in the thick-layer photo plates.

Frenkel immediately got involved in the uranium splitting studies, and, with the quickness so characteristic of him, created the first quantitative theory of this phenomenon. He submitted his first paper («Electrocapillary theory of the heavy nuclei splitting by slow neutrons») to JETP (it was received by the editors on 16 April 1939), and the English version — to the «Journal of Physics», a journal published in English in the Soviet Union. Simultaneously, one copy of the English manuscript went to Bohr, and another to Hill in the United States. At J.I.'s request, Hill completed a brief summary of this article and passed it on to the editorial board of the «Physical Review», where it was published, in the form of the letter to the editors, on 15 April 1939.[23]

In the Niels Bohr Archives is a letter J.I. Frenkel wrote on 12 March 1939 to Bohr.[24] We quote several paragraphs from it which show how the news of the discovery of uranium splitting was delivered to Leningrad.

«Dear Professor Bohr,
About the end of February we learned for the first time of the discovery of the new type of splitting of the U-nucleus (from Joliot's paper in C.R. and somewhat later from the American «Science and News Letters»). A few days after that I worked out a theory of this process, which turned out to coincide in its essential points with that proposed by Frisch and Meitner (in a few words about the decrease of surface tension owing to electric charge), and especially to your letters to «Nature» and «Phys. Rev.»

As I have worked out in quantitative detail the main points which may be lacking in the papers of other authors, I am sending my paper to the new Soviet physics journal (which is to replace the «Physik. Zeitschr. der Sow. Union» and the «Technical Physics of the USSR») — the «Annales Physicaux» which will still be published by our Academy of Sciences. I am afraid however that the paper will appear there after a great delay (although I have been told that the first issues for this year are at press already). In any case, I think you will be interested in reading my paper before its publication. I am accordingly sending you a copy.

---

23) Physical Review, v. 55, p. 987, 1939 («On the splitting of heavy nuclei by slow neutrons»). This paper was included in 1979 in the collection dedicated to the 40 year jubilee of the Hahn and Strassmann paper: «40 Jahre Kernspaltung: Eine Einführung in die Originalliteratur». Herausgegeben von H. Wohlfarth. Darmstadt: Wiss. Bücherges., 1979, S. 108–110.

24) I express my gratitude to Dr. H. Levi for sending me a copy of this letter.

I am planning to investigate a few other questions, hitherto unsolved, on the existence or nonexistence of a stable non-spherical equilibrium shape for the heaviest nuclei, on the kinetics of splitting after excitation, and especially on the capture of some electrons by the two daughter nuclei.»

The note of Frenkel's published in the «Physical Review» presented the basic formula describing the energy balance in the excited nucleus and the quantitative explanation of the splitting of the nucleus by neutrons. It showed that at a certain ratio between the surface and electrical energy of the nucleus (see below), its spontaneous splitting may occur, and it also derived the instability condition. The same note discussed the stationary non-spherical (ellipsoidal) shape of heavy nuclei and pointed out that in their case one should expect the anomalously large quadruple moments of nuclei. Both these assumptions were soon confirmed experimentally.

The detailed calculations were given by J.I. in a JETP paper. There he demonstrated that the main features of the phenomenon may be understood with the help of the liquid charged droplet model; the instability of the nucleus then proves to be caused by its surface oscillations. The same idea was independently suggested and developed several months later in a paper by N. Bohr and J. Wheeler.[25] «In these two papers by Frenkel and Bohr/Wheeler the foundation of the modern theory of heavy nuclei splitting was formulated for the first time.»[26]

We would like to stress one more point before leaving our brief historical survey and proceeding with the exposition of Frenkel's work. The possibility of heavy nuclei splitting was mentioned in a paper by the German physicist Ida Noddak (1934)[27], and in a paper by Fermi and his co-workers on the transuraniums which had appeared a little earlier.

When constructing his theory of electrocapillary splitting of heavy nuclei by neutrons, J.I. proceeded from the following very transparent analogy. Let us imagine droplets of a common fluid, such as mercury, which are so familiar to anyone who has broken a thermometer. These droplets «as agile as mercury», rolling over the surface of a table or simply on the floor, merge together. Under the total volume conservation, this merging corresponds to the total energy decrease, because the surface of the formed (spherical) drop, and the extra surface (capillary) energy along with it, is less than the sum of the surface areas of the droplets before they merged. If one now starts to charge mercury droplets (with charge of the same sign), then at the beginning the trend to merge will be preserved, but there will be no more merging after some point. On the contrary, the mercury drop whose charge

---

25) A footnote in this paper (Phys. Rev., v. 56, p. 426, 1939) indicates that the main formulae derived in it coincide with those obtained by Frenkel, as one sees «from the manuscript sent by him».

26) Ya.A. Smorodinsky and I.E. Tamm. «Works by Frenkel on the Theory of Electrons and Heavy Nuclei. Coll. of Selected Papers of J.I. Frenkel», v. II, AN SSSR Publ., Moscow-Leningrad, 1958, p. 458. (in Russian).

27) I. Noddak. Angewandte Chemie, Bd. 47, S. 653, 1934. On page 654 the author says: «It seems to be possible that under the neutron bombardment of heavy nuclei these nuclei split into large parts which are isotopes of the known elements, but not the neighbours of those bombarded.»

(or its potential proportional to it) exceeds some value, splits into two droplets. The explanation is that, besides the capillary (intermolecular) forces, electric repulsive forces enter the game. J.I. used this fact as an analogy to explain the behaviour of droplets in the case of nuclear liquid recognized by an extremely high density and the value of the surface tension. If one assumes these droplets-nuclei to be of a spherical shape (to which experimental data testified), then it is easy to calculate the energy balance for the splitting of such a drop into two. The energy of the initial nucleus, less the rest energy of the nucleons constituting the nuclei and the bond energy, which are conserved under the nuclear liquid splitting, is

$$W = 4\pi R^2\sigma + \frac{3}{5}\frac{Ze^2}{R},$$

where σ is the surface tension, R is the radius and Ze is the charge of the nucleus. The first term corresponds to the capillary energy, U, of the nucleus, and the second one to the electrical energy, E (this electrostatic energy is calculated accounting for the distribution of charge over the volume, and not the surface, as in the case of charged mercury, of the nuclear liquid).

If, to simplify the situation, one assumes that the maternal nucleus breaks into two equal parts, then their energy is

$$2W_1 = 2\left\{4\pi R_1^2\sigma + \frac{3}{5}\frac{(Ze/2)^2}{R_1}\right\}.$$

Thus the energy balance for such splitting process reads

$$\Delta W = W - 2W_1 = 0.37\frac{(Ze)^2}{R}\cdot\frac{3}{5} - 0.26\cdot 4\pi R^2\sigma,$$

and the obvious from the volume conservation condition relation $R_1 = 2^{-1/3}R$ is used in the expressions for W and W_1

Thus the value $E/U = 0.70$ corresponds to the condition $\Delta W = 0$ (reaction energy equals zero). If one introduces the values $\sigma = 10^{20}$din/cm, $R = r_0 A^{1/3} = 1.5\cdot 10^{-13}A^{1/3}$ (here A is the integer equal to the number of nucleons in the nucleus, and $2r_0$ is the mean distance between the neighbouring nucleons) and sets $Z = A/2$, then for the atomic weight of a nucleus capable of an electrocapillary splitting ($\Delta W = 0$) one ends up with the value $A^* = 87$. Bearing in mind the roughness of the approximations made, one must admit that the calculation correctly reflects the fact that instability is characteristic of heavy nuclei.

The condition fulfilled does not, however, necessarily imply the splitting. To cause in reality the decay of a nucleus, it is useful to transfer some energy to it. In the case of uranium nuclei the role of this «activation energy» carrier is played by neutrons «adsorbed» by the surface of the nucleus; the «adsorption heat» is just the activation energy for this nuclear reaction.

Evidently, a spontaneous splitting of heavy nuclei happens when the activation energy, instead of the energy of reaction ΔW, turns to zero. We denote the corresponding value of the atomic weight by A^{**}. Frenkel suggested the following extremely simple method to estimate A^{**}. Consider a spherical nucleus of radius R and charge Ze and let us mentally subdivide it into two halves (hemispheres). The capillary forces bonding them together each equal $2\pi R\sigma$ (any part of the nucleus is «fastened» to the other one, just like a drop of water hanging from the end of a faucet; the drop equator plays here the role of the faucet's hole). On the other hand, each half is repulsed from the other one due to its charge $Ze/2$. If we consider the distance between the halves of the sphere to be equal to its radius, then the force of the electric repulsion is $\frac{(Ze/2)^2}{R^2}$. Setting these two forces equal, we find the conditions under which the hemispheres «separate» without any external actions, i.e., with zero activation energy. The corresponding calculations for uranium ($Z = 92$, $A = 238$) give $Z^2/A = 35.5$. In reality, as is shown by Frenkel and Bohr & Wheeler (by other, more rigorous means) the activation energy turns to zero when factor Z^2/A is 1.6 times larger than the repulsion energy.

The model discussed demonstrates that heavy nuclei rest on the verge of stability. Frenkel's paper published in JETP solved the problem of the stability of heavy element nuclei (treated as spheres uniformly charged over their volume) relative to small deviations from a spherical shape. The corresponding theory leads to the formula for the oscillation frequency of incompressible uniformly charged fluid about its unperturbed spherical form, and at zero charge the formulae tend to the classical Rayleigh expressions for capillary oscillations of an uncharged droplet.

In the work by Bohr and Frenkel on the splitting of heavy nuclei, the possibility of their spontaneous decay was indicated. This effect was discovered by two young scientists, G.N. Flyorov and K.A. Petrzhak, I.V. Kurchatov's staff members, at the PTI in 1940 (by this time the research of W. Libby had been completed which shows that in case such decay is possible, its half-life exceeds 10^{14} years). Their study and the theoretical one that appeared in its wake, by Khariton and Zel'dovich (on the calculation of the chain reaction of the uranium decay) were, it seems, the last essential landmarks on the path to the freeing of the intra-atomic energy which Russian physicists introduced before World War II.

In concluding the survey of Frenkel's studies on nuclear physics, I would like to return to a period long before 1939, about 15 years earlier. In 1925 in Leningrad a book by well-known German science popularizer H. Günter was published under the title «Technical Dreams».[28] Its editor, Frenkel, wrote a small appendix entitled «Does intra-atomic energy exist?» Frenkel closes this appendix with the following words: «By joining hydrogen atoms to make helium, the excess energy of the hydrogen atoms must be released. Obviously, if we could initiate such a synthesis,

---

28) G. Günter. Tekhnicheskie mechtaniya, Leningrad, Seyatel', 1925. Ed. by Prof. J.I. Frenkel (in Russian).

any worries about energy would be eliminated. Though the energy we discuss would be only 0.8 per cent of the «fictituous» [equal to mc^2 — V.F.] energy which is a dream of many, it is real, and we can at least muse about possessing it.»[29]

One should not place too much importance on such a prediction; had Frenkel himself done so, he would have published it in some specialized magazine. But it seems to be worth mentioning that this guess was made several years before the publications of F. Houtermans & R. Atkinson and H. Bethe. It testifies more to the fact that Frenkel had pondered the subject long before the cycle of his published work on nuclear physics (1936–39).

As in previous chapters, we are going to dwell upon Frenkel's teaching activity.

In the period from 1931–1940 the whole course of theoretical physics, which had partly been written earlier, was published.

In 1933, a Russian edition of the first volume of «Wave Mechanics» appeared. When, a year before, J.I. started working on the second volume, the book had already been accepted for publication by the State Technical-Theoretical Publishing House (GTTI). To save time and to avoid the translation of the second volume from Russian into English, J.I. wrote this part also in English (keeping in mind its publication at the Oxford University Press). The simpler task of the «reverse» translation from English into Russian was completed by someone other than Frenkel; this work was done practically at the same time as the writing of the corresponding chapters.

Also in 1933 the first part of «Statistical Physics» (classical statistics) was published, and in 1935 — the «Analytical Mechanics». In 1934 and 1935 both volumes of «Electrodynamics» (considerably expanded versions of the German edition) came out. Finally, after a break, in 1940 J.I. Frenkel published «Theoretical Mechanics», in which was included some material from «Analytical Mechanics» and «The Course of Vector and Tensor Analysis with Applications to Mechanics».

Together these books form the first course of theoretical physics published in our country. Therefore it is difficult to overestimate their role in the development of generations of Soviet physicists.

How did J.I. manage to cope with this gigantic task? After all, he did not stop for a moment his research work, or giving lectures and consultations …

T.A. Edison was once asked how he had managed to do so much. He replied: «Very simple by eating half as often, sleeping half as much, working twice as long». That was also the way Frenkel lived, with the exception, however, of Edison's first rule: Frenkel worked and worked without any break or rest.

True though, each summer he went with his family to the country near Leningrad. In the 1930s the favourite place of rest was «Zhelezo»[30], a resort

---

29) See also my papers in «Archives for History of Exact Sciences», 1974, v. 13, # 1, pp. 1–26, and in UFN, 1994, v. 164, # 4, pp. 352–356.

boarding house on the bank of the Luga river, about a hundred kilometers from Leningrad. The Frenkels' friends, — the Semenovs, Talmuds[31], Frederiks', and Kharitons also vacationed there.

We here quote a passage from a letter Frenkel wrote to his parents (adding the lines concerning the work during the rest to those about rest).

«Zhelezo, 17 July 1935
... I do not do any exercise or even go for a walk. Strange as it is, I have not got enough time: until dinner I regularly work, and sometimes do the same also upon returning from the «children's hike» [in Zhelezo J.I. went boating each day with a company of children — V.F.]. Today, for instance, I wrote about 15 pages in-quarto.

... Yesterday played music a lot, until one of the women living in the neighbourhood asked me to stop, complaining of a headache. I'll have to go to the forest to play music, there I might probably find a less exacting audience ... I have had no time for drawing yet. Perhaps I will be able to draw only after the book is finished. I hope that I'll manage this by about the 25th of July. Then I shall come again to Leningrad, for a couple of days, probably, to submit my book to the editors. Unfortunately, even then I shall not be completely free: there will remain two papers, and the editing of the translations. Cannot remember a summer I was absolutely free from literary duties. Nevertheless I am satisfied with the stay at Zhelezo, both for the family and myself.»

In the 1930s V.A. Fock (from 1928 to 1933), L.V. Rozenkevich, M.P. Bronstein, S.V. Izmaylov, T.A. Kontorova, S.N. Tkachenko, O.M. Todes, and N.L. Pisarenko all worked at the Chair of Theoretical Physics of the Polytechnic Institute headed by Frenkel. J.I. invited, moreover, almost all Leningrad theoreticians to give lecture courses there, both regular and elective (he himself also gave lectures at Leningrad University and other institutes from time to time). An independent group of mathematical physicists headed by Professor G.A. Grinberg belonged in the 1930s to the Chair of Theoretical Physics as well.

For all Physics and Mechanics Department third-year students, starting with the academic year 1936/37, Frenkel taught a course on the structure of matter. He thought that this course would serve as an introduction for those who would attend later specialized courses of theoretical physics, and, on the other hand, it would conclude the acquaintance of students specializing in mechanics and technology with modern physics. The students of various physical divisions (electronics and the physics of dielectrics) also attended lectures on theoretical physics (statistics, electrodynamics, quantum mechanics, electron theory of solids).

---

30) Meaning «iron» in Russian.

31) David L'vovich Talmud was a known physical-chemist, corresponding member of the Academy of Sciences.

Along with this, lectures on elasticity theory, theoretical mechanics, physical kinetics, thermodynamics, etc., were offered by the department (in 1939 and 1940 Frenkel himself was engaged with theoretical mechanics).

I will now give several examples of imaginative explanations and model representations from Frenkel's lectures.

Why is there no resistance to shear in liquids? Having posed this rhetorical question, Frenkel asked the audience, which included future professor A.G. Samoylovich, to provide him with a match box and demonstrated how it resists the attempts to push its parallel surfaces against each other. But then he took out the inner box with matches, so that only the empty case remained, and repeated his attempt. The shear occurred immediately and without any effort; the case changed its shape from a rectangular parallelepiped to an inclined one. «Liquid does not resist shear just because there are voids in it,» explained J.I.

When giving a comparative description of solid (crystal) and liquid (amorphous) bodies, Frenkel said (and later wrote in a popular science article): «If the number of neighbours of a paving block is uniform all over the paved road, and in this sense it can illustrate a crystal, then a cobblestone road, whose stones have a varying number of neighbours, may serve as an illustration of fluid. In a liquid there exists an even vaster space for the atom motion than in a solid. Under melting of a crystal its volume increases 3 to 5 or even 10%. This increase in volume leads to a substantial animation of the transition process of atoms from one equilibrium position to another one nearby as compared to what takes place in crystals where atoms are clasped almost to the limit. This model can to some extent be compared to an overcrowded tram car. If just one or two passengers get off, the others obtain a much greater freedom of motion.»[32]

Here is how Frenkel, in addition to giving theoretical estimates, explained the process of vacancy formation and melting of a solid (as recollected by Academician B.G. Lazarev): «Imagine a brickwall from which bricks are being taken out in various places: vacancies are appearing («holes» in the literal sense!). Up to some degree of the «hole density» the wall will still stand, but at some critical point it breaks down suddenly; that is, the crystal melts.»

Professor Ya.E. Geguzin recalled a lecture Frenkel gave in the spring of 1939 in Khar'kov. Frenkel had just finished his paper on electrocapillary splitting of heavy nuclei. To make it clear how a neutron «starts up» the splitting reaction, he used the following analogy. He drew on the blackboard a water faucet with a drop hanging from it. Without any external influence it rests at the equilibrium state, attached by capillary forces. But one flick of the faucet (and J.I. flicked his drawing), and the drop breaks away.

Many more such examples could easily be given.

---

32) From a paper based on a lecture given by Frenkel in 1947 in Moscow. See J.I. Frenkel. In: «Na zare novoy fiziki», Leningrad, «Nauka», 1970, pp. 373–4. series «Populyarnye proizvedeniya klassikov estestvoznaniya», (in Russian).

At the Theoretical Physics Chair meetings the members discussed course curricula and the methodology of exposition of some of their parts, the distribution of the teaching load, the research work of graduate and postgraduate students and the programmes of the postgraduate examinations. (An unusual question asked at one meeting was formulated as follows: «Is it ethical or not for a male professor to invite his female postgraduate student to the cinema?» Along with the affirmative answer to this question (of which of course the professor whose actions had caused the discussion was sure), the opposite point of view was supported with equal passion. When the debates proved irreconcilable, Frenkel was appealed to as an arbiter. With a smile, he pronounced his sentence: «If the girl is pretty, then it is ethical!»)

Along with the teaching, research work was also done at the Chair. Frenkel and his closest co-workers at the Polytechnic were mainly investigating mechanical and electrical properties of solids, developing a statistical theory of liquids, and applying the theory of nonlinear oscillations to the problems of turbulence.

A high specific weight at the Chair of theoretical physics was with the preparation of graduate and postgraduate students. In the 1930s Academician I.Ya. Pomeranchuk, professors V.B. Berestetsky, F.F. Vol'kenshtein, Yu.N. Obraztsov, V.S. Sorokin, I.M. Shmushkevich and others, mostly pupils of Frenkel's, studied at the Chair.

During almost all of the second decade of his teaching activity, Frenkel, just as he had done in the 1920s, most frequently gave lectures to a comparatively small audience comprising a study group of usually not more than 30 people. Of course, this promoted direct contact with the listeners.

Probably everyone is familiar with the experience of being a spectator at a theatre run by a good and nicely balanced company of actors. One understands that the performance watched is an extraordinary event for the viewer but a normal, perhaps routine, one for the actors: they have already given this piece and are going to repeat it many times in front of other spectators. But the mastery and inspiration of the actors causes a sensitive spectator to yield to the illusion that the performance is being given for him personally.

A good and experienced lecturer can read his lecture dramatically, following not only the spirit, but also the letter of the lecture notes (often written by himself), and a student cannot escape the feeling that everything is being said for the first time here and now, that all the formulae are being born at the chalk point. The mastery and, especially, enthusiasm of the speaker make a greater impression the more sincere he is: the reproduction of this or that result may even become exciting again for the lecturer himself. His excitement is akin to what a musician feels when unfolding before the listeners a subtle and clear melodic tapestry created by a composer's genius. Such lectures are easy to listen to and to remember.

J.I. Frenkel was a lecturer of just that extremely rare kind. Many of his students used to say that «Frenkel is thinking aloud,» or «is creating directly during a lecture,» and his listeners joined him in this creative process. Presumably it is more correct to speak about «improvisations on a given theme». Let us recall a letter Frenkel wrote on 18 November 1930 from America. He wrote that he was trying to give his lectures about Russia in a new way each time, so as not to bore himself. The same goes for his lectures on theoretical physics. By his own admission, Frenkel never could make himself prepare them in minute detail. He outlined only a general scheme of exposition without rehearsing to the end the derivation of any formula — and with this outline he went to the audience. And, as he recounted, this scheme was usually prepared in ten minutes while walking from home to the university.

In his lectures Frenkel was able to expose a complicated theory with an exceptional simplicity. And often, by the same token, to show that something which seems simple and obvious is in fact very complicated.

Here are several traits characteristic of Frenkel, the lecturer, recalled by his students from different periods.

At the start of the first lecture of the course given for electrophysicists, J.I. referred to some results from the theory of elasticity but, as it happened, this theory had not yet been introduced to the students, and the majority of them were not acquainted with it. «Right,» said Frenkel, «then perhaps I'll start with elasticity.» And he immediately began discussing it as if he had prepared beforehand exactly this lecture.

Sometimes Frenkel would introduce a new lecture by explaining that he was going to present the material from the previous one in some other way. At first, the new exposition would appear to be quite different, yet it would be an entirely new line of argumentation suddenly leading to the old result.

From time to time, when Frenkel obtained a formula by a method new even to him, he would remark: «Look, how interesting! How come I didn't manage it earlier!»

Before the lectures, as well as during them, questions were usually asked. Periodically he was asked to repeat the derivation of some formula demonstrated in the previous lecture. This often caused a characteristic misunderstanding: the derivation he gave always differed from the previous one, and one had to remind J.I. in which way the formula in question had been proven last time.

I am far from willing to represent Frenkel as an ideal lecturer. Theoretical physics is a difficult discipline, and, of course, it sometimes happened that the audience could not follow him: lost in in his science, J.I. sometimes forgot the level of knowledge of his listeners. The loss of contact with the audience is felt rather quickly by an experienced teacher. «Is it not all clear to you?» Frenkel would ask in such cases, and he would try to make the difficult point of the lecture clear to the listeners. Several people recalled one case when in reply to the above question the following was heard: «J.I., it is not even clear to us what exactly one should understand here at all!»

The students of Frenkel's from the 1930s said that once, at the end of a semester, Frenkel came to the next lecture on thermodynamics and said that he was supposed to finish with his course that day. He did not manage it, however, and the finale was postponed to yet another lecture, and then again to the next. As a result, the students asked him: «J.I., and do you know what thermodynamics is? It's a science with three beginnings and no endings!»[33]

Another incident went like this: after a rather cumbersome chain of arguments, Frenkel obtained the result and wrote it up. Then he went away from the blackboard and, inclining his head, looked at it and said something like: «Now we have to change plus to minus here, remove the cubic root leaving the first power and then we'll get the formula known under such and such name. You please check the derivation at home, find where I was mistaken, and get the correct result.» (In such cases students joked that Frenkel had given a lecture on experimental mathematics.) And what is really surprising is that such incidents did not bother his listeners at all, such was the relationship between the lecturer and his audience.

From the student's point of view, each lecturer is characterized not only by his manner of presentation but also by how he grades examinations. A bad lecturer may be «nice» about the exams, and a good one may be really «harsh» about them; one can go over all the possibilities here.

So, how was Frenkel as an examiner? He seems to have been the first teacher at the Physics and Mechanics Department of the Polytechnic to give exams on lecture notes.[34] Anyone having attended his course of lectures had to report to the exam or test with an outline of the lectures composed by himself. J.I. would look through these notes and pose some questions; the examinations had the form of a scientific dialogue.

Here is a characteristic example of the informal attitude typical of Frenkel in regards to the examination procedure, based on his good knowledge of his students. J.I. leafed through the outline of one of them and was evidently satisfied with it. «Well,» he said, «tell me now what question you would like to answer best of all?» The student said he was ready to answer any question Frenkel found proper to ask him. «Then let us proceed like this,» said Frenkel smiling. «I'll put my finger into your notes at random and you will tell me about what I am pointing at.» He closed the note-book, put his finger inside and found the indicated place to be blank. «If this is the case, I won't ask you anything,» said Frenkel with a laugh and gave the top grade to the student.

In a notebook from 1939 or 40, a list of students from one of the Physics and Mechanics Department study groups is to be found. Next to one of the names there is a note: «excellent», and, in parenthesis, «in advance». Presumably, it was a good student who had passed the examination somewhat below his abilities. Frenkel's former students said that he really sometimes gave the grades «in advance», and, with only rare exceptions, this «debt of honour» was always paid to the examiner.

---

33) In Russian one says «3 beginnings of thermodynamics» instead of «3 laws of thermodynamics».

34) J.I. pointed to Lagrange who gave exams simply by opening his own treatise and questioning students on specific topics.

The Physico-Mechanical Faculty of the Leningrad Polytechnical Institute's staff. From left to right (first row): A. Lurie, P. Tartakovskii, A. Ioffe, Ya. Frenkel, F. Knorre, E. Nikolai, M. Frank, G. Grinberg, L. Prokofieva; (second row): (?), V.V. Skobel'tsyn, L. Sena, I. Palei, Yu. Terminasov, A. Tudorovskii, I. Melnikov, L. Freiman, B. Gorelik. Leningrad, 1939.

«We all went to J.I. well prepared,» recounts Professor L.A. Sena who graduated from the department in 1930, «but I would have died of shame if I had received a bad grade from him!»

A.F. Ioffe was the organizer of Soviet physics, a national figure. Knowing this, one has to be really surprised by the enthusiasm with which he implemented his plans to «transplant» physics outside Leningrad. The best, most gifted scientists left the PTI and served as «condensation centres» for new institutes created on the PTI sample both in Leningrad and other cities of the Soviet Union.

In 1927 Teplotekhnichesky Institute (the Institute of Thermal Technology) was hived off the PTI; in November of 1930 the Khar'kov Physical-Technical Institute was opened, with the Tomsk one preceding it a little, and in 1933 the Physical-Technical Institute in Dnepropetrovsk was founded.

In October 1931, the Leningrad Electrophysical Institute headed by A.A. Chernyshov and the Leningrad Institute of Chemical Physics headed by N.N. Semenov detached themselves from the PTI. In the beginning of 1932, the Urals Physical-Technical Institute in Sverdlovsk was organized.

Ioffe sent the most capable staff members, ones mature in talent if not in years, to lead and direct the activity of those young scientific institutions. A.I. Leipunsky, I.V. Obreimov, K.D. Sinel'nikov, L.V. Shubnikov, and theoreticians D.D. Ivanenko, L.D. Landau and L.V. Rozenkevich went from Leningrad to Khar'kov. It was in Khar'kov, where Landau worked from 1932 to 1937 (i.e., before his departure, to the Institute of Physical Problems in Moscow), that the first representatives of the Landau school of physics appeared. To Tomsk went P.S. Tartakovskii, to Dnepropetrovsk went G.V. Kurdyumov and B.N. Finkelstein, and to Sverdlovsk went S.V. Vonsovsky, Ya.G. Dorfman, I.K. Kikoin, B.G. Lazarev and others.

All these institutes needed their own theoreticians. We have already spoken about the strong theoretical group at Khar'kov. However, at first things did not go so well elsewhere, and it was Frenkel who really helped a lot in this respect.

Sparing no efforts, he spent much time and energy on these seedlings of Soviet physics. He headed the theoretical department at Semenov's Institute and organized a seminar there, and he frequently visited Odessa, Kiev, Dnepropetrovsk, Sverdlovsk and other cities of our country.

Among J.I.'s papers is a letter from the Rector of the Tbilisi University to him which shows both Frenkel's influence on the development of physics in the country and also the public's positive opinion of him, even at that early date.

«Tbilisi State University, 24 February 1936
To Professor J.I. Frenkel

Recently the Research Institute of Physics was organized at the Tbilisi State University. The new institute and especially its theoretical department are in strong need of highly qualified guidance, so your visit to Tbilisi would be very useful to us. You could perhaps give a number of consultations and provide the theoretical group with several topics for future studies. We hope that your visit would constitute the onset of a closer scientific and organizational relationship with you, which would be of great value to us.»

J.I. received quite a number of such letters and always responded in the affirmative, believing it his duty as a citizen to help the new institutes. One should add that J.I. returned home from such trips always freshened and full of new strength and ideas. He exhibited a real thirst for such travel, having an inexhaustible curiosity about new people and places. Here are several excerpts from his letters of this period:

«Odessa, 25 May 1933 [to his parents].
... Having given the students a lecture entitled «A solid in a liquid and a liquid in a solid», I returned to the office of Kirillov [Professor E.A. Kirillov, Director of the Physics Institute at Odessa University — V.F.] where, along with solid pastries and liquid «Narzan» mineral water, I found things falling in neither of these categories, such as letters ... I am supposed to weigh anchor on the 30th to look at the Dneprokombinat before coming to Dnepropetrovsk.»

«Khar'kov, 18 May 1934 [to E. Hill].
... I have just arrived here to attend a conference on theoretical physics in which a number of foreign physicists will participate. Among them is Bohr who has been our guest in Leningrad for 10 days. He is a charming man, and I greatly regret having missed him at our previous meetings. The conference will cover

Kharkov Conference on Theoretical Physics, May 1934. From right to left: Mrs M. Bohr, N. Bohr, Ya. Frenkel (in open coat).

various topics from nuclear problems to those relative to the solid state. I intend to speak about some parts of my latest theory of superconductivity (which is still unpublished and rather qualitative in character) and about some advances along the lines of Born's modified classical electrodynamics ...

I shall come back to Leningrad about the middle of June. After the close of the Khar'kov conference I shall proceed to Odessa where I shall attend another conference on semiconductors and solid rectifiers. Then I shall stay in Odessa for a couple of weeks giving lectures at the university and trying to work a bit in a quiet atmosphere (that of Leningrad is anything but quiet for me).

I recently received an invitation to give 6 lectures at the Institut Henri Poincaré in Paris on a subject of my choice. I have accepted the invitation and plan to go to France at the end of September (via England, where a conference on nuclear problems will be held at the beginning of October), returning right at the beginning of November ...»

«Khar'kov, 18 May 1934 [to his parents].
... The conference starts tomorrow. Unfortunately, today I was not able to prepare my presentations because of a headache.

Bohr, with whom I traveled from Moscow to Khar'kov, is extremely content with what he has seen. He was especially impressed by the penitentiary work colony of GPU near Moscow (nothing similar exists in the West!). He is going to visit us again this fall. He is generally very pro-Soviet, and one can only feel sorry that he has not been to our country before.»

«Odessa, 25 May 1936 [to his parents].
... For the first three days of my stay in Odessa I will not be burdened at all with work for the Institute, having limited myself to a small consultation on the 23rd. Today, for a change, I sat around from 10.30 a.m. until 3 p.m. In the next few days, after finishing the consultations, I am going to go into an intensive lecturing phase (3 lectures for the university, one report at the Institute Colloquium, and 1 lecture at the Odessa House of Scientists). I simply do not know how I'll manage all these in 4 or 5 days. Maybe I'll stay in Odessa for another day. From here I'll go to Dnepropetrovsk, the last stop on my Ukrainian tour, to attend a conference on solutions.

... Yesterday I was absolutely free; until 5 p.m. was working in my hotel room (wrote a paper completed a long time ago) ... Now I shall return home and do my studies until late at night.»

On 13 December 1929, in the Grand Physics Auditorium of the Leningrad Polytechnic Institute, the first of three public debates on the nature of electric current took place. The debates were held between the electrical engineer V.F. Mitkevich and J.I. Frenkel, and one cannot apply to them the standard diplomatic formula of «they were held in an atmosphere of complete mutual understanding»; more likely, exactly the opposite occurred. A.F. Ioffe, P.S. Ehrenfest, D.A. Rozhansky, I.V. Mestchersky participated in the discussions.

If one is to judge by their titles, the talks dealt with the nature of the electric current. But actually they ended up as a discussion of more general questions, in particular the problem of ether and its role in the electromagnetic theory. Einstein's theory was appropriately appreciated and used by all progressive scientists, and they had abandoned the concept of ether. However the idea of ether continued to live among a rather numerous group of physicists, including, for instance, J.J. Thomson. The German physicist P. Lenard, known for his conservative views, had even «quantified» ether and introduced its two types: the usual ether and the «primary ether» (Uraether). Commenting on these delusions, J.I. said: «The incapability of a substantial part of mankind to convert to atheism, and of a substantial part of contemporary physicists to reject the seeking of ether, are of the same source and breed: they rest on traditional and inert modes of human thought.»[35]

Ehrenfest spoke at both the first and second debates. He started his speech at the second one with the words: «It could never happen in Europe, nor in America that four thousand man-hours would be so vigorously spent on such a complicated question, as happened here, and this circumstance alone attracts me

35) J.I. Frenkel. «The mysticism of the world of ether». Vestnik znaniya, # 6, p. 426, 1925 (in Russian).

to you. Secondly, I learned a surprising amount both the other time and today, and from all who have spoken here, without any exception. I did not always understand, and I did not fully agree with all that was said. Sometimes I protested deep down inside, but this is of no consequence since I learned something from everybody, and therefore I understand why you were content to concentrate on this for so many hours. Now you will probably become angry with me if I say that a good physicist seldom goes into philosophy, and then only when nothing else is left for him, and, unless I am not mistaken, he always does it extremely poorly. And this scares me out of my wits ... Last time I was discredited and several of my friends called me a typical European compromiser, because I had tried to say that I did not understand the difference: it is my conviction that Mitkevich and Frenkel said the same. I would like to correct this unfavourable impression, and therefore today I will be an anti-compromiser.»[36]

Ehrenfest's role as an anti-compromiser resulted in the unfavourable evaluation of ideas developed by V.F. Mitkevich. Ehrenfest energetically supported Frenkel's arguments.

We conclude the story of these emotional debates with a quotation from Frenkel's last speech at the third discussion on 14 March 1930, in which he analyzed the concept of far-range interaction (these thoughts were further developed in his report on the analogy method in physics) and outlined the contemporary state of electrodynamics.

«I stated in my previous speech that any «understanding» is relative, that the process of understanding may be compared to the process of solving an equation when one value, which we call the unknown, is expressed in terms of another one, which is assumed to be known. When we, however, try to express this supposedly «known» value in terms of something else, then we find ourselves at a dead end, because it cannot be reduced to something else. We reduce the more complicated things to less complicated ones, and these less complicated things then cannot be understood. We just have to reconcile ourselves to this. Far-range action represents something not understood. This lack of understanding, nevertheless, is in no way evidence against the admissibility of this notion for physics. If this viewpoint is convenient, if it provides the possibility of describing simply physical phenomena, then there is no need to reject it, to regard it as not clear enough, to not be able to imagine a phenomenon according to this point of view. The viewpoint of an intermediate medium is reduced to either exactly this far-range action, if the medium is a set of separate particles, or, if the medium is imagined to be continuous, as is sometimes suggested, to an even worse, not better, than that in which the viewpoint of far-range action rests. We recognize no continuous material medium. Each material body which we meet in nature consists of separate particles, and it really is impossible to imagine a continuous medium since we have no example in our world as we know it. The question here is not about how

36) «The nature of the electric current». VEO, Moscow-Leningrad, 1930, p. 101 (in Russian).

easily we can imagine and understand the fundamental premises of the theory of electromagnetism; the question is about to what extent the picture developed on the grounds of this concept is closed and simple. I think that the picture obtained on the basis of electrons interacting at some distance aside from any intermediate medium [i.e., ether — V.F.] is actually the simplest sketch of electromagnetic processes ... The conclusions of the electron theory do not change a bit if we eliminate the concept of ether. The only change is the modification of our notions, and this is really a change for the better in the sense of simplifying our world picture. Our picture of electromagnetic phenomena before this was clouded by the ether concept, which was not necessary. By removing the idea of ether, the picture of electromagnetic phenomena reduces to one of motion and interaction of electrons through emptiness.»[37]

The example of these disputes on the nature of electromagnetic phenomena shows that a physicist dealing with general rather than applied problems of his science sooner or later has to face questions of a philosophical nature. This happened, naturally, with Frenkel. On the 16th of December 1929, three days after the first discussion, J.I. made a presentation at PTI seminar on dialectical materialism PTI under the title: «The theory of cognition of physics». From a short note published in the Polytechnic Institute newspaper «Tovarishtch» («Comrade»)[38] we learn that one of the listeners suggested «sending Frenkel into industry, since he regards mathematics as a classless science.» (It is true though, the author of this article notes, that this speech was not taken seriously by anyone and earned a «laugh from the audience».)

In the second half of the 1930s of which we have been speaking, and at the end of the 1940s, which we will describe later, critics assessing Frenkel's philosophical views referred almost exclusively to one talk he gave at the 8th Physical-Chemical Conference held in Leningrad in November 1931.

From the letters Frenkel sent from the United States to his relatives we see that, in his lectures on Russia there, he talked about all the good that had occurred in his motherland in the 1920s. He described, in particular, the unheard-of growth of education, the elimination of unemployment, and the successes in industry. These positive changes appeared even more impressive when discussed against the background of the depression beginning in the United States, and fascism rearing its head in Germany.

Upon returning home Frenkel had a fresh view of the Soviet reality and considered it to be not only his right, but also his duty to strive to eliminate everything that, in his opinion, was inhibiting the development of physics, the science which gave the principal meaning to his life.

I am not going to go into a detailed description of the history of the confrontation between Soviet physicists and those who opposed them with all the support

37) Op. cit., pp. 112–114.

38) Tovarishtch, # 42 (128) of June 10, 1930 (in Russian).

of the then party apparatus. I will only note that to the cohort of anti-quantists and anti-relativists there also belonged several serious scientists who truly believed that everything beyond the framework of classical physics was devilishly evil. Among them were the academician V.F. Mitkevich, physicists N.P. Kasterin and A.K. Timiryazev, and some of their disciples. Also holding these conservative views was a group known as the vanguard, or the philosopher-marxists as they called themselves. The latter group included first of all M.B. Mitin and A.A. Maksimov.

The polemics took place in the pages of «Pod znamenem marksizma»[39] (PZM; its present title is «Voprosy filosofii»), popular science journals, and major newspapers, not to mention various conferences.

It was in one of the PZM issues, in a paper by Maksimov, that I read that in 1931 at the physical-chemical conference, «J.I. Frenkel allowed himself a number of purely idealistic statements. The statements were made so aggressively and, at the same time, were so much in opposition to the views of the communists participating in the conference that it was decided to call a meeting of the VKP(b)[40] faction to deal with it. In order to double check which were J.I. Frenkel's true views on philosophy and his actual attitude to dialectical materialism, he was invited to the faction meeting, where he delivered some comments which left no doubt as to what Frenkel's attitude toward dialectical materialism really was.»[41] Several quotations from Frenkel's speech at the VKP(b) meeting are given below. First, however, I will discuss a letter Maksimov sent to Ioffe. Maksimov wrote: «You, Academician Ioffe, would like to promote all the idealistic twaddle of J.I. Frenkel [he is referring here to Frenkel's «Wave Mechanics» — V.F.] to the rank of the laws of nature and to speak sarcastically about those who would not. However, it would be better if you, Acad. Ioffe, would think a little about whether a man pretending to play a leading role in Soviet physics can be permitted to declare Mach's inventions to be natural laws. Instead of showing grief at such a sad state of affairs, Acad. A.F. Ioffe not only speaks sarcastically, not only proclaims idealistic blabbering to be «laws of physics», but pronounces Frenkel himself as a materialist and quantum mechanics (in his, Tamm's and other such Menshevik idealists' interpretation) to be a particular implementation of materialism.»

This is more or less the style and manner of Maksimov's whole paper. Having carefully looked through the proceedings of several physical-chemical conferences I, however, did not find any «criminal» statements by my father. He spoke about rather spezialized questions, such as the theory of adsorption, the conductivity of metals, and the physics of electrolytes. Nevertheless, after Maksimov's paper appeared, the quotation about Frenkel's report in 1931 started to wander from one

39) «Under the banner of Marxism».

40) Vsesoyuznaya Kommunisticheskaya Partiya (bol'shevikov), i.e., the All-Union Communist Party of Bolsheviks.

41) A.A. Maksimov. «On the physical idealism and its defense by acad. A.F. Ioffe», PZM, 1937, # 11–12, p. 171 (in Russian).

publication dedicated to the philosophical questions of physics and their interpretation to the next.

G.A. Savina (at the Archives of the Russian Academy of Sciences, Moscow) kindly provided me with the verbatim record she had found in Maksimov's papers on the meeting of the party group of the third physical-chemical conference to which Frenkel was invited after his report (the record was excluded from the conference proceedings). Some fragments of this record are found below. The chemist S.A. Balezin chaired the meeting (later, in the war years, he worked on the State Defense Committee).

Frenkel: I am eager to speak before the conference but not to take back my words, as I do not reject them.

Shvetsov: I don't know how one could separate the questions of methodology and politics, and Philipp Frank considers you to be a Machist.

Frenkel: ... Frank knows me for my works on physics, and I am interested in gnosiological questions. I think that the theory of dialectical materialism is not a final product of a human mind that may satisfy all reasoning mankind. Independently of how it is necessary to the foundation of socialism. The dialectic method has no right to claim the leading role in science. The relation between philosophy and science is similar to that between the being and consciousness. I consider our policy to be extremely harmful when it forces the youth and scientists to adopt the views of dialectical materialism. This does not mean that I deny there is any truth in dialectical materialism.

Balezin: So what is your philosophy?

Frenkel: I am hardly a baby in philosophy. What I read in Lenin and Engels cannot replace my gnosiological views. This is my opinion, and I will not reject it. I consider it my duty to make amends for any damage effected by my speech, but will not denounce my views.

Balezin: Is it possible to conclude that dialectical materialism hinders the development of science, in particular natural science?

Frenkel: Yes, at least in the form in which it is now dogmatically taught. The dialectical method you are forcing into the natural sciences is the Hegelian dialectical, and it is not necessary for physics.

Kedrov: Your speech goes against all the decrees of the Party, against all the decisions of recent scientific conferences. How could one combine your views of philosophy with those of politics?

Frenkel: My views are an impartial analysis of real facts, contrary to the preconceived opinion with which you, dialectists, approach every phenomenon. You put an equal sign between antidialectical opinion and anti-Soviet spirit. I give myself up to the Soviet power, but I do not admit diamat [dialectical materialism]. For me and my scientific activity the Party resolutions on the role and significance of dialectical materialism are not obligatory. That is why I am not a Party member. I do not consider it necessary for me just to drift downstream or to say something and think quite the opposite, as some people do. Your policy I regard as erroneous. Only I, as a friend of the Party, pronounce openly that I do not acknowledge dialectical materialism, while many specialists are afraid to say this and vote formally for your suggestions. I think, I hope, that the higher ranks of our Party will soon abandon their view of dialectical materialism as one of the dominating values.

Shevtsov: For us Marx, Engels and Lenin are the founders of the dialectical method in natural sciences. What is your attitude to Engels and Lenin?

Frenkel: What I have read in Lenin's and Engels' treatises has far from delighted me. Neither Lenin nor Engels are authorities for physicists. The book by Lenin is a sample of fine analysis but reduces to the firm establishment of truisms which are not worth fighting for. You put yourself in a false position with Lenin: as soon as you leave Lenin, you are just left with nothing. Your philosophy is a reactionary one; I hope that the Party will soon be convinced of this [Frenkel refers to Bukharin as an example]. I personally, as a Soviet man, cannot agree with an opinion that harms science. There cannot be proletarian mathematics, proletarian physics and the like. ... As an honest man, as a Soviet one, I am saying this to you frankly and openly.

Balezin: I think the question is exhaustingly clear. Let us not delay Professor Frenkel any more.»

One can state that the conference speech was Frenkel's reaction to certain philosophers' dogmatic approach to the interpretation of new physics theories. To understand the reasons for such a reaction, we quote from a paper by I.E. Tamm entitled «On the work of Marxist philosophers in the field of physics» published a little later: «The principal evil lies in the fact that the vast majority of representatives of Marxist philosophy working here in the field of physics and related disciplines are simply ignorant of the present state of science ... Certainly, the questions [of new physics — V.F.] are difficult. One cannot just storm them in a direct attack. To acquire a good understanding of them one has to learn a bit, to get acquainted with the actual material through more than popular brochures and papers; one has to think the issues over. But it is not such a big deal; no special talent is needed to do this, but it is absolutely necessary to do certain work, and not just a small amount ... Without a doubt, those who equate the philosophy of dialectical materialism with some of its unsuccessful adepts are wrong. Unfortunately there are many scientific workers who do so. Given this, one can understand their attitude but can in no way justify it ... Therefore we can only hope that there will soon appear new scientific workers who can help those few worthy ones in the arena of the methodology of physics, mathematics and natural sciences by combining the arms of dialectical materialism they possess with an understanding of the basics of the sciences to which these arms are applied.»[42]

42) I.E. Tamm. Pod znamenem marksizma, # 5, pp. 220–21, 1933 (in Russian).

The other source for the critique of Frenkel on the part of the philosophers about whom Tamm wrote about was the first volume of «Wave Mechanics» published in Russian in 1933 (the second volume appeared a year later). I shall not dwell again here on specific features of this book since it was done previously, though briefly. Other physicists were «criticized» along with J.I.

In this respect one has to admire the courage with which Abram Fyodorovich Ioffe came forward to defend Soviet physicists. In his paper «On the situation of the philosophical front of Soviet physics» he supported theoreticians Frenkel, Tamm, Fock, Landau, and Bronstein, as well as some experimentalists such as Vavilov and others.

Everyone who knows the history and present state of Russian physics associates the names cited with the galaxy of our outstanding scientists who brought world recognition to scientific research in our country. For A.A. Maksimov, however, these people were instead «the compact group which has fallen under the influence of idealism and which tries to identify itself with the collective of Soviet physicists as a whole.»[43]

Ioffe wrote: «Maksimov, [A.K.] Timiryazev and some others ... proclaim the whole dialectical process of the development of contemporary physics and the synthesis of wave mechanics achieved on this basis to be overall idealism, and all the Soviet physicists to be servile imitators of the West. This is an undeserved libel which cannot be left unanswered. In writing off the theoretical physics of Fock, Frenkel, Tamm, Mandelstam, Landau and their pupils, we would erase all Soviet theoretical physics, leaving no trace ... Is it true that Soviet physics is nothing, that it has no ideas of its own? The theory of solids and metals by Frenkel, the theory of the photoeffect by Tamm, Landau's generalization of the indetermination principle and his theory of diamagnetism, the theory of the atom by Fock ... have all these not influenced Western physics and led it in new directions?»[44]

Ioffe then wrote that the Western physicists who had not only created the brilliant new theory, quantum mechanics, but who also sympathized with the Soviet state, were also being attacked by the dogmatists and vulgarizers of science. He stressed that these attacks were directed against quantum mechanics itself. «At the prompting of comrade Maksimov, some people now think that dialectical materialism (and not its unsuccessful representatives) forbids quantum mechanics and the theory of relativity. On the contrary, dialectical materialism rejects that idealistic mist which, along with these theories, some bourgeois idealists are trying to force upon us. Such thought is a harmful delusion, and it should be pushed aside as soon as possible!»[45]

The most damaging feature of the critique about which Ioffe wrote was that it had a strongly pronounced political subtlety: the physicists named were called

---

43) A.A. Maksimov. Pod znamenem marksizma, # 6, p. 52, 1937 (in Russian).

44) A.F. Ioffe. Same journal, # 11–12, p. 135, 1937 (in Russian).

45) Op. cit.

«Menshevik idealists», etc. Here is a very typical example of this. Tamm, when speaking at the March 1936 session of the Academy of Sciences about the report in which Mitkevich continued to insist that magnetic field lines were material, remarked that to discuss the properties of the latter was just as senseless as to inquire about the colour of a meridian. His critics used this sentence as proof that Professor Tamm was opposed to the red colour of our meridian (in this case literary meaning was attached to the political colouration)!

Yet one more small but characteristic episode. The students of the Physics and Mechanics Department were issuing their own tabloid «Experimentator» («Experimentalist»). They invited Frenkel to participate in one of its 1935 issues. He responded readily and wrote some jocular aphorisms. After that an article appeared in the institute's newspaper «Industrial'ny» written absolutely seriously and even with a good bit of wrath, under the title «Pearls of idealistic wit». The author reproduced some of J.I.'s aphorisms (for instance: «Vacuum-phobia is an illness similar to hydrophobia, i.e., rabies; so be wary of the physicists and philosophers infected by it»; in this we easily hear the reverberation of the recent disputes on ether), two of which especially outraged him. What were those phrases?

Frenkel had written: «A physical theory is similar to a suit tailored for nature. A good theory is like a good suit, and a bad one just hangs unevenly,» and then generalized: «A theoretical physicist is like a tailor». This joke provided J.I.'s critic with an excuse to say: «So, from Professor Frenkel's viewpoint the theory is «sewn» for nature and put on it when ready — this is just what is called idealism. This is the break between an idea and reality, mentality and nature. The meaning of these aphorisms is that it is not nature that provides laws for us, but we who issue laws for nature.»[46]

So what was Frenkel's own attitude toward this kind of critism? I think he was distressed not because such critique was targeted at him, but because people listened to it in those years, and he feared that it would stay in the minds of many hundreds of readers and disorientate them.

I recall that one day when Frenkel said he could not take such accusations seriously, one of his friends lamented on the light-mindedness which his reaction revealed. «Well, yes, perhaps I'm light-minded,» responded J.I. with a spontaneous joke, «but please do think about what that means: it means my mind works easily. Do you really prefer hard-mindedness?»

Publications by S.I. Vavilov, Ioffe, Tamm, Fock and others did finally, after a certain delay, affect the situation: as of the end of 1938 the unsound accusations that Soviet physicists were idealists stopped. Among those who had rebuffed Maksimov and the like in the press, in the «Pod znamenem marksizma» (PZM) in particular, I did not find the name of Frenkel. It occurred to me that he himself might have decided in this case to revert to silence. Such caution was not in his nature. Presumably, I thought, there had been special reasons for him to remain silent.

---

46) «Industrial'ny», April 19, 1938 (in Russian).

And indeed, about three years ago the previously mentioned staff member of the Archives of the Academy of Sciences, G.A. Savina, informed me that, while sorting out Maksimov's papers, she had discovered a letter Frenkel wrote to the editors of PZM and his article in reply to Maksimov. Having read this reply, I understood why it had not been published. I would like to present parts of it here.

The paper is written in an antagonistic spirit. Referring to the known (Lenin's) definition of the materialistic view, according to which the world exists independently of our consciousness and can be more and more accurately imagined over the course of the scientific development, J.I. wrote:

«From the point of view of this criterion I, as well as any other physicists who have not gone out of their minds, am and always have been a hundred percent materialist. In fact, a physicist, both the experimentalist and the theoretician, dealing with the experimental study and theoretical explanation of phenomena of the surrounding world, simply cannot help being a materialist. He is at the very least a «spontaneous», unconscious materialist.

«Maksimov writes in his article that «as a rule, political reaction goes hand in hand with reactionary philosophical views». I consider it my duty, however, to indicate that the views of this group [Maksimov and his colleagues — V.F.] are astonishingly similar to those of the group of reactionary German physicists led by Professor J. Stark. Stark, having been authorized by the Fascist party to represent it on questions of physics, proclaimed the theory of relativity to be a Jewish invention and saddled one of the creators of quantum mechanics, Heisenberg, with the stigma of «Weisse Jude» [«White Jew»]. Like their ideological parents who regarded Jews as the cause of the coming revolution, the modern German fascists considered the revolution that started in physics 40 years ago, and which is continuing its progress now, to be a result of the Jewish conspiracy both inside and outside Germany.

«In their attitude to modern physics, Mitkevich, Timiryazev, Kasterin, and Maksimov along with them, differ from Stark, Lenard, E. Gehrke and other representatives of German obscurantism only in the fact that they replace the term «Jewish» with the term «idealistic».

«As for myself, Ioffe and a number of other physicists whom Maksimov tends to suspect of a counter-revolutionary spirits ... I hope that citizen Maksimov had not meant to accuse me personally of being politically reactionary. In such a case I would be forced to institute legal proceedings against him for a scurrilous libel.»[47]

Frenkel's reply was forwarded to Maksimov; it contains marginal notations by the latter which now would only cause bewilderment and laughter. But October 1937 was not at all a time for laughter! I am sure that other archives kept, and perhaps still keep, some specific responses to Frenkel's rebuttal and his speeches in defense of the new physics. On one occasion such a document was found in his

47) The complete paper «Reply to A.A. Maksimov» is published in: «Readings in the memory of A.F. Ioffe, 1989». Leningrad: Nauka, 1990, pp. 53–60 (in Russian).

personal file in the Archives of the Academy of Sciences. I will give it in full but will omit the name of its author, because he has long been not among the living and, as his co-workers told me, had an excellent reputation during the last years of his life. So let us write off his statement as one of the «youthful mistakes» which tragically characterize that period of time. Here is the document:

«I know him [Frenkel — V.F.] as a man essentially anti-Soviet and as standing for the enemy's ideological positions. The former Scientific secretary of the physical group comrade Divil'kovsky (member of VKP(b)) and comrade Maksimov from the Institute of Philosophy of the Academy of Sciences of the USSR must know him well. The latter has to possess stenographic materials characterizing his ideological positions. From the words of Divil'kovsky I know that Frenkel, in a private talk with him, stated that dialectical materialism was a juggling of facts, that Lenin made a mistake in turning to the questions of dialectic materialism, that the heads of young people were being littered with it in vain, that it was time to reject this harmful fashion et cetera, et cetera.

«He surrounds himself with pets, «geniuses», dubious people. One cannot point out the communists trained by him. May 19, 1939 Signature»

The history of the «Reply to Maksimov» helped me to clarify yet another puzzle over which I pondered for many years. Extant and recently published letters physicists wrote to the Procurator's Office and other institutions contain appeals for reconsideration of the arrests of their colleagues. I can name several such people: M.P. Bronstein, Yu.A. Krutkov, P.I. Lukirskii, V.K. Frederiks; I can add the recently published uncredibly courageous letters by Kapitza to Stalin and Molotov in the defense of Fock and Landau. I was surprised by the absence of my father's signature. A simple explanation of this was provided by Professor D.D. Ivanenko. In a letter he informed me that Frenkel had been the first to start collecting signatures for such letters, sparing neither time nor effort (such was, in the words of Ivanenko, his own case just after his arrest in 1935). But to put Frenkel's signature on such documents would have only reduced the probability of positive action: in «competent institutions» it would have been regarded negatively.

Ours is the century of science, which has become more and more involved in everyday life and has entered each house. This process is naturally reflected in literature: plots are more and more frequently based on a scientific conflict, and creative people of science figure as heroes.

The importance of science had been understood long ago by scientists and writers. So in Leningrad in the early 1930s a group of writers (V.A. Kaverin, B.A. Lavrenyov, S.Ya. Marshak, K.I. Chukovsky, Yu.N. Tynyanov, M.M. Zostchenko and others) established a good relationship with the scientists (A.F. Ioffe, N.N. Semenov, J.I. Frenkel, Ya.G. Dorfman, P.P. Kobeko, D.L. Talmud and others). Their meetings were held under the slogan: «Tell millions of people about the glorious victories of Soviet science!»

A discussion was held on scientific creative work where Frenkel delivered a lecture on the subject. He began his presentation by saying: «Scientific and artistic creation usually are opposed to each other as stiff brainwork, on one hand, and the free flight of creative fantasy, on the other. This opposition is, however, absolutely erroneous, since fantasy plays as important a role in the process of scientific creation as it does in art. Scientific activity is, of course, also associated with work of a mechanical or collecting nature, such as calculation, accumulation of data, and so on. But this work is only auxiliary, similar to stylistic work in literature or the finishing touches in painting. It gains some sense and value only in relation to this or that idea which always comes as a product of the creative fantasy of a scientist.»[48]

A physicist «is fed» experimental material, but latent laws and properties of the field are not given directly to the researcher. He has to unveil them. However, to conduct the search one has to know where to look. It is necessary to suppose or have a suspicion of the existence of certain effects and trends; one has to work out some conjecture. Based on the analysis of the most fruitful conjectures, J.I. put forward the principles of analogy and simplicity as the guiding criteria of «hypothesis construction». «Having exhausted the method of analogies, a physicist trying to sort out new material does not at all revert to logic, but mounts the horse of his fantasy.» Frenkel concluded this part of a lecture by saying: «A new theory is better, the simpler its form and the broader its content as far as the amount of experimental material covered and, so to speak, the larger its efficiency coefficient. A physical theory with high efficiency is called beautiful and provides an aesthetic delight. This is also true for its mathematical formulation, as well as for purely mathematical theories and proofs. The notion of «elegance» is as relevant and valuable to a mathematician or physicist, or any scientist in general who not only observes phenomena but interprets them, as it is to a poet, musician or artist.»[49]

The second part of Frenkel's lecture was dedicated to the question of the artistic exposition of scientific data. He discussed the same question in another speech at the meeting with writers in the winter of 1934. In Frenkel's opinion, the aim of writers whose work has some «scientific flavour» is to help the readers get involved in scientific creation, to understand its drama and poetic nature. As was characteristic of his lectures, J.I. purposefully overrode his hand (which makes them not only clearer, but, unfortunately, more vulnerable if considered formally). «Until now,» Frenkel said, «literature has dealt mainly with the study of people's interrelations; nature has only served as a background, filling in those boring pages which would be omitted by an impatient reader. At present, when we are approaching the exposition of the relations between man and nature, intra-personal relationships may occupy the background» ... «A writer is a victim of the old

48) «Literaturny Leningrad», # 13, November 6, 1933 (in Russian).

49) Op. cit.

Ya. Frenkel lecturing his seminar in the Physico-Technical Institute in the mid-1930s.

A caricature of Ya. Frenkel done by artist Antonovskii and published in 1934 in *Literaturny Leningrad.*

attitude toward natural sciences. A man who graduated from either law school or a philological department was regarded as highly and broadly educated, while the math and natural and physical science departments were thought to produce only limited specialists ... When we speak about the scientific theme in literature, we have to speak about the expansion of the knowledge of the writers themselves. One of the issues of the first stage of the combined work of writers and scientists should thus be, naturally, the learning of science.»[50]

A.F. Ioffe said many interesting things at the meeting with writers. He explained that many scientists love, unsurprisingly, good detective novels. They are attracted to such novels by the search for truth, the selection of facts and, finally, guesswork. Conjectures happen to be true and false. «But», Ioffe said, «when we describe our conclusions, we depict the matter as if we understood immediately and had never been mistaken. In fact, we are deceiving the reader ...»[51]

Frenkel was interested in books with scientists as protagonists. There were very few such books about in the 1930s (or in the 1940s), and they were mostly about medicine and medical doctors. He had a high opinion of Paul de Crui who wrote about real scientists; he very much liked «The Citadel» by Archibald Joseph Cronin and «Arrowsmith» by Sinclair Lewis, whose characters were fictional.

In July 1938, J.I. was slightly ill. He wrote to his elder son: «... I am still writing to you from my bed. I have used my time lying here for intensive reading. The day before yesterday I read «Struggle with Death» by P. de Crui, and yesterday, with equal pleasure, a French detective story. Took it in one gulp, like a cold drink on a hot day, and after that I grimaced with disgust, feeling the drink turning sour.»

Perhaps it would be appropriate to speak here about Frenkel's literary tastes. He was not one of those people who spend the major part of their free time reading, which one can see from his letters. J.I.'s favourite poet was Alexander Pushkin. In this respect, as in many others, his taste was, so to say, traditional. He knew by heart many of Pushkin's poems and had read his letters several times. Frenkel liked some poems by A. Blok very much, though he remained absolutely indifferent to the other verses by this author (the majority, in fact). He would always say that a poet should be judged by his best verse; the lesser work should be ignored. He judged other kinds of creative work, in particular, scientific creation, in the same way. And that also was how J.I. related to people. There is a typical phrase in one of his letters: «one should be indulgent to their weaknesses and value their merits, which most people do not completely lack» (he was appealing to his addressee to be patient).

In June 1949 my father wrote to me from the Baltic lands: «The weather here has become stable. These days I do a lot of walking (near Maiori), go boating, swim a little, write my book occasionally, practically do not play music, and do a little painting ... The day before yesterday we were the guests of Marshak, who is vacationing here, or, rather, works a lot and very successfully. He read us his own poems as well as some newly translated ones, to our great delight. Tomorrow we shall visit him again to say good-bye.»

---

50) «Literaturny Leningrad», # 9 (31), February 23, 1934.

51) Op. cit.

Ya. Frenkel with an easel. Maiori (near Riga), 1949.

Frenkel and S. Marshak had met back at the onset of the 1930s. After the war J.I. did Marshak's portrait in oil and saw him rather frequently. He admired

Marshak's translations of Shakespeare's sonnets, especially the very lyrical # 77 («The glass will show thee how the beauties wear ...») and # 104 («To me, fair friend, you never can be old ...»), and the tragic and wrathful # 66 («Tired of all these, for restful death I cry ...»).

I have already mentioned that J.I. never knew the joy and distress of collecting; in particular, he did not collect books. There were three bookcases in his study: one of them contained issues of «Physical Review» and JETP, and the other two held books on physics, mostly gifts from the authors, and some fiction. Sometimes the former and the latter were mixed, as G.A. Grinberg observed. He noted that, side by side with the «Orlean Virgin» by Voltaire there happened to be a book by a very famous theoretical physicist. It was known that this scientist dealt with nothing except theoretical physics, which is why Grinberg wittily remarked, to the great pleasure of Frenkel: «This proximity, however, in no way threatens the Orlean Virgin!»

Several books were found directly on J.I.'s desk, among them four volumes of the prewar edition of the collected works of Il'f and Petrov, and «Julio Churenito» by Erenburg; these books, as well as «The history of One Town» by Saltykov-Stchedrin, Frenkel frequently reread.

I recall the pleasure with which my father read «Zheleznaya volya» («Iron Will») by Leskov, of which an enormous number of copies were published during the war: it was perhaps the only book the Frenkels bought during those years. Also in the war years, in evacuation, my parents «in one gulp» reread «Voyna i mir» («War and Peace») by Leo Tolstoy; there was a long queue for this book, which happened to be in the possession of one of their friends. Father often recalled Tolstoy's famous saying that we love people for all the good we did for them and hate them for all the evil we brought upon them.

When leafing through Frenkel's letters and reading his opinions of people, one finds them all to be very nice, kind; J.I. did not stint his praise. One finds practically no sharp evaluations of anyone in hundreds of his letters. This, of course, does not mean he had no understanding of people, though his excessive trustfulness, of which we have already spoken, made one sometimes, if only for a moment, suspect that this possibility could not be eliminated.

My mother, after my father's death, remembered that he had told her once: «You know, I notice that I am extra lenient in the estimation of the scientific achievements of those physicists with whom I do not sympathize, while sometimes I am unsoundly severe in regard to the others, whom I do love. The reason is that I am afraid that my personal relationship will influence, to some extent, my opinion of the work, and thus I find myself being unnecessarily extreme.» Another time, when he was told by one of his acquaintances that he gave the impression of being a person who liked having others play dirty tricks on him, Frenkel tried to explain even this as an overt example of the same manifestation.

Frenkel highly valued Soviet classics, both prewar works («Konarmiya» («Mounted Army») and stories by Babel', «Tikhiy Don» («And quiet flows the

Don») by Sholokhov, «Razgrom» («Rout») by Fadeyev, stories by Lavrenyov, «Khozhdenie po mukam» and «Pyotr Pervy» («Travel in Sufferings» and «Peter the Great») by A. Tolstoy) and postwar ones («Zvezda» («Star») by Kazakevich; «V okopakh Stalingrada» («In the Trenches of Stalingrad») by V. Nekrasov, etc.).

Among the works by Hemingway published during Frenkel's life his favourites were the novel «To Have and to Have Not» and the series of Nick Adams stories.

J.I. started to read more in the last years of his life, when his health deteriorated and his beloved active ways of relaxing now tired him. It was in these years that he read almost all the memoirs of Anatole France. He considered «Les Dieux ont Soif» to be one of France's masterpieces. Generally, J.I. read a lot about the history of the French Revolution; he also knew Russian history well. From his American letters we know the interest with which he read «The Rise of American Civilization». He was a connoisseur of the history of the American Indians and liked to talk about their customs. One he liked was that the orators of one Indian tribe uttered their speeches standing on one foot, to ensure the brevity of the exposition. Talking about this custom, J.I. remembered a similar Jewish parable. A young man had been asked to state «the whole essence of the Torah» while standing on one foot. And the youth was able to do it by just saying that one simply should not wish upon his neighbour anything which he would not wish upon himself.

Here is another Indian custom that Frenkel remembered. The Native Americans always kept with them a tightly sealed vessel with some admirable aromatics. When something extraordinarily pleasant happened in their lives, and only then, they opened the vessel and inhaled the aroma, to preserve the impression. A familiar aroma of perfume, a voice that you know, all these could, in J.I.'s opinion, suddenly recall very strong memories which had seemed completely buried. (Yu.L. Sokolov remembered a brief remark of J.I.'s about scent memory. On a white night they had been walking in Leningrad; the jasmine was blossoming, and Frenkel said: «The aroma of jasmine is pleasant, but seems to be something infinitely common and naive. It reminds me of a worn melody.»)

The Oxford University Press regularly sent Frenkel advertisements for books to be published, proposing he order some of interest. And once, after the war, a packet with Jim Korbett's «Cumaon man-eater» arrived. It turned out that Frenkel had ordered this book from England, and he confessed with a smile that he had dreamed of becoming a tiger hunter; so the Korbett book was a debt paid to the memories of his childhood. J.I. read the book nonstop. Then he wrote to one of our publishing houses that in his opinion it should be translated. Soon Frenkel received yet another of Korbett's books, «Leopard-man-eater from Rudraprayag», accompanied by a letter from the author who presented the book to J.I. This gift was Korbett's response to Frenkel's letter (both these and two more books by Korbett were published in Russia later, at the end of the 1950s and the beginning of the 1960s).

In 1951, during the last fall of his life, J.I. read for the first time, with delight, «The Forsyth Saga» by Galsworthy, and reread almost all of Dostoevsky's works. The book which he read on the last evening of his life was a collection of stories by Ol'ga Forsh «Pod kupolom» (about the French Academy, a good supplement to the noteworthy pages A. France and A. Daudet wrote on the subject).

In his letters Frenkel often spoke about his musical tastes and exercises. When not on vacation, he seldom picked up the violin. But, once having started, he improved his performance rather quickly, approximately within three days of playing for an hour or two. Frenkel usually started his «postwar» programme with the slow and pensive «Siciliana» by J.S. Bach, the «Slavonic Dances» by Antonin Dvořák or the miniatures by Fritz Kreisler and, proceeding gradually with the more technically demanding pieces, turned to the concertos of Antonio Vivaldi and J.S. Bach, the «Devil's Trills» by Giuseppe Tartini or the violin opuses by S.S. Prokof'ev.

In 1950 Nikolai Nikolaevich Davidenkov[52] celebrated his 70th birthday. Frenkel decided to paint his portrait. For about two months, Nikolay Nikolaevich went to J.I.'s each week and sat for his portrait. After the painting session both the artist and his model switched to a musical programme, leaving the study for the sitting-room. Davidenkov went to the piano, and Frenkel tuned his violin. They played sonatas by Edvard Grieg and L. van Beethoven and, in particular, Frenkel's favourite sonata by C. Frank.

Of the Russian composers Frenkel liked N.A. Rimskii-Korsakov very much; he played the violin arrangement of «Zolotoy petushok» and admired the choirs from «Kitezh». He liked the «oriental tone» of many Rimskii-Korsakov melodies; the oriental colouring in general he considered as very characteristic of Russian music.

The last time my father and I attended a concert at Maly Zal (Small Hall) by the Leningrad Philharmonica was on 10 January, two weeks before his death. It was an evening of chamber music; and the programme included Brahms' trio for piano, violin and French horn (the piano and violin were played by L. Oborin and D. Oistrakh this evening). After the concert we went out into Nevsky Avenue. Father began to whistle the main theme of the trio. Then he said: «What a brilliant composer! I don't know why R. Rolland was so angry with him?!» [Rolland, the famous French novelist, had expressed a great dislike of Brahms in his novel «Jean-Christophe».]

Getting an enormous aesthetic pleasure from music, Frenkel as a physicist considered it paradoxical that the progress of technology produced no influence on the techniques of musical performance. By means of radio and television music can enter houses many hundreds and thousands of kilometers away from a hall

52) N.N. Davidenkov, Academician of the Ukrainian Academy of Sciences (1880–1960), was a specialist in the physics of metals and strength and one of the first staff members of the PTI.

where an artist stands in front of a microphone and «draws a horse tail across ox tendons».

J.I. listened with great interest to the Termen-vox invented by L.S. Termen at the PTI. But Termen's musical device continued to require a highly skilled performer. He greatly enriched the sound range of expression, and due to solely this simple, good technique instead of virtuosity, the first performers earned the Termen-vox and its inventor broad recognition at the end of the 1920s and early 1930s.

It is not difficult to imagine Frenkel's interest in a musical instrument which combined all the advantages of Termen's instrument with the solution of the performance problem, thus «overthrowing» virtuosos. I am speaking about the variophone of E.A. Sholpo.

On 8 March 1941 Frenkel arrived in Moscow specially to attend the Supreme Attestation Commission (VAK) where the question of awarding E.A. Sholpo with a scientific degree for his invention was to be discussed. Frenkel, who was Sholpo's official expert reviewer, explained the basic principle of the instrument in the following way:

«If you put a row of absolutely transparent images on a [dark — V.F.] film and then make the film move with a known velocity against the photo element, you will obtain a row of electrical pulses. Their number will be equal to the number of light spots. If the number is above 30 pulses per second, you will have a sound whose frequency corresponds to that of the signals. This is the simplest method of obtaining artificial sound. The sound intensity will depend on the current which may be altered easily. The sound timbre depends on the shape of the apertures. By perforating rows of apertures of different shapes you can obtain any timbre, reproducing both existing musical instruments and also those unknown to contemporary music.

«How are spots or signs recorded on the film? There is a special device called a variophone ... The possibilities of musical performance associated with Sholpo's instrument are infinite. I have had the pleasure of listening several times to pieces performed by Sholpo's instrument and must state that I have never heard anything remotely similar to it, even at an orchestral performance ...»

In this book some of Frenkel's drawings are presented to give an impression of his artistic style. In dozens of his albums hundreds of such sketches and landscapes are kept. (To prove that his artistic creations were popular, J.I. boasted, with a smile, of how several of his portraits and landscapes were stolen from his apartment during the war.)

As for landscapes, he was mainly attracted by the capricious and rapidly changing colours of the evening sky. He drew in his albums clouds with sheaves of sun rays storming through them, and marked the different colours with words. After that, from memory, he transposed what he had seen onto canvas. Each vacation, especially one spent near the sea, brought dozens of such oil landscapes.

The Billiard Marker. Oil portrait by Ya. Frenkel, 1949.

In the Hermitage Frenkel loved above all Rembrandt Hall. Commenting on the colouring of later portraits and pictures by the great master, J.I. wondered about its origin. He then put forward the idea that perhaps it is related to the way a crystalline lens tends to turn with age into a kind of yellow filter. Later on, from a conversation with people acquainted with physiological optics, it became clear that the crystalline lens has a trend to turn with age sooner to a blue filter (and Frenkel was stuck by the question: does this not explain the blue period of Picasso?). I want to stress from this example how organic Frenkel's very physical view of the surrounding world was, in particular, to the world of art.

In the 1930s at the Physical-Technical Institute the staff members would often gather to spend an evening relaxing together.

Frenkel, as well as his colleagues from the PTI, liked these evenings. There were also musical items in their programme. Talented musicians are not difficult to find among physicists. Famous Western physicists, Planck, Born, Heisenberg, Sommerfeld and Ehrenfest were brilliant piano players, and the violin is inseparable from our image of Einstein. Otto Hahn had a beautiful voice. There were some musicians among the PTI staff members as well. L.S. Termen, already mentioned as the inventor of the world famous electronic musical instrument, the Termenvox, should be among the first on this list. The Academicians N.N. Andreev and N.N. Davidenkov were good pianists, J.I. played the violin and L.S. Vagin, the cello.

Along with the performances of amateur musicians, humorous pieces occupied a significant place on the agenda of these evenings. Back in the 1920s a special journal «Physikalische Dummheiten»[53] was published at the institute in which the PTI scientists teased each other. P.P. Kobeko performed well as the master of ceremonies. Witty scripts for puppet performances were written by M.P. Bronstein. J.I. often participated in the humorous parts of the events, posing usually as a lecturer, and he was always a success. The representatives of the first PTI generation recalled his talk in the Ehrenfest style (Frenkel himself wrote the following about talk: «At the Institute, a small soiree had been organized where the All-Union Congress of physicists was humorously dramatized. I took an active part in this stage version and, above all, mimicked Ehrenfest with great success».) Once Frenkel imitated himself delivering a report. With an absolutely serious appearance he told some funny science-like rubbish whose effect was amplified by the fact that he used the most exotic letters of cyrillic for writing the formulae (which usually is never done). The improvised presentation concluded with something like the following phrase: «Introducing the previous expression into this formula and neglecting its right-hand side, we easily obtain the known result of quantum mechanics which, however, turns out to be wrong.»

His other lecture, «On the theory of dances», J.I. had prepared carefully, that is, he had written it from beginning to end. Its effect was partly based on the similarity and interplay of some Russian words, in particular, «spin», which just means spin, and «spina», which means «back» in English. After describing dances in the spirit of the classical Newton mechanics Frenkel, combining it with «emotion dy-

53) «Physical Stupidities».

namics», created the quantum theory of dances. There is an appropriate prohibition principle in this theory which states that only two partners may move along the same quantum orbit, and a firm condition is that their spins (backs!) are oppositely directed. This postulate of the theory, as well as the others, was demonstrated on the stage by the lecturer's assistants, i.e., a pair of dancers. The inadmissibility of the dance with the partners' spins directed identically was demonstrated in the same way. After the quantum theory of dances, in the spirit of Bohr, Frenkel proceeded with the wave theory. Since a dance is not just a body motion but is associated also with some movement of a soul, that is with agitation, one was able to speak about the wave theory of dances.[54] In the wave theory of dances, the concept of a quantum orbit was removed as incompatible with free will. The path of a dancing pair became absolutely indefinite, and its position at this or that instant of time could be defined only in terms of probability theory.

J.I. concluded his report with the following words: «An enormous thanks for the discovery of the wave theory of dances and, in particular, to its experimental testing, goes to the scientists of the PTI who have been working in this direction since last year with all the power of their feet.»[55]

P.L. Kapitza began his memories of Rutherford with the following episode. At the Cavendish Laboratory a grand meeting had been held, dedicated to the memory of Maxwell. Maxwell's disciples gave their recollections of him. Rutherford asked Kapitza about his impressions of these memories. Kapitza answered that the lectures were very interesting but he was surprised that everybody had only said good things about Maxwell, presenting him in the form of, so to say, sugar syrup. And he would prefer to see Maxwell as a real, live man, with all his human features and drawbacks. Kapitza recalled:

«Rutherford laughed and said that he ordered me to tell future generations, after his death, about what he was really like. Rutherford said this in a semi-jocular manner, and I also laughed.

«And now, after his untimely death, I am speaking about him and would like to fulfill his behest. But when I start to draw his image in my mind as I would like it to be presented to you, I see that his death and the time which has passed since parting with him have absorbed all his petty human shortcomings ... Now I really do understand those of his disciples who spoke then in Cambridge.»[56].

I have often pondered over these lines from Kapitza's lecture on Rutherford while working on this book, and when talking to people who knew my father closely, I have even asked them to tell me especially about his faults.

---

54) In Russian the words «wave» and «agitation» have the same root and are pronounced similarly.

55) This report, «The quantum theory of dances» is published in: «Fiziki prodolzhayut shutit'», ed. V. Turchin. Moscow: Mir, 1968, pp. 281–86 (in Russian). Several other jokes of J.I. Frenkel's are also found there.

56) P.L. Kapitza. «Zhizn' dlya nauki. Lomonosov, Franklin, Rutherford, Langevin». Moscow, Znanie, 1965, p. 35 (in Russian).

My father himself liked to say that people's faults are the analytical continuation of their merits (which echoes an English saying).

Frenkel had a very broad range of interests; he was, according to W. Ostwald's classification, a romantic type of scientist. This means that he did not work for all his life on one or two problems, as is characteristic of classical scientists. But this peculiarity of the latter is as much a drawback as an advantage, because the distance between purposefulness and goal-orientation and narrow-mindedness, is just as large (or small, depending on your opinion) as the distance between broad-mindedness and being a little bit scatter-brained. Kapitza once told Frenkel: «You would be a genius if you published 10 times less than you do!», a very characteristic hyperbole!

In a letter J.I. wrote to Hill on 5 October 1931 he commented on one of his unsuccessful (in his opinion) papers: «Well, well, I have been a bit hasty in publishing the matter. But on the one hand I have always been hasty in publishing rubbish — as if it weighed too heavily on my shoulders — and on the other hand, I have usually been too slow in publishing good things.»

It is interesting to note that E.V. Shpol'sky, recalling P.P. Lazarev, wrote: «He considered it necessary to publish his results as soon as possible since, in his opinion, an unpublished result affected the scientist's mentality and hindered his further work.»[57]

The trait Shpol'sky indicated is very characteristic of scientists of this bent. Ostwald stressed that the other such trait is the speed of mental reactions. And J.I. provided an example of a «romantic» scientist. He could catch instantly the essence of a phenomenon and immediately interpret it.

In his article on Frenkel's scientific legacy Ioffe wrote: «One can pose the question: why did J.I. Frenkel sometimes publish his studies in cases when he had not been able to find a complete and rigorous solution to a problem? One may think that here a teacher, a leader of a school, manifested himself in J.I.'s behaviour. He passed on relationships noticed by him, uncovering a new corner of the physical picture. It was as if J.I. were saying: here is a thought with a healthy core; try to develop it, for it is a promising topic for a theoretician. Such papers by Frenkel stimulated the thoughts of his official and unofficial pupils and led them to penetrate the mechanisms of physical processes.»[58]

If we return to the list of «analytical continuations of the merits» of Frenkel, we have to mention his surprising trust in both people and information. His wife, Sarra Isaakovna, joked: «It is boring to pull your leg: you believe instantly anything you are told.» This trust by J.I. taught him some sad lessons which, however, he forgot very soon. He was a «hopeless optimist», as his friends called him, and not because optimists enjoy an easy life. But it was easier for him to live because

---

57) E.V. Shpol'sky. UFN. v. 27, # 1, p. 3, 1945 (in Russian).

58) A.F. Ioffe. In: J.I. Frenkel. Collection of Selected Papers, v. II. Moscow-Leningrad: AN SSSR publ., 1958, p. 18 (in Russian).

he was an optimist. He often laughed at some home-bred sceptics, saying: «They know about how to make themselves comfortable: if their gloomy predictions do not come true, then they are fine. Otherwise it is even easier for them, knowing they have proven to be prophets.» He valued highly the great skeptics Voltaire and France because he suspected an enormous love for life behind their scepticism.

We learn about another feature of J.I.'s nature from a letter Sarra Isaakovna wrote to a friend: «... what you are writing about Yasha's scrupulousness makes me happy ... He feels that everybody is always glad to help and serve him, and he considers that he has the right to accept everything since he is always attentive himself in every respect. So please do not encourage him and thus spoil many years of my educational work. I keep telling him in capitals that he must not just appeal to anybody for anything, that at present everybody has enough worries of their own [the letter was written during the war — V.F.]. However, I know he is incorrigible!»

When searching for Frenkel's shortcomings, people usually recall that precision was not among his merits. He was always late — for a train, for a seminar, for a lecture — and in this respect presented a target for good-spirited bantering. Once the Director of the Polytechnic Institute was informed of such an example of Frenkel's lateness for a lecture (it happened around 1949–50). The Director demanded from Professor Frenkel, in a way suitable perhaps in the case of a guilty student, to present in writing the reason for his lateness. Frenkel wrote a report: «In reply to your inquiry I inform you that I was not late for my lecture since it had not yet started before my arrival.»

However, Frenkel's habit of being late contradicted his reputation for doing everything he had promised without fail, within the limits of his responsibility. His notebooks are covered with notes regarding requests and commissions. And the mark «done» stands against each of them.

He had a very good memory, though often could not remember names well. This professional memory never failed him, and his encyclopedic knowledge of physics was recognized by everybody. Once after the war, one of Frenkel's colleagues mentioned to him a paper published in one of the foreign physics journals. It turned out that J.I. knew nothing about it. «In this case,» the man told Frenkel, «I have to change my opinion of it. I considered it to be just a regular paper. In reality, I see it is absolutely outstanding: in the many years of our acquaintance this is the first paper to remain unnoticed by you!»

Chapter 7
Years of War

On Sunday, June 22, 1941, J.I. woke up early as usual, had breakfast and started working in his study. A preexam consultation had been scheduled for 1 p.m. at the Polytechnic Institute; it was the end of spring examinations, and Frenkel with his family would be going on vacation afterwards. The resort place had been chosen and the train tickets even purchased long before. But then the radio brought terrible news about war: for several hours the fight had already been thundering on the western borders of the USSR. A deep gulf emerged in the life of every Soviet citizen, and everything, even recent happenings, became distant: «this occurred yet before the war».

In the morning of June 23, Frenkel went to the recruiting point of his Vyborgsky district in Leningrad. He appealed to be accepted as a volunteer in the Red Army, but his name was known in Leningrad. The District Military Commandant said to him: «Professor Frenkel, think, where would you be of more use? At the front we would be able to make use of only your hands. Your knowledge is needed much more.»[1]

With the beginning of the war the work of a number of the PTI laboratories was reoriented towards military subjects. Frenkel managed to get connected with the scientists of these laboratories, consulted the researchers from there, and tried to invent something by himself. One of the topics was prompted by the city's Anti-Aircraft Defense, to which Frenkel had reported with a proposal of service. There were «white nights» in Leningrad, but the fall was not at all far off. And then the problem of the blackout of the city would arise. And in July and August on dark enough evenings and nights, tramcar flashes broke the blackout.

On July evenings J.I. would stand for a long time at the last tram stop across from the Polytechnic Institute observing the dazzling flashes and pondering the frequency of their appearance. He figured out how to eliminate the flashes by conceiving of a way to keep the tram's brush contact and the wire from disconnecting. Later on this work was published[2]. It is significant that even in such a strictly technical calculation problem Frenkel adhered to his usual way of exposition: the article is written in vivid language and the ideas are illustrated with analogies ...

The other question Frenkel got interested in was the problem of the armour-piercing ability of materials, their interaction with bullets and shells. A group

---

1) Recently Professor I.I. Zuckerman, in the article «My teacher», dedicated to the memory of G.A. Grinberg, recalled how he had entered the People's Emergency Volunteer Corps at the very beginning of the war: «My decision [to enroll — V.F.] was directly influenced by what I had been told in the Dean's Office [of the Physics and Mechanics Deptartment of the Leningrad Polytechnic, where he had just completed his third year — V.F.]: among the enlisted volunteers were Jacov Il'ich Frenkel and Georgiy Abramovich Grinberg.»

2) JETP, v. 12, 1942, p. 171 (in Russian).

of PTI researchers had suggested some interesting ideas about diminishing of the armour-piercing action (striking force) of projectiles by mounting one or two special grating screens on the front part of a tank hull. The screens would provide the striking projectile with a transverse displacement before it hit the armor; now it would hit at some angle, sometimes even tangentially, and thus the force of the impact would be smaller. The same screens might also cause an early shell-burst, which was desired, since the armor would better resist shell-splinters. It is interesting to note that, according to the Marshall I. Konev's memoirs, during the storming of Berlin, the tank hulls were «shielded» by sheets of thin tin, with the same objective: they «knocked off» the front strikes of the Faust-cartridges. Such a simple tool proved to be very efficient!

In the beginning of 1942, Frenkel presented his calculations related to these and similar questions in the form of reports entitled «On the resistance of fibrous media to the motion of a bullet», «On the bending of the shell or bullet axis moving in a medium with high resistance», and «Statistical theory of the turn of a projectile (or bullet) penetrating a grating perpendicular to its trajectory».[3]

Before the war Frenkel had been more or less regularly consulting some works on determining aircraft velocity and the formulation of principles for its measurements with the help of the cockpit instruments.

M.S. Sliozberg, a staff member of one of the special Leningrad institutes, recalled that in 1939 Frenkel did a presentation at his institute on the design of a radar system to locate planes. After the war began, J.I. returned to this problem and improved the initial project. Calculations related to these ideas were carried out at the PTI in a tight collaboration with the members of Yu.B. Kobzarev's laboratory who dealt with the pulse radars.

On 1 October 1941, Frenkel wrote from Moscow to his brother, Vladimir Il'ich (V.I. Frenkel was mobilized from the very start of the war and served as a military medical doctor in the 23rd Army of the Leningrad front until the war ended): «... I went to Moscow on the order of S. Kaftanov, a representative of the State Defense Committee. I had written him about my Anti-Aircraft radio invention and asked him to summon me from Kazan', and the engineer guiding the work at the plant, from Leningrad, to transfer the whole work to Moscow. I don't know what will come of it. Until now I have had no information about the progress of the work from Leningrad. Possibly, the plant has been evacuated.»

At the conference in Moscow where Frenkel's results were reported, Professor S.E. Khaikin was present. He later recalled that it was the electrodynamic calculation of the detection system suggested by Frenkel. The latter consisted of a high-frequency continuous transmitter and several optimally positioned receivers. One was able to determine the coordinates and velocity of the target by the Doppler frequency shift of the radio signal reflected from the moving plane.

---

3) Archive of the A.F. Ioffe PTI, file # 24 (in Russian).

The chess party. Ya. Frenkel and his brother, Dr. V. Frenkel. Leningrad, in the mid-1940s.

One should, however, state that these ideas were not developed at that time, for a number of reasons. First of all, the efficiency of the detection is proportional to the airplane's velocity, which was relatively small in those years. Moreover, the continuous regime of operation required a lot of power (and, correspondingly, powerful generator lamps in the meter and decimeter wavelength range), as well as special efforts to detect a weak reflected signal against the background of the strong transmitted one. Finally, a serious obstacle to implementing Frenkel's detection system was the necessity of rather large computational work, in particular, to eliminate the so-called false targets. With the action rate of the mechanical computation devices used then, this kind of work could hardly have been carried out during a given short period of time. Therefore progress in the development of radar technique in the war years was associated with research on pulse radio-location. (It is interesting to note that the idea of synthesizing pulse and continuous operation regimes is now implemented in long-range radio-location detection stations.)

At the end of July, 1941, a part of the staff and equipment of the PTI laboratories was evacuated to Kazan'. J.I. and his family left Leningrad with the second PTI train on 23 August. This was one of the last trains which actually left Leningrad; two days later all railway connections between the city and the country were cut off by the advancing German forces. The train went along the northern route, with frequent stops. I recall one stop at Mga made at night in response to the alarm signal. A German reconnaissance plane dropped a light-bomb attached to a parachute which slowly descended over the station, flooding it with blinding light. Very soon Mga was cruelly bombed; we were told about it by those who passed it in the morning on the already restored rails and caught on with the PTI train. And it continued its slow progress, via Tikhvin, Cherepovets, Vologda, Yaroslavl', and Murom.

In Kazan', where the train arrived on 2 September, a majority of the physics institutes of the Academy of Sciences had been concentrated by this time, both the Leningrad and Moscow ones. Almost all of them were situated in the building of the Kazan' State University. One of its wings was occupied by the PTI, Institute of Physics Problems, P.N. Lebedev Physics Institute. Several of the PTI laboratories had the Ethnography Museum at their disposal. Here, in two tiny nooks, the theoreticians settled themselves. One of the nooks was occupied by Frenkel's staff members (B.I. Davydov, T.A. Kontorova, Yu.N. Obraztsov, and his postgraduate students K.S. Shifrin, A.I. Gubanov, O.B. Firsov and others). The other one, with hardly enough room for a desk, was for J.I. himself.

Both walls of a kind of a corridor formed by museum shelves were covered by pictures of North-American Indians which everybody going to Frenkel's nook looked at. They were dressed in many-coloured costumes of various tribes; there were also exhibits of the weapons and domestic life of Eskimos, and other Native American tribes. In winter of 1941 one of the institute staff members was lucky to get some unground rye. He ground it with the help of stones exhibited in the museum.

Starting about the spring of 1942, Frenkel engaged himself with work on a book in which he wanted to gather and develop all his results concerning the kinetic theory of liquids. He spent the major part of his time on this work. He worked at home, not in the small communicating room occupied by him and Sarra Isaakovna, but in the laundry placed at his disposal by the landlady. The laundry was half-dark: the dim light oozed through a narrow window cut in the log wall. J.I. wrote on a sheet of veneer in his lap which replaced a desk. He had no books, as always. However, all the inconveniences were outweighed by one factor: silence.

From a letter J.I. wrote to his brother on 9 July 1942.

«It seems that I have not written to you for about a week and a half, if not more. The reason is that I'm too much involved in the writing of my book which, in any case, I must finish by the beginning of August. I can spare practically no time for anything else, and, also, all my interest is temporarily concentrated on it. Will it be needed by anyone besides myself? If everything goes right, and I am absolutely sure of this, then yes, the book will be needed. Anyway, there is nothing better and more useful I would be able to attend to. Along with the book, I work with my postgraduate student two times a week; from time to time I consult the students and give exams; even more seldom I consult the members of our institute. Recently I've started a portrait of O. Shmidt. I am satisfied with it, also because it provided me with the opportunity to get closely acquainted with Shmidt, with whom I sympathize in many respects.»

O.Yu. Shmidt was the Director of the Institute of Theoretical Geophysics of the USSR Academy of Sciences. Frenkel got to know not only him, but the studies carried out at his institute as well. J.I. also told Shmidt about the passion of his youth to investigate the Earth's magnetism. Perhaps exactly these conversations gave rise to a long period when Frenkel turned to the problems of geophysics.

The portrait of Shmidt was not the only one J.I. painted in Kazan'. Not long before Frenkel had completed the picture of academician A.N. Krylov, Kapitza's father-in-law. When the portrait was ready, Frenkel, who had always liked Krylov and admired him, as everybody did, asked him to accept it as a gift. And J.I. was really surprised when Krylov categorically declined this offer! «Alexey Nikolaevich, it means, you do not like it?» asked Frenkel with distress. «But, Jacov Il'ich, on the contrary, I like it very much!» «Then what's the matter?» «The problem is, I cannot accept such a valuable gift from you, I could only buy it». «But with your permission, Alexey Nikolaevich, I have all my life made my living with a pen, but never with a brush!»

Probably, a simple wish of Krylov's to help Frenkel was hidden behind this unexpected proposition. The winter of 1941–42 was very hard, cold and brought real hunger.

I recall two episodes characteristic of this period. Starting with the spring of 1942, when the Kazanka river, which crossed the whole city, had freed itself from ice, river shellfish appeared at the city markets. Various dishes were prepared from them, pepper or other spices being added to cover their rather unpleasant flavour. Mrs. Frenkel once made a paste out of these shells. «It seems it's good», she said to her family with a touch of uncertainty. And J.I. responded in the following way: «You know, we are ready to eat this, but please don't assure us it is tasty!»

Another time, after a tiny Kazan' breakfast, Sarra Isaakovna recalled how they had breakfasted once before the war back in Leningrad. That other day J.I. had been in a bad mood and had said he was not satisfied with the relatively simple menu. Mrs. Frenkel had started to defend herself by naming the dishes on the table. «But you would also tell me that we have bread, butter and sugar!», exclaimed J.I. This recollection served as a good dressing to the Kazan' breakfast!

Here are some more letters J.I. Frenkel sent to his brother during the war years.

On the Kazan'-Sverdlovsk train, 19 January 1943.
«My dear chum! I have just been informed about the breaking of the Leningrad siege. My warmest congratulations to you on this occasion! The day is not far when the railway connection of our city with Moscow, with the east and north, will be restored again. The day is not far when we will rejoice over a final victory in the fight against the damned fascists.

... I committed myself to go to Sverdlovsk for 1 week a month. The Kazan' situation, with its everyday hardships and worries, rather tires me both physically and, especially, morally. In a week or ten days of travel I become completely relaxed and return to Kazan' with a new store of spirit.

... I am seldom at home. Go away in the morning, about 9 a.m., to the institute and come back from there by 9 p.m., as hungry as a wolf (in between I only get the so-called «academic» breakfast). During the last 2 or 3 months thoughts have started to cross my mind again, especially during my Sverdlovsk detours, when I relax about the domestic and other Kazan' worries and problems.»

On the Kazan'-Moscow train, 19 October 1943.
«My dear brother! I've hardly come back from Moscow with Sarra when I am heading back again. This time I'll stay in Moscow for only three days and then go to Gor'ky and Sverdlovsk.

... What is driving me, would you ask, over the world? Financial difficulties, in the first place, as well as my own restlessness. I had been invited to Gor'ky long ago (to the university).

... On my previous visit I got an invitation from Shmidt to work at his institute (of theoretical geophysics). I wanted to do it very much. I even thought about transferring from the PTI to Shmidt's institute and, in this case, about going to live in Moscow. Have decided, however, not to force all these changes and to take some work from Shmidt only as a part-time job.

As for Sverdlovsk, I am going there probably for the last time, to complete my old obligations.

Later on I am going to visit only Moscow, first from Kazan' and then, with the help of God, also from Leningrad. All of us in the Kazan' evacuation hope to return to Leningrad in the spring of 1944.»

«Moscow, 13 February 1944.
My dear chum! Yesterday I received your letter of congratulations. Thank you for your best wishes. Have to state that I don't yet feel the burden of my years. I spent my fiftieth birthday[4] in an overcrowded third-class car on the way from Gor'ky to Moscow. I stayed in Gor'ky for 4 days, during which has been developing an intensive lecturing and consulting activity.»

«Moscow, 28 July 1944
... I engaged myself with putting in order my English book which has been submitted to the All-Union Society of Cultural Relations with Foreign Countries. They promised me to send it with a courier of the Ministry of Foreign Affairs by plane in few days. 2–3 days more, and I'll be done with the «backward» translation in Russian of the added sections for the Russian edition; after that I'll consider these «twins» to be completely dealt with and will attend to other matters.

Here in Moscow I've recently been offered an interesting and honourary job at the Council on Radio Location (my first war invention is devoted to it). This work

4) J.I. Frenkel's 50th birthday was on 10 February 1944.

does not require me to live permanently in Moscow but provides the possibility of frequently visiting it. The question about whether we shall go to live in Moscow or shall await the possibility of returning to Leningrad is still not resolved by me either way.

I congratulate you, my dear, on our succes on the Western Front. Hope those on the Leningrad front will follow them soon.

It seems to me that the hour of the final rout of the fascist filth is really close now. Mussolini is already finished.»

In his biography (1946), J.I. wrote: «Though I dealt with a large number of questions related to various branches of physics, physical chemistry and geophysics, the fundamental contribution provided by my works is to the problems associated the structure of matter, especially the structure of liquids and solids.»

It is thus not surprising that «Kinetic Theory of Liquids» was the most successful of all the books written by Frenkel. He himself was also of this opinion.

It has already been mentioned that J.I. wrote the book under difficult conditions. The following lines are in the preface, dated July 1944: «I have written this book at a time ill suited to the pursuit of purely academic tasks. I may be excused for this by the consideration that, when victory had become certain, it seemed advisable to revert to topics of pre-war and post-war interest. In recalling the state of mind and the conditions in which this work has been carried through, I wish to express my debt of deepest gratitude to the men and women of my country who, in ranks of the Red Army, have heroically and victoriously struggled for the salvation of our life and civilization against the ruthless forces of the Nazi barbarians.»[5]

Even since Archimedean times, the trend has established itself in physics to bring liquids and gases closer together. From the purely superficial side this was explained by the fact that both liquids and gases are shapeless, or more precisely, they take the shape of the vessel containing them. But the volume of liquids does not depend on the vessel's volume, which makes them closer to solids. Furthermore, the density of liquids differs little from that of solids, and this means that interatomic distances change little under melting. The same follows also from the fact that the latent vaporization heat of solids is 30 to 40 times lower than that of liquids.

Liquids resist not only compression (as gases do) but expansion as well: it is a remarkable property established back in 1911 by the direct Meyer experiments. (He filled a vessel to the brim with mercury at the same temperature and soldered it. Upon cooling the vessel size decreased less than the mercury, and, nevertheless, it was expanded without breaking from the walls; the break occurred at pressures of about 100 atmospheres. If not for the pressure, it would have withstood even larger stretching stresses.)

---

5) J. Frenkel. «Kinetic Theory of Liquids». Oxford Univ. Press, 1946, p. 3.

The transition from the solid to liquid state is accompanied not by the drop of heat capacity, as one might expect from the point of view of the «gas-like» liquid; on the contrary, it causes to a certain extent its rise (for one-atomic solids heat capacity is 6 calories per gram-atom, by the law of Dulong and Petit, and gases have half that value). Frenkel concludes from here that the thermal motion of particles in liquid changes relatively little and preserves, in particular, the oscillatory character near their equilibrium positions. This conclusion is one of the most important among those forming the basis of Frenkel's theory.

X-ray analysis (and, to the same extent, the experiments on electron and neutron scattering in liquids) demonstrated that liquids keep elements of order in their structure. X-ray diagrams of liquids obtained for the first time by Stewart, as well as the whole set of X-ray studies carried out most intensively in the USSR by V.I. Danilov, testified to the existence in liquids of a short-range order under the loss of the far-range one. X-ray pictures of liquids, being essentially different from those of monocrystals, are in many respects similar to the X-ray patterns of polycrystals consisting of many monocrystals «pressed» into a solid.

The above facts are so meaningful and transparent that one should only wonder how they could have been ignored for such a long period of time, and how, by Frenkel's expression, «the gas end was considered to outweigh» the liquid state, which actually occupies an intermediate position between the solid and gas states.

The revision of this point of view is tightly connected with the name of J.I. Frenkel. His study did not move too far in the direction of a solid, but managed to see features of liquids in solids and vice versa. This is the basis on which the kinetics of the liquid state rests.

One should note that the unification of the notions of hardness and fluidity had been performed back in the 1870s in the particular case of the theory of elastic relaxation developed by Maxwell. But Maxwell's theory, which stated that the same force may cause in a solid (amorphous) body both a reversible elastic deformation and an irreversible viscous flow, was a purely phenomenological one. Only the concepts developed by Frenkel of the character of thermal motion in solids and liquids provided a molecular-kinetic foundation for this theory. From the viewpoint of the a «nomadic» nature of atoms' movement in liquid (and solid) bodies, it becomes clear that if the time of an external force action is smaller than the «settled» lifetime, τ, of a particle (see the above formula for it), then the body behaves like a solid one. In the opposite case the body starts to flow in the direction of the applied force, i.e., in the dominating direction of the atom «jumping» from one (temporary) state of equilibrium to another.

Frenkel's book was published in 1945 by the Academy of Sciences. A year later its translation appeared in England. It was reprinted many times after the death of the author: in the United States in 1955 and 1968, in Germany in 1957. It also constituted the 3rd volume of the Selected Papers of J.I. Frenkel issued by the Academy of Sciences of the USSR (1959), and in 1975 the Academy published it in the prestigious series «Classics of Science».

The very fact of the multiple printings of Frenkel's book shows that his ideas about the nature of the liquid state have not grown old. Naturally, they are not the last and final word of science. The theory of liquids is far from being completed and continues its development. But an exceptionally broad range of phenomena gain their exhaustive explanation within the framework of J.I.'s concepts and are described by the formulae derived by him: future oblivion hardly threatens them.

There was a lot of work to be done by Frenkel by the end of 1943, when the last page of the «Kinetic Theory of Liquids» was written. He had to translate the book into English. Everyday work at the PTI consumed considerable time: in 1943 Frenkel dealt with the viscosity theory, mechanical properties of solid bodies, and the theory of oxide film formation on the surface of metals.

Glimmers of peaceful life then appeared in the country: sessions of the Academy of Sciences, meetings and conferences were held again in Moscow. J.I. continued to consult researchers at the Physical-Technical Institute, and his relationships with the physicists of Sverdlovsk, especially S.V. Vonsovsky, were deepening.

As early as in the beginning of 1942, J.I. headed the Chair of Theoretical Physics at Kazan' University and simultaneously started lecturing there. Along with this he became interested in the studies of a young Kazan' physicist, Evgeny Konstantinivich Zavoysky (later a member of the Academy). I.E. Tamm wrote that J.I., when discussing with Zavoysky his work, «... in particular, noted that some experimental curve obtained by Zavoysky in the study of paramagnetic absorption in solutions, was a branch of the resonance curve,[6] and delivered the theory of relaxation losses associated with paramagnetic resonance phenomena in both solutions and solids. It is known how widely these phenomena are now used in physics, as well as in chemistry.»[7]

The role of J.I. Frenkel in the discovery made by Zavoysky is similarly outlined by A.F. Ioffe: «Defects of Frenkel are known under this name all over the world. But few people know that one of the most significant discoveries of solid state physics, the paramagnetic resonance by Zavoysky, is due to a scientifically sound guess of Frenkel's»[8].

By this time Frenkel's interests leaned more and more in the direction of geophysics, and, starting in 1944, he began to work intensively in this field. The Institute of Theoretical Geophysics, within the frame of whose activity this work was mostly carried out, had been reevacuated to Moscow back in 1943. J.I. started going there regularly from Kazan'.

---

6) This very curve was discovered later by E.K. Zavoysky with the broadening of the frequency range — V.F.

7) I.E. Tamm. UFN, v. 76, # 3, p. 424, 1962.

8) Quoted by Ioffe's paper in: «J.I. Frenkel. Memoirs. Letters. Documents». — Leningrad: Nauka, 1986, p. 17 (in Russian).

About the same time Frenkel established relations with the All-Union Institute of Aviation Materials. Academician S.T. Kishkin, one of the leading members of the institute, wrote about it: «Jacov Il'ich knew that our institute is an experimental one in the field of material studies, and we all knew that J.I. was the most distinguished theoretician in metal physics. Unification of the experimental abilities of our institute and Frenkel's theoretic potential proved to be most fruitful. That is why a new laboratory was set up, organized specially for J.I. Naturally, Frenkel was appointed to head the laboratory.»

One should say that J.I. had thought for a long time about opening his own experimental laboratory and had many times spoken about it in the PTI. Therefore he started to work with passion at the new institute, together with a small collective of physicists and physical chemists.

Frenkel's work there, as well as his advances in geophysics, are described in the next chapter.

In the spring of 1944, Frenkel returned to Leningrad. On 5 April he wrote to his co-worker T.A. Kontorova, who was staying in Kazan': «It is not without a certain surprise that I have written at the head of this letter the name of our native city, which during the last two and a half years has seemed to be something distant and inaccessible. Today at 4 p.m. I saw it again ... I've seen only two or three houses destroyed by bombs (one near the Main Post Office, the other on the embankment, near the House of Scientists). Perhaps in some cases the inner destruction is hidden from a superficial view by boarding covering the empty eye-sockets of windows. Despite these covered windows the city is still very beautiful, especially the banks of its channels. Today I walked along the Griboedov channel. Your house is safe, but the windows are mostly without glass.

«What strikes me is the variety of damage to the windows. In one and the same house very often some windows are unharmed, while the others are destroyed. In some houses luxurious mirrors are as they were, and next to them are the others covered with boards, like blind people covering their unseeing eyes with black patches. And nevertheless the city remains astonishingly beautiful. Anyway, it pleases the eye with its architecture, with the gratings of its channels, with the Neva River, and with its sunset sky against which unforgettable silhouettes of the Petropavlovskaya Krepost', Burse and Rostral'nye Columns show.

«People are distressfully scarce. Old people and teenagers are absolutely absent. There are many children below 12. They drive their scooters and girls jump over pavements covered with chalk drawings, all under the caressing rays of the spring sun, as if there were no war, siege and hunger!»

During the years of siege theoretical physics in Leningrad suffered severe losses. One of Frenkel's closest co-workers, N.L. Pisarenko (1905–1942), was killed in the war. Another talented theoretician who died, E.E. Lysenko (1909–1945), had also worked with Frenkel before the war, first at the Polytechnic, then at the LPTI. When the war started, Lysenko (just like Pisarenko) had turned to the theory and practice of magnetic screening of ships, worked at the Black Sea Fleet in the group of I.V. Kurchatov, and went to the front from there. He was killed in Romania on the very eve of the victory in May of 1945.

In 1939, after graduating from Leningrad University, G.S. Zavelevich (1916–1941) came to J.I. at the PTI as a postgraduate student. A very gifted theoretician, he practically finished his dissertation in the second year of his postgraduate work. In June, 1941, he volunteered for the front and was killed in the first months of the war near Leningrad.

The main body of staff of the PTI Theoretical Department stayed with the institute in Kazan' until February, 1945. Frenkel's family was also there waiting for reevacuation. J.I. visited Kazan' from time to time, and in Leningrad he resumed his teaching at the Polytechnic. The Voeykov Main Geophysical Observatory became the Leningrad base for his geophysical studies.

Chapter 8
Years 1945–1952

In the festive month of May, 1945, in Moscow and Leningrad meetings of the anniversary session of the Academy of Sciences were held. Formally they were devoted to the 220th anniversary of the Academy, but every participant regarded them as a celebration of the return to peaceful life and to the normal international scientific relations broken by the war.

The participants were present at the unforgettable victory parade in Red Square in Moscow, having seen with their own eyes the fascist banners thrown down at their feet.

At the session Frenkel again met his friends from abroad, F. Joliot-Curie, P. Auger, and I. Langmuir, and he hosted them in his house in Leningrad.

Foreign delegates were greatly impressed by all they saw in our country, the main contributor to the victory. We can see this from excerpts from a letter by M. Born:

«Moscow, 27 June 1945.
My dear Frenkel, I am writing this letter to express my gratitude to you for your exceptional kindness and attention. There is no need to tell you that 2 weeks spent in Russia are a great event for us. We were welcomed with magnificent hospitality and saw a lot of brilliant things in the fields of science, art and so on. I was very glad for the chance to exchange some ideas concerning physics with you and other physicists. I would like to thank all of them, but as it is impossible, I ask you to pass on to them my heartiest greetings and gratitude.

I learned many things in this country. However, I fear that my presentations were not good enough. In any case, you know what we are doing, as well as I know what is happening with you.

My cordial regards to Mrs. Frenkel. With the best wishes,
M. Born.»[1]

During the session Frenkel was honoured, along with other scientists working at the Academy, by the Labour Red Banner Order; two years later his «Kinetic Theory of Liquids» obtained the First Grade State Prize. So Frenkel's scientific service received official recognition.

However, even at that time there appeared the first hints that a change in social policy was occurring, one that became apparent in the middle of 1946. The first gusts of the cold winds of change reached Frenkel soon after the anniversary session. In December of 1945, he was called to the director of the Polytechnic Institute, Professor P.L. Kalantarov, a recognized specialist in electrical technique. The invitation happened to be related to the session of the Academy of Sciences.

1) The letter was passed on by Frenkel to the newspaper «Polytechnic» and published there on 3 September 1945.

In June, 1945, one of the members of the American delegation asked Frenkel whether his wife could purchase (or, as it used to sound then, to «get») several pairs of silk stockings. The reason for the request was that in the United States during the war silk had been something like strategic material used for parachutes, so silk stockings were in shortage. I don't know how, but my mother, S.I. Frenkel, miraculously organized the purchase. The American, in his turn, wanted to be somehow of service to J.I. Now, a little before this, Frenkel's watch had been stolen. Good watches are very necessary for a man always living on a tight schedule. J.I. asked him to buy a good wristwatch, adding that he would compensate the expenses from the modest sum of money remaining in England from the royalties for those of his books published there. The matter was settled, and Frenkel forgot the whole story. But Professor Kalantarov later showed J.I. a copy of a letter from V. Kemenov, the Chairman of the All-Union Council for International Cultural Relations (ACIC), to S. Kaftanov, the then-Chairman of the Committee on the High Education Affairs.

V. Kemenov wrote to S. Kaftanov on 10 December 1945: «Professor M. [as we shall call him — V.F.] from the United States attended the anniversary session of the Academy of Sciences. Before leaving the USSR he was approached by J.I. Frenkel, professor at the Leningrad Polytechnic Institute, with a request to buy for him in the United States a wristwatch. Frenkel said that he had «some money» (something like an honorarium due to him). Professor M., having bought the watch, passed it on to the Consulate General of the USSR in San Francisco to send it to Professor Frenkel, who would pay the price of $ 56.35. The Consulate General put an inquiry into the Ministry of Foreign Affairs in effect as to what money Professor Frenkel had meant, and since further delay in the reimbursement of the watch purchase was impossible, it was forced to pay the sum to Professor M. (from the account of the ACIC).

«Drawing your attention to the inadmissibility of such behaviour for a Soviet scientist, I am asking you to take the necessary measures. ACIC Chairman, V. Kemenov»

Kaftanov forwarded a copy of the letter to Kalantarov, who called for Frenkel (and gave him a copy). Everything moved into full swing! My father was reproached for this wristwatch for a very long time, no less than five years in fact.

Why do I not publish the name of Professor M.? The reason is that either Kemenov or Kaftanov, or perhaps Kalantarov, sent yet another copy of the letter to PTI Director Ioffe. So J.I. had to have the corresponding conversation also with Abram Fyodorovich too, though this time in absolutely friendly tones. It has come out that Professor M. insisted on the immediate payment of the small sum stated in the letter, which caused all the fuss. Ioffe asked Frenkel: «Didn't you know M.?» «But yes, I knew him well!» «So how come you did not know he enjoys the glory of a great miser?»

I think, nevertheless, that people from the Consulate General, and not Professor M., proved to be impatient. It seems that the simplest way to handle the matter would have been to write to Professor Frenkel himself. But no, it was easier to mount «proceedings» against him.

The absurdity of this foolish story becomes clear with the help of a «reductio ad absurdum», a known method in mathematical proofs. Let us imagine a meeting of the scientific council of the American university where Professor M. worked to consider his personal affair. Accusation: servility to a Soviet scientist and an antipatriotic action in the form of the request from his colleague from the USSR to buy several pairs of silk stockings for his wife. Even accounting for the «witch-hunting» which soon began in the United States, it is simply impossible to imagine.

The so-called struggle against cosmopolitism and servility began to gain force at the end of 1946. Physicists were accused of uncritical acceptance of quantum mechanics and the theory of relativity, and of an overstated appraisal of their value and of the role of Western scientists in their development. Publication by Soviet physicists in Western physics journals, encouraged in the 1920s, was declared an unpatriotic practice (after the fact!). Frenkel was accused again of idealism, just as he had been in the 1930s. He was a beautiful target on whom to paste the label of servility to Western science. His books and papers were published in Germany, England, America and France. His «Wave Mechanics» and «Electrodynamics» had appeared for the first time abroad. For the most eager «critics» of J.I. the reason for this was absolutely clear. The publication of these, as well as other books and papers in Germany and England, testified to the fact that Frenkel was in a hurry «to help the Americans use the achievements of Soviet science in the interests of monopolistic capitalism», no more and no less! This phrase is taken from a paper by a staff member of the Leningrad Polytechnic Institute named M. Kuz'min[2]. The quotation is not at all taken out of context but corresponds completely to the content and spirit of this libelous note. Frenkel's reply to Kuz'min's article, addressed to the «Vestnik vysshey shkoly» is found among J.I.'s papers, which had been refused for publication. In it he denounces the mendacity of these essentially political accusations, which are stated in an inadmissibly vulgar tone.

Let me give yet another example of the then-popular critique. In a paper of some V. L'vov («Literaturnaya gazeta», November 24, 1948) the following idea was put forward: «Recent gigantic achievements of physics in the field of the release of atomic energy, the study of cosmic rays, etc., have all been made not so much due to, but in spite of, the present state of the theory.» Why has it happened? Because «for two decades the Copenhagen masters and a number of Soviet professors servile to them keep atomic physics in the tight stays of «complete» and «closed» quantum mechanical descriptions.» Frenkel wrote in his reply to the paper L'vov had submitted to the newspaper, «It seems that «Literaturnaya gazeta» can hardly serve as a podium for the discussions of special problems of physics, the more so as the latter are performed by such absolutely incompetent people as comrade L'vov, and in so unfair a way ... Instead of adressing the principal

2) Vestnik vysshey shkoly, 1948, ## 2–4 (in Russian).

consideration of idealistic accretions of foreign quantum physics and the clarification of their essence, the papers by L'vov turn society aside from the point in question, disorient the reading masses, and actually form an attempt to discredit Soviet physics and to throw it half a century back.»

In the manuscript which probably was the basis of a talk Frenkel gave at one of the meetings dedicated to the philosophical problems of modern physics, he wrote: «Only very limited or extremely arrogant people can find idealistic concepts in my «Wave Mechanics» ... These people are incapable of understanding those metaphoric expressions which I had intended for more educated readers. For instance, when speaking about quantum statistics, I made use of such expressions as «sympathy» between electrons with opposite spins and «antipathy» between those whose spins are oriented in parallel directions, I did so not because I am inclined to anthropomorphic images but, on the contrary, because the latter are absolutely alien to me and hence are absolutely safe. I recommend that people who may fall into temptation by reading such anthropomorphic metaphors not read serious books ...

«I personally do not consider it necessary to write my books in a dry and formal language; I carefully etch everything in order to enliven and to promote the better acquisition of sometimes dry material. The right to use metaphors is not the monopoly of poets; scientists must also be able to use them.»

I would like to note that, in defending himself against the accusations of «servility to Western physics», Frenkel of course did not to any extent deny his admiration of the genius of quantum mechanics' creators. This feeling was the greater because the majority of them were known to Frenkel not only as outstanding scientists, but as excellent, progressive people as well.

At one of the meetings of the Polytechnic Institute Scientific Council, Frenkel, rejecting the accusations of servility, wittily remarked that the publication abroad of a book by a Soviet author proves, from the point of view of formal logic, just the opposite of servility (and the sin of publications abroad served as a refrain for all the accusations against J.I.). In fact, those are being «servile» to a Soviet author who translate and publish his works in other countries; this is just as absurd a statement as its inverse.

It seems that at the same meeting another episode took place. One of the speakers said something like this: «Professor Frenkel denies his servility to, and admiration of, the West. But here is a characteristic fact: even his handkerchiefs are marked in Latin letters!»

How should one react to this «direct evidence»? J.I. took out his handkerchief and blew his nose into it, accompanied by the sympathetic laughter of the major part of the Council, still retaining some sense of humor.

Here is one more example illustrating the level of the criticism of Frenkel at the Polytechnic. The director of the institute, one K.N. Shmargunov, accused him of derision of the Soviet kolkhozes (collective farms) on the grounds that J.I. had used the term «collectivization of electrons» in his book «Introduction to the Theory of Metals». But the director's wrath was aroused more greatly by the other term Frenkel introduced, namely, «forced collectivization of electrons», or «collectivization under pressure». Frenkel meant, of course, that under very large pressure applied to a dielectric, a phase transition dielectric — metal occurs. Electrons stop being possessed by particular atoms («masters») and begin to be free, that is, to be a property of the collective, i.e., the crystal as a whole. It is hardly necessary to remind the reader that in the beginning of the 1930s, the joining of private farms into collective ones in Soviet agriculture was usually executed under severe pressure (and repressions), which gave rise to Shmargunov's «accusation».

One has to state, with regret, that it was at the Polytechnic Institute itself, where Frenkel had nurtured so many future scientific researchers and engineers, that the fierce company against him was organized in the latter half of the 1940s. The next stage (in the series of trials already mentioned), especially hard for J.I., was the revision of the Theoretical Physics Chair of the Institute by a representative of the Ministry of Higher Education, an Associate Professor P.E. Zrebny. J.I. was almost boiling when he described fresh impressions of the talk with him. Some of Zrebny's questions, according to J.I., were of a provocative nature, but I remember father assuring us that he had not fallen for this provocation. It was in the winter of 1948.

I think that M.A. Bulgakov could not suspect how long the life would be of a phrase «Manuscripts do not burn!» from his famous novel «Master and Margarita» published at the end of the 1960s, long after the death of its author. These words were really destined to acquire wings! I recall them now to proclaim: «Verbatim records do not burn either!» But then, in 1948, my father did not indicate that such records of his conversation with Zrebny had been taken ... More than 40 years had passed when my brother, Professor S.Ya. Frenkel, received at his disposal a folder with yellowed sheets of typed text. On the first page it was written:

«Verbatim Record
of the talk between the representative of the Ministry of Higher Education Associate Professor Zrebny and the Chairman of the Theoretical Physics Chair of the Leningrad Polytechnic Institute, corresponding member of the Academy of Sciences of the USSR, Jacov Il'ich Frenkel, January 7, 1948. The talk was held in the presence of the Party Committee Secretary of the Leningrad Polytechnic Institute, VKP(b) member, Deputy Dean of the Physics & Mechanics Department and VKP(b) member, Chairman of the Dialectic Materialism Chair of the Institute.»

There are 34 pages in this document. I am going to quote here some of Zrebny's questions and Frenkel's answers to them, not arranging them in any particular order but just following the verbatim records.

«Zrebny (Z): What has been provided for the socialist people's economy by your best works? How much was saved, or really what particularly is being constructed by them or where are they directly used?

Frenkel (F): Theoretical physics is much farther from industry even than from experimental physics. It is of great importance for metallurgy. I consider those branches which deal with these questions to have brought much profit to our industry. I can judge about this, for instance, because many people come for help here ... As it is, theoretical physics is not directly related to technology or industry.

Z: Does your theory of liquids somehow help the construction of socialism?

F: I think yes, it does. Were this not the case, I would hardly have been awarded by the Stalin Prize for it.

Z: How do ideological questions stand at your Chair? How is the struggle carried out for the fulfillment of the TsK VKP(b)[3] rulings on ideological questions? ... How do you state the question of the party line in science, and questions of dialectical materialism in your course?

F: We do not touch on the question of the party line in science in our lectures. Philosophical questions of physics also are not treated. The questions of dialectical materialism, if touched on at all in the lectures, are considered only occasionally, without any lure to the opinions forming its essence. I do not think that physics' explanation of the questions of physics can contradict materialism. One has to teach students to think in the spirit of materialistic principles, but the dialectical philosophy itself should develop on the grounds of facts accumulated by science.

Z: Philosophy is not severed from physics, and philosophers often are better equipped to sort out questions of physics than physicists themselves: take treatises of the Marxist classics, for example.

F: Allow me to have my personal point of view on this problem.»

Later, Zrebny, trying to interpret one of my father's papers, stated that among concepts developed in it there are some which are «directed against dialectical materialism».

«Frenkel: Then it is expedient to change some points of dialectical materialism.

Zrebny: Perhaps it is the structure of theoretical physics that must be reconsidered, and not Marxist philosophy.

F: Philosophy should develop, not stagnate, and it can only be developed on the firm ground of facts revealed by science. Most of the people who dealt with the questions of the quantum theory, had not neglected these facts.

Z: One must take the works of Marx, Engels, and Stalin as a basis [for the work on textbooks — V.F.]. This is the spiritual source for the development of all sciences.

F: I need particular results and their implications, not merely the general principles. I must base my theories not only on principles, but on facts as well.

Z: I give an example regarding the role of dialectical materialism. The Red Army gained the victory over the Germans because its strategy and tactics were

---

3) Central Committee of the All-Union Communist Party (of Bolsheviks).

Marxist ones. And we, armed by dialectical materialism, will also celebrate victory in physics, and idealism will lose».

The criticism of Frenkel's views and works continued to grow. On 17 December 1948, at the initiative of the Ministry of Higher Education, the decision was taken to call an All-Union meeting of physicists. The agenda was formulated in the following way: «The present state of the science of physics and the problems of improving the quality of physicists' training». By March of 1949, 32 sessions of the Organizing Committee had been held. Their minutes have been retained. Professor A.S. Sonin published detailed excerpts from them in 1990 in the «Priroda» magazine (issues 3–5). The Chairman of the Organizing Committee A.V. Topchiev (later Academician and Vice-President of the Academy of Sciences of the USSR) said then: «I think that our meeting must be like the session of VASHNIL[4]; we have to hold it at a high ideological level.»

In the archives of the Frenkel family a folder is kept on which I wrote in 1953–54, while sorting out father's papers, «Persecutions». There is a typed manuscript of the report he delivered at the Organizing Committee meeting. According to the title of the All-Union meeting, the report is mostly dedicated to the problems of quantum mechanics, statistical physics, and the relationship between physics and mathematics. But at the beginning of the report Frenkel returned to the 1931 physical-chemical Conference (see above).

What was, in my opinion, the essence of this conflict? Note that in this period J.I. did not fight at all against the materialistic approach to natural phenomena. More than that, he indicated that he himself agreed with this approach. Frenkel's speech was directed against the intellectually limited people, ignorant of physics, who had allowed themselves to teach the specialists as mentors, «from above», that dialectical materialism (reduced by them to the known triad) was the necessary and sufficient means for finding the solution of all, the most difficult, problems of contemporary science. They had thus politicized physics, forcing their dogmas, which, by the way, obviously contradicted the standard postulates of dialectical philosophy, suggesting the development of its own concepts, on grounds of experimental data. The next thing against which Frenkel protested was the requirement of canonizing some philosophic works, by Engels and Lenin first of all, such as Lenin's «Materialism and Empiriocriticism», and «Dialectics of Nature» by Engels. J.I. read Engels after the war and, as far as I remember, did not find in this book the ways to overcome the difficulties facing physics. In the 1880s Engels was infinitely far removed from these difficulties, the new stage of physics developments having been started after his death in 1900–1905 by Planck and Einstein. J.I. also regarded Lenin's book critically, rightly considering it to be mostly polemical, though containing interesting pages and hypotheses. The latter are even more surprising since Lenin never considered himself either a physicist

4) This is the sadly famous session of the V.I. Lenin All-Union Academy of Agricultural Sciences (VASHNIL) where T.D. Lysenko and his sympathizers smashed Soviet genetics.

or a philosopher. Frenkel used to repeat one phrase he ascribed to Lenin: «Nothing so promotes the discrediting of an idea as its reduction to absurdity.» Of my father's comments on the «Dialectics of Nature», I remember only his indignation over Engels' evaluation of Newton («the inductive ass»).

A.S. Sonin, in the series of articles in «Priroda», provided a detailed exposition of reports made at the Organizing Committee by S.I. Vavilov, M.A. Markov and J.I. Frenkel. In the discussion of Vavilov's report, V.F. Nozdrev of Moscow University said: «To fight various bourgeois traces in our Soviet science, it is first of all necessary to completely destroy cosmopolitism as the «theoretical foundation» of all ideological perversions and wanderings, as well as to expose such concrete carriers of these manifestations, rootless cosmopolites as Academician Kapitza, Professors Frenkel, Markov, and the like ....» «Physical idealism is exactly the chain which binds scientists to the chariot of imperialism. Unmasking the obscurantists, open philosophic idealists, we must expose their service to the reactionary camp of idealistic physicists».

A.S. Sonin writes further: «It became the «bon ton» at the Organizing Committee meetings to smash Frenkel's «idealism»; only rarely did a speaker not afford himself this pleasure. But perhaps the most vociferous critic was physicist P.E. Zrebny, Associate Professor at the Light Industry Institute; he devoted his whole speech to attacking Frenkel.» Later on every speaker considered it his duty to curse M.A. Markov. «But everyone understood that he was not the principal victim, but Frenkel. Frenkel also understood this,» wrote Sonin.

The result of the discussion of J.I.'s report was formulated by A.V. Topchiev: «You know,» he said addressing Frenkel, «all the party decisions taken in recent years, you know the discussions ... Therefore we ask you to admit that your mistakes are serious ones ... I have not heard you saying: «My critics are right; I was mistaken».»

But Frenkel held his position: «To summarize, I would like to say the following. At the meeting I will eagerly, without any enforcement, criticize the views I espoused in my «Wave Mechanics» and repeated in other books, indicating, however, that they were not my views but those of the creators of quantum mechanics, and which I had not been able to oppose at all then and so included without any criticism. This was my fault, and I admit it completely. But it was caused by a rather naive relationship with foreign science.»[5]

The sense of the last sentence in the quotation was, I think, that in the years when «Wave Mechanics» had been written, the successes of this new outstanding branch of physics were so impressive that it was too early to speak about any constructive criticism of it.

As it is known, the meeting on physics was ultimately never held; people say that a telephone call broke through to Kaftanov two or three days before its announced beginning, and he was advised to cancel it. According to the most

5) Quoted in a paper by A.S. Sonin, «Priroda», 1990, # 5, p. 96 (in Russian).

Ya. Frenkel during his lecture in the Physico-Mechanical Faculty. Leningrad, 1950.

probable version, the then powerful I.V. Kurchatov came into action. He, a brilliant physicist himself, knew well all the persecuted scientists and understood what a harmful role the arranged analog of the August 1948 VASHNIL session could play.

The struggle with «the servility to the West», cosmopolitism, and philosophical idealism, however, was not terminated, but was continued at the «local», instead of the All-Union, level. At the beginning of 1950 this became the level of the councils and the Party organizations of the PTI and Polytechnic.

In the «Persecutions» folder there are many typed versions of Frenkel's speeches. One can see there a number of corrections made by various people to whom my father had previously shown the texts. I recognize the handwriting of my brother, who was considered the philosophy expert among us, Father's colleague T.A. Kontorova, and finally A.F. Ioffe. All this was extremely hard for my father; the question of whether it was worth trying to persuade him to «repent» was not at all a rhetorical one, since all this harmed his health. But we, his relatives, were scared by much more horrible prospects.

For instance, in «Persecutions» a statement J.I. made to the PTI Party Committee is kept. On the pages of this document dedicated to a physics discussion of the indetermination principle, Frenkel wrote: «I was guilty, in some cases, of using inaccurate terminology, absolutely inadmissible in a discussion of philosophical questions. While explaining a series of general questions concerning quantum mechanics I used inaccurate formulations, metaphors which could be interpreted erroneously, that is, literally, by a number of readers (especially those who are not physicists). I mean, for example, such unfortunate expressions as the freedom of volition of an electron, as well as phrases about the absence of determinism for elementary events, i.e., about the violation of the causality principle, etc., associated with this metaphor.»

A psychological motive for J.I.'s behaviour suggests itself again, and it is revealed by the last part of the previous quotation. He was convinced his action was correct, not dangerous. But various people, some of them respected by him, tried to make him believe the opposite: that the metaphors he used to animate and clarify, imaginative remembering of the text were unacceptable for textbooks because they would divert young people from the materialistic view of nature. And now, in the period of sharpening of the ideological struggle, it became absolutely inadmissible, harmful, and it supported our ideological and political antagonists.

In short, Frenkel, the great Soviet physicist, suffered a real hounding during a period of unsteady health. It was, as I imagine, some kind of a Kafkaesque scenario: serious physicists, contributors to science, accused Frenkel of sins which, of course, they themselves did not regard as sins at all. Naturally, among his «critics» there were some really limited people, ignorant of physics, who were convinced of his idealism; there also were pure careerists who gained pseudoscientific and political capital from the «fight against idealism».

It was the task of the sitting PTI Party Committee to convince J.I. Frenkel of the objective harm done by his idealistic statements, allegedly occurring above all in «Wave Mechanics»[6]. The PTI Party Committee demanded that J.I. write a statement for the central press («Pravda» or «Izvestiya») admitting all his ideological mistakes and denouncing his statements made in 1931 concerning the dialectical method. Frenkel presented a version of such a statement which was not considered repentant enough and was not accepted by the Party Committee; they suggested he write another one.

I recall only 3 or 4 public statements (but not in the press, of course!) in defense of Frenkel, in particular, those made by the Polytechnic Institute professors A.A. Gorev, L.N. Dobretsov, and M.P. Kostenko and Associate Professor A.R. Shul'man. Even when one declined to speak at those of meetings, silence was then taken by everybody as a positive action.

---

6) It is appropriate to cite two responses to J.I.'s «Wave Mechanics». The first one is from Richard Feynman and was expressed in a private letter to me and then included, with his permission, in a paper about my father. Here is the testimony of the greatest contemporary theoretician: «May I add that your father's great books on wave mechanics were a strong and valuable influence on me when I was a student». I heard about the other response just recently. Recalling the year 1941 and his many days' evacuation way to Ashkhabad, A.D. Sakharov wrote: «We spent ten days in Murom waiting for the train. These days happened somehow to be very fruitful for me scientifically: reading Frenkel's books on quantum («wave») mechanics and the theory of relativity, I understood a lot all at once.»

Frenkel was especially distressed by the action taken against him by his junior colleague, Professor D.N. Nasledov. In June, 1950, the latter published in the newspaper «Leningradskaya Pravda» an article in which he repeated the accusations made by him and other physicists (a pity, but Ioffe was among them). Here is the end of this article: «Soviet physicists, like all Soviet scientists, understand and appreciate the care of the Party, and the care of the greatest scientist of our epoch, our leader and teacher, Comrade Stalin. And we fully comprehend that it is exactly for the reason that we are guided by such genius of humanity as is Iosif Vissarionovich Stalin, that Soviet scientists and technicians have succeeded in so rapidly creating new and advanced technique ... To him, to the leader of peoples, to Comrade Stalin of great Party of Bolsheviks we are grateful for those unusual conditions for the development and growth of Soviet Science which now exists in our country.» It is difficult to imagine now that such words were then accepted as the obligatory norm for such publications.

In 1973 I was preparing a paper about my father for some of the American journals on the history of science[7]. It came to my mind that American physicists could have responded to the 1945 anniversary session of the Academy of Sciences (as their English colleagues had done). It seemed appropriate to me to present their evaluation of my father's scientific contribution. To find it, I looked through several thick volumes of the American periodical bibliography of that time. They were equipped with various helpful indices, in particular, with the name index. I was going to look up the names of the members of the American delegation there, but habitually looked under the letter F as well. To my surprise, I found under my father's name a reference in the famous journal «Science News». In its issue of 16 September 1950, a small item was published under the title: «USSR Physicist Rebuked». It went like this: «The Russians have slapped down their top theoretical physicist for being a «speaker of idealistic assertions» who has a «negative attitude toward dialectical materialism and in his writings sometimes acted as a loud-speaker for the opinions of bourgeois physicists». This latest victim of the «Marxist-Leninist theory» is Dr. J. Frenkel, recognized as a top man in theoretical physics and quantum mechanics. Dr. Frenkel's work could conceivably have been in connection with the development of the Russian A-bomb, although American physicists who know and admire his theoretical work decline a guess as to that.»

Further in the note I found a brief exposition of the Nasledov publication in «Leningradskaya Pravda», with a sharply negative assessment of it; the magazine indicated that obviously something was being prepared in Soviet physics similar to what Lysenko did in biology. It added that information came from a reprint of Nasledov's article in the other American magazine, «Physics Today». In the September 1950 issue of this journal, Nasledov's paper was published in English

7) V.J. Frenkel. «J.I. Frenkel (1894–1952): materials for his scientific biography». Archs. Hist. Exact. Sci., 1974, v. 13, # 1, pp. 1–26.

without any comments, since, perhaps, the editors had soundly assumed they would be superfluous.

In 1973 Nasledov was still alive and continued to work at the PTI as the head of a large laboratory. I often met him at various seminars and meetings. At first I wanted to «nail» him, that is, to present him with a xerox copy of the paper, one more addition to his «list of scientific publications», and the addition published in a prestigious foreign journal. At that time publication in such journals was not at all regarded as shameful. But Nasledov was already old, looking poorly, and I rejected this method of «revenge».

At the end of the summer of 1946 Frenkel received a letter from Professor P. Shillp inviting him to participate in the collection of papers «Albert Einstein: Philosopher and Scientist» which was going to be published to celebrate the 70th birthday of the creator of the theory of relativity. J.I. was especially enthusiastic because he had started to ponder over ways to overcome some of the fundamental difficulties which then faced quantum mechanics. The title of the paper, as is seen from the correspondence retained, was changed several times. First it spoke of «Matter and field», then of «Field and matter in Einstein's theory». The final version chosen was «Remarks on the field theory of matter». The English historian of science Dr. P. Hoch sent me a copy of the telegram Frenkel sent to Einstein on his birthday, 14 March 1949: «Accept my heartiest congratulations, deepest admiration, best wishes. Shall send you my paper in Russian with English translation next June. Frenkel.»

By this time the campaign against cosmopolitism and servility to Western science was in full swing. Frenkel was convinced by his colleagues to protect himself somehow before sending the article abroad. He made the following decision: since V.M. Molotov had helped him in 1944 to bring the manuscript of «Kinetic Theory of Liquids» to England, he appealed in a letter to him to help once more with the mailing of the paper for the Einstein collection.

One should say here that Father would probably have liked the lines by Boris Pasternak: «You have to live without archives and worry with the manuscripts.» He never kept copies of letters, even important ones he sent, partly because he practically always wrote by hand. However, he stored his colleagues' letters to him, though all his prewar correspondence with Einstein (several letters), Ehrenfest, Bohr, Pauli, Dirac, Brillouin, and Joliot perished during the siege of Leningrad, and after the war, correspondence with his foreign colleagues soon ceased to exist. The few things I managed to collect later were either obtained from archives, both Soviet and foreign, or directly from the addressees of Frenkel's letters. Of course, it probably is possible to find the letter to V.M. Molotov in the Archives of the Ministry for Foreign Affairs, but I remember its contents well since I read it before it was sent.

Molotov sent the letter to his deputy, A.A. Gromyko. One evening, when all of us were gathered in my father's study, the telephone rang and a lady from the intercity switchboard announced that Moscow was calling Professor Frenkel.

Then another lady, Gromyko's secretary, connected Father with her boss. I do not remember my father's exact words but do remember that he, after hanging up, said with satisfaction: «Well now, everything is all right. No objections to sending it!» My mother wanted him to be specific: «Did he say this exactly?» Father answered: «He said the following: «Professor Frenkel, the paper is yours and you have the right to use it as you wish. We, however, would not advise you to publish it».»

We started trying to convince father that this was really an indirect refusal, but he emphasized the first part of the conversation («you have the right»), held his position, and became angry with us when we continued to defend our, to us rather obvious, interpretation.

Mother then went to the other room where the other telephone was located and, without speaking a word to Father, called his junior comrade at the PTI Professor S.Ye. Bresler. Having quickly described the situation, she asked him to come immediately, despite the late hour. Semyon Yefimovich lived relatively near us and soon arrived, to the surprise of my father. He did not conceal my mother's phone call. I remember his phrase: «Oldie (this is what he called father, who was 55 then), you just went crazy!» J.I. was talked, with difficulty, into not sending the paper, and he gave his word on it.

He kept his word, of course, but was very annoyed by this. It is also regrettable and painful for me now, when I look at this luxuriously printed volume (from the well-known «Modern Philosopher's Library»), because there is no J.I. among its galaxy of authors!

However, the article Frenkel wrote for the Einstein birthday volume served for him as a stimulus to a whole series of works on quantum field theory (or, more precisely, on the field theory of matter). He had time to publish some of them in Soviet physics journals, but the better part remained in his archives: the paper «Remarks on the field theory of matter» itself, a similar text in Russian entitled «Causality principle and the field theory of matter» planned for publication in «Uspekhi fizicheskikh nauk», and several of its versions.

Frenkel first espoused his views on these questions in a lecture given for a public audience at the Academy of Sciences in Leningrad in January of 1947. We shall not dwell on a detailed explanation of Frenkel's arguments which were, as he himself often stressed, only initial rough ideas. They are far from the construction of the final theory, indicating instead the need for such theory to be developed. We note, however, some of the principal points of his «Einstein» paper.

J.I. analyzed the origin of the causality relationship between physical phenomena and stated that in classical mechanics, contradicting the common notion of a chronological sequence of the cause and its consequence (J.I. called this notion an «everyday» one), they are simultaneous. The laws of classical mechanics, which usually treat force as the cause of motion, conflict with the everyday notion of causality. For instance, in the case of the motion of a particle with mass m in an electrical (or gravitational) field, the motion equation

$$m\frac{d^2\vec{r}}{dt^2} - k\frac{\vec{r}}{r^3} = 0,$$

where k is some constant, does not include the force as either the cause or the consequence of motion. This fact corresponds to the circumstance that the force in the cases described by this kind of equation (for example, the motion of the Earth around the Sun) cannot be measured in the direct, static, way. In classical electrodynamics, where the notion of the instant propagation of interaction is absent, equations for the scalar and vector potentials of the field allow for both advanced and retarded solutions. Usually the latter is preferred to the first, which is, as J.I. remarked, largely just a tribute to our «everyday» image of the natural, chronological sequence of the cause and its effect. Nevertheless, for a finite spatial domain the solution to the electrodynamics equation may be constructed, with equal success, using both advanced and retarded solutions (and taking into account the initial and boundary conditions).

Frenkel treats in particular the question of the gravitational field. Since it determines, according to Einstein's theory, the space-time geometry, Frenkel regarded the attempts to unify the electromagnetic and gravitational fields into unified field theory as doomed to failure, just like the attempts to quantize the gravitational field. «The last fact,» he wrote, «seems to be natural, because the gravitational field is of macroscopic, not microscopic, character, specifying only some frame for the description of physical events in space and time, while quantization deals only with microscopic processes in material fields.»

The ideas Frenkel suggested in his «Einstein» paper did not lead to success. The main road to it was laid about the same time in work by Feynman, Schwinger and Tomonaga. We have included here some excerpts from his articles guided by the following arguments. First, they present a picture of attempts to overwhelm the difficulties physics faced at the end of the 1940s. Second, one can find in Frenkel's scientific practice (although in less «global» examples) similar attempts to find a way out of a dead end. In such cases he was not always sure that the chosen way (or even the direction) would lead to the target. For instance, participating in 1937 in a discussion of a report by D.V. Skobel'tsyn, J.I. noted: «Since nobody suggested any explanations of the anomalies observed by Skobel'tsyn, I'll risk inventing one which may easily prove to be incorrect but provides a means for grappling with the problem ...»[8] To come forward with a kind of «catch» and with its help to approach the truth, or to see some gleam from it, even if the catch would be subsequently criticized, is how Frenkel understood his task[9].

---

8) A more detailed description of Frenkel's ideas is given in the Russian edition of my book about J.I. Frenkel, pp. 427–29.

9) J.I. Frenkel. Speech in the discussion of a lecture by D.V. Skobel'tsyn. Izv. AN SSSR, ser. fiz., 1938, # 1–2, p. 101 (in Russian).

At the end of the 1940s J.I. started to work on the book «Theory of Cognition of Physical Sciences». This is its plan (a long preface to the book and the beginning of the first chapter are retained in Frenkel's archives).

Plan of the book

I. Resemblance and distinction

1. Cognition and recognition. 2. Physical analogies. 3. Formal (mathematical) analogies. 4. The structure of a physics theory and its relation to the experiment (direct and indirect experiments). 5. Relationship between a mathematical theory and the physical reality.

II. Cause and consequence

1. Coexistence and the sequence of causes and implications in mechanics and electrodynamics. 2. Reciprocity of causes and consequences and the concept of stability in physical phenomena. 3. Instability in the phenomena of everyday life. 4. Anthropomorphicity of causality and its replacement by governing laws. 5. Determination and randomness in statistics and mechanics (statistical laws; randomness).

III. Variation and constancy

1. Invariants and conservation laws. 2. Invariants and relations (Remark. Elimination of subjective points by means of invariance or symmetry (depersonalization)). 3. Covariance and relativity. 4. Variety and uniformity. 5. Understanding as a reduction to constancy and uniformity.

IV. Macroscopic and microscopic considerations

1. Microscopic theories of macroscopic phenomena. 2. Macroscopic theories of microscopic phenomena. 3. Models and the nature of matter. 4. Discreteness and continuity. 5. The role of the observer and action quanta.

Frenkel concluded the preface to the intended book with the following words:

«. . . One would be able to recall a very large number of memoirs by different mathematicians and physicists of the end of the previous and the beginning of this century dedicated to gnosiology, which influenced, to a large extent, the development of contemporary mathematics and especially physics. This influence was not always positive and, in some cases, delayed the progress of science. For instance, such were the attempts of W. Ostwald, and partly of E. Mach and P. Dughem, to force on physics a phenomenological view of physical facts and to reject the search for a molecular-kinetic mechanism concealed in them. Usually, however, the philosophical études of mathematicians and physicists were at least harmless, if not beneficial.

«On the grounds of the conviction that the repetition of such attempts have to lead, in the long run, to the creation of a correct gnosiology, which would further the development of scientific cognition of both the surrounding world and our own thought, I dare to issue this book, the fruit of my reminiscence and studies I started in my youth, which have until now found only a fragmentary expression in a few oral presentations. I hope that my example will impel some of my colleagues who tried to analyze their creativity and to establish the corresponding trends and tendencies to share the results of this analysis. And a comparison of such notes delivered by representatives of different scientific specializations would promote the construction of the comparative gnosiology of sciences and would help to reveal the general laws of motion of a scientific thought.»

Several other ideas of J.I. concerning the relationship between physics and mathematics, expressed in some of his printed and, mostly, oral reports, adhere to the above ones.

In the paper «Theoretical physics in the USSR over 30 years» Frenkel devoted the introductory section to the definition of the place occupied by this discipline in the system of physical sciences and to its specific features. He wrote: «The broadness of a physics education often gives ... the theoretical physicist an advantage over a mathematician in solving complex physical problems of mathematical character; lacking the appropriate understanding of the essence of phenomena under consideration, a mathematician is usually inclined to look for an exact solution where a simple approximate one would do, the latter being obtained by means of a reasonable, i.e., not distorting the sense of the matter, simplification of the problem.

«Thus, if the inability to carry out physical experiments is the negative characteristics of a theoretician, the positive one is a wide encyclopedic knowledge of the questions of physics combined with a sufficient mathematical armament.

«Depending on the ratio between these two factors, a theoretical physicist may either approach an experimentalist or a mathematician. Theoreticians of the experimentalist type deal mostly with various particular physical or physical-chemical questions agitating experimentalists, with whom they are in close contact. Theoreticians of the second, mathematical, type are usually involved in abstract questions or in the solution of special mathematical problems associated with phenomena quite clear in their essence but demanding exact quantitative description. This latter field is usually recognized as a separate specialization called «mathematical physics». Its representatives are mathematicians rather than physicists: their attachment to either this or that category is conditional and depends on what questions they deal with, either purely mathematical or of physics.»[10]

J.I. Frenkel belonged to the theoreticians of the «experimentalist» type. It does not, of course, follow from this that his work bore a qualitative character. However, he always considered a fresh physical idea to be the principal point which could serve as a source for further, more detailed, research exploiting the mathematical arsenal.

In this respect the words of J.J. Thompson were dear to him; they provide an epigraph for J.I.'s whole creative life: «Of all the services that can be rendered to science the introduction of new ideas is the very greatest.»[11]

10) J.I. Frenkel, UFN, v. 33, # 3, p. 295, 1947 (in Russian).

In the 1930s some younger theoreticians were inclined to regard the lack of «powerful» mathematical tools in a number of J.I.'s papers, along with the fact that the ideas presented were not elaborated, as disadvantages. In their opinion, this drawback often annihilated the very idea of the paper. One of these theoreticians recalled: «We even had a somewhat lenient term «frenkelism», for a joined notation of semiqualitative papers. Now, with the years passed, I understand that we were not right and that the main thing is new ideas, for they feed science. Our former statements may be excused by our young age then.»

This confession reminds me of a joke told by my father. A fifteen-year-old boy tells his pal about his parents: «You know, they've become much cleverer in recent years.»

Frenkel wrote his papers in a very simple and clear manner. Max Born wrote to Frenkel in March of 1947: «It was most kind of you to send me your paper on «Fission». I read it with great interest, although I am no expert in nuclear physics. You always write so clearly and simply that it is easy to follow»[12] It is interesting to compare this positive and favourable assessment by Born with a witty remark by Einstein (made in connection with reading a treatise by Aristotle which was written, in Einstein's words, in a very heavy and meandering way): «The majority of people have a profound respect for those words which they are not able to understand and regard an author they do understand as being superficial. This is a striking manifestation of a limited mind.»[13]

While preparing this book, I went again into the family archives and discovered there a record of one of my conversations with M.F. Deygen (Kiev, April 1964). He spoke about his meetings with J.I. and recalled one episode directly related, it seems, to the questions of theoretical physics and the role played by mathematics in it.

In 1938, Deygen was a student and attended Frenkel's seminar at the PTI. Once a speaker presented a paper on neutron scattering which was saturated with mathematics. Deygen soon got lost and shifted his attention to the seminar «entourage»: questions, reaction of the audience, the speaker and especially J.I. It seemed to Deygen that Frenkel had become a bit bored with the mathematics. And, in fact, J.I. soon said to the speaker: «Well, enough of mathematics, let's proceed with the result!»

The speaker, who was a bristly and abrasive man (J.I. called him a hedgehog), answered resentfully: «And how can one manage without it? Or can you, J.I.? Then here you are, here is the chalk, please draw the curve of the scattering cross-section!» «Well then, perhaps I really can!» Frenkel went to the blackboard and plotted the curve. Turning to the audience and the reporter, he asked: «Correct?»

---

11) Quoted from: Amer. J. of Phys., v. 19, # 7, 1951.

12) St. Petersburg Division of the AN Archive, record group 953, inventory 1, file 134.

13) K. Seelig. Albert Einstein. Moscow, Atomizdat, 1963, p. 176.

The speaker confessed, a little sheepishly: «Correct!» «Now you see, one can manage without mathematics!»

Here, A.B. Migdal exclaimed from his seat: «J.I., let us give him a break; he worked so hard!» He was supported by A.I. Alikhanov: «Allow him another 20 minutes.» Frenkel, of course, agreed.

«What a fantastic intuition J.I. had!» Deygen concluded.

Among Father's papers are preparatory notes for one of his lectures from the end of the 1940s. Frenkel wrote: «Mathematics can give us, in a reworked form, only what we ourselves put into it. To get new physics results, it is necessary, either consciously or not, to put into the «mathematical meat-grinder» new physical ideas, at least in an unfinished form ... Physical problems may be solved only by means of physics ...

«An unhealthy passion for formal mathematical apparatus, a formal approach to physics theory brings more damage than profit by tempting physicists to be satisfied with cheap mathematical trophies and to forget about the true essence of the problems considered».

In a manner of speaking, Frenkel objected to «shooting guns at sparrows». However, he kept the inclination for mathematics he acquired in his young years and was up-to-date on the news in the mathematical community. Professor G.A. Grinberg wrote: «I remember, for instance, that J.I. once decided that his knowledge of group theory was insufficient and that he should study the two-volume treatise by Schreier and Sperner dedicated to it. He read these books as fast as most people read fiction, turning over pages quickly and practically never going back to parts he had already read. That was enough for him to grasp the material completely.»[14] We would like to note here that J.I. Frenkel was the editor of the Russian translation of the well-known Van-der-Waerden book on group theory[15] published in 1938. Perhaps in the course of this work he turned to the book mentioned by G.A. Grinberg.

In his last years Frenkel dreamed about finding time to write a textbook on higher mathematics in which the basics of calculus and the theory of differential equations would be illustrated with problems of physics.

After starting to work at the Institute of Theoretical Geophysics Frenkel became interested in the study of seismoelectricity. He was interested in the work of Soviet geophysicist A.G. Ivanov, who had discovered in 1939 that the propagation of vibrations in soil was accompanied by the appearance of a potential difference between electrode probes installed in the ground. To explain this effect, Ivanov used the Helmholtz and Smoluchowski theory, according to which the pressure

---

14) In: «J.I. Frenkel. Memoirs. Letters. Documents». Leningrad, Nauka, 1986, p. 53 (in Russian). Probably Grinberg meant the book: O. Schreier und E. Sperner. «Einführung in die Analytische Geometrie und Algebra», Bd. 1 u. 2. Leipzig, Teubner, 1931 (in German).

15) «Die Gruppentheoretische Methode in der Quantenmechanik». Berlin, Springer, 1932 (in German).

difference between two points in wet soil (caused in the situation considered by the propagation of longitudinal waves) gives rise to an electrical potential difference proportionate to it. A generalization of the corresponding formula accounting for effects associated with periodic compression and rarefaction of the solid soil core and surrounding water, as well as for other factors, led J.I. to his theory of seismic and seismoelectrical phenomena. The formulae of this theory (in contrast to those of Helmholtz and Smoluchowski) were in a good agreement with the experimental data and, in particular, with the dependence of the effect value on the degree of the soil's porosity and with the frequency of elastic vibrations. The formulae also explained directly the dependence on the pore radius. We would like to stress here one very characteristic feature of this work. It can be illustrated best by the concluding lines of J.I.'s paper.

«Let us finally note that solid metal bodies may also be treated as two-phase systems similar to soil, in which the role of the rigid carcass is played by the crystal lattice formed by free (collectivized) electrons. From here it follows immediately that propagation of longitudinal acoustic oscillations in a metal body has to be accompanied by electrical effects similar to those observed under the propagation of seismic vibrations in soil. This question needs special consideration which I will undertake in the near future.»[16]

After the research on seismoelectricity J.I. started a series of papers on atmospheric electricity. Work in this field goes back to that of B. Franklin and M. Lomonosov. In particular, it was Lomonosov who discovered the so-called electrical phenomenon of good weather, that is, the presence of an electrical field in a cloudless sky. «Thunder and lightning», with which the studies of atmospheric electricity had started, were treated after the Lomonosov discovery as disturbances in a «quiet» electrical field in the regions of good weather.

As was noted in a survey of J.I. Frenkel's studies of geophysics, his service was, particularly, «that he established a close relationship between cloud electricity and the fields of the cloudless atmosphere and pointed out the primary role of clouds in the creation of the atmospheric electrical field.»[17]

Frenkel dealt with questions of atmospheric electricity from 1944 until the end of his life; two of his last papers on atmospheric physics, written in 1951, appeared after his death. In 1949, J.I. published a short monograph («Theory of Atmospheric Electricity Phenomena», in Russian) in which the results of his papers on this topic were collected and generalized.

To conclude this brief survey of the geophysical cycle of Frenkel's works, we shall describe briefly his paper dedicated to the explanation of the origin of the

---

16) J.I. Frenkel. Izv. An SSSR, ser. geogr. i geofiz., v. 8, p. 150, 1944 (in Russian). The effect predicted by J.I. in this article is now known as the acoustoelectric effect and is studied intensively both in theory and experiment. The first publication on it is dated 1955.

17) V.V. Bazilevich and K.S. Shifrin. «Works of J.I. Frenkel in geophysics». In: «J.I. Frenkel. Collected Papers», v. II, 1958, p. 515 (in Russian).

Earth's magnetic field. An acquaintance with it allows the reader to imagine the road J.I. went along to construct the theory of the Earth's magnetism, a problem he had been interested in back in his youth.

His work on seismoelectricity led J.I. directly to the questions of our planet's structure and of the existence of a liquid metal core whose presence may be detected by seismic wave propagation.

On the other hand, while learning about the physics of the atmosphere, Frenkel encountered the S. Chapman theory developed to explain day by day variations in the Earth's magnetic field. Electron and ion flux moving in this field are directed, in the process of their motion, into opposite directions according to the signs of their charges. Electric current associated with this effect causes a magnetic field which reduces (according to Chapman, who assumed electrons and ions went downward under the gravitational force) the basic magnetic field of the Earth.

Frenkel first demonstrated that, because of very small heat sources (radioactive elements may play this role) in the Earth's core, convective currents may appear there. If the core is situated in an external magnetic field, then, with a certain direction (type) of this convective motion, the secondary field associated with it would lead to the growth of the external field. In this way the Earth's magnetic field can increase to the observed values. The mechanism is similar to the self-excitation mechanism of the dynamo machine: J.I. treated the Earth's metallic core as a kind of self-exciting turbogenerator and called his theory the «turbogenerator» theory of the Earth's magnetism. In the common dynamo, some part of the power is set aside to overcome the friction between the parts of the machine; in the case of the «Earth turbogenerator» this part is used to overcome electrical friction, that is, turns into Joule heat. The energy supporting the magnetic field is that of the radioactive decay.

Frenkel concluded his work with a quantitative formulation of the above ideas. Recently these ideas and calculations became widely recognized and further developed, in particular by English geophysicist E.G. Bullard. The theory may be extended, by the way, to the origin of proper magnetic field of the sun and stars, as J.I. himself noted. At present this point has been completely confirmed.

The third direction of Frenkel's postwar activities concerns his work as a scientific head of the All-Union Institute for Aviation Materials. There, along with the solutions to certain special problems, J.I. dealt with both applied and more general questions. He worked mainly on molecular physics, metal physics, hydrodynamics, friction and, in this connection, on the interaction of liquid drops with a solid surface. We can supplement this list with research on suspension electrical conductivity and metal corrosion.

Frenkel paid attention, in particular, to the motion of droplets compressed by two surfaces and arrived at the idea of a drop-shaped bearing[18]. In the cases when the compressing forces were not very large, the use of such bearings would

18) Zhournal Tekhnicheskoy Fiziki, v. 20, p. 1927, 1950 (in Russian).

be, in Frenkel's opinion, highly profitable as compared to the usual metallic ball bearings.

Publication of this series of papers by Frenkel and his staff members started in 1948. They studied the trends of the liquid droplet's motion on a solid surface, which was not at all a purely academic problem; it was related to the fundamental technical problems of friction and lubrication. With the increase in the inclination of the solid surface, at some critical value a drop starts to roll down the surface. This rolling Frenkel compared with the motion of a caterpillar or a tractor. A droplet rolls down by flowing over its back end to the front one. If it moistens the surface well, then it leaves a film on its way, a «plane» (two-dimensional) jet. This plane jet is as unstable as the usual three-dimensional one studied by Rayleigh in his classical works. And, like the latter, it is fragmented into smallish droplets. «These droplets», Frenkel wrote, «extending along the direction of motion of the primary drop and forming a kind of punctured trace on the surface, are well known by everybody; they are especially clearly seen on the windows of a car or train speeding through rain.»[19]

The three areas of research mentioned do not exhaust the range of Frenkel's scientific activity after the war. It had, as always, surprisingly many facets and at first glance could even seem too diversified. But if one looks more attentively at it, this variety evidences a common ground, a uniformity in the approach to problems.

Frenkel was interested in the problem of atmospheric precipitation, droplet electrification. From this he passed on to the general problem of the electrification of pulverized liquids and solids. Investigations of pulverization turned his thought to the action of an electric field on a liquid (electrolyte) jet: in the paper dedicated to this question (written with G.P. Vager) he at first sharpened, and then developed further Rayleigh's classical results concerning the stabilizing influence of an electrical field on a vertical liquid jet. It turns out that the stabilizing effect takes place when the fields are comparatively weak, while strong electrical fields cause an enhanced fragmentation (pulverization) of the jet. It is of interest that, in explaining the jet fragmentation effect, Frenkel referred to his theoretical work on the electrocapillar splitting of heavy nuclei (1939), applying the formulae obtained there to the effect being studied. Recall that when N. Bohr (with J. Wheeler) did his classical studies of the nucleus splitting, he had at his disposal the first published paper on experimental measurement of liquid surface tension by means of research on jet surface oscillations.

J.I. loved riddles and word games, starting with funny puns and children's verses. I remember that once before the war he told me: «Perhaps you are very strong; well, show your muscles!» Puffing my cheeks, I bent my arm. Father asked the adults observing the scene: «What has just happened?» The answer

---

19) Op. cit.

was: «The arm was bent and the muscles flexed.» Very satisfied, he corrected: «On the contrary, the muscles flexed and then the arm bent!»

One can find such things in J.I.'s scientific work as well. When at the end of the 1940s he decided to write a book on higher mathematics for physicists, he wanted to start it with the notion of the integral that he regarded as more visual than the derivative. For his book on theoretical mechanics he started not with statics, as is usually done, but with dynamics. Statics, as a particular case of dynamics, concludes the book. One could easily increase the number of examples of this kind.

It is not without cause that in an oral parody of Frenkel his imitator explained vibrations of a string in the following way: «Let us imagine a particle trapped in a one-dimensional potential well. It is described by the famous Schrödinger equation. Introducing the inverse square of the wavelength, instead of the difference between the total and potential energy over the square of the Planck constant, we arrive at the equation for the harmonic oscillations of a string.»

Now, many years after the appearance of quantum mechanics, one frequently observes young students of physics learning certain fine details of optics or classical mechanics with the help of analogies from quantum mechanics.

Turning again to the list of questions Frenkel addressed in the last period of his life, we also find here work on nuclear physics (magnetic properties of nuclei, their γ-spectra, specific features of heavy nuclei fission), on the surface tension of metals and amorphous bodies, on the theory of solid solutions, and on biophysics (the relaxation theory of hearing and electrical parameters of dispersed systems). Between 1945 and 1952 he published more than 50 works.

Starting with the first prewar enrollment, about 150 people a year entered the Physics and Mechanics Department of the Leningrad Polytechnic. During the 3rd and 4th years of their study, they took a complete course in theoretical physics (with the exception of mechanics), namely: electrodynamics (for all the students), statistical physics and quantum mechanics (for the students specializing in physics). The 5th year students attended lectures on theoretical physics which were addressed to particular groups, the topics corresponding to the group's specialization (theory of atomic collisions, theory of nucleus, electron theory of solids and so on).

The style of the lectures Frenkel delivered to a large postwar audience of about 150 people differed essentially from those given to comparatively small groups. J.I. performed practically no «experiments», as in the 1920s and 1930s (or perhaps carried them out with such mastery that nobody noticed). Maybe the reason was that he was teaching standard courses, though he added to the «canonical» text new applications of general principles of quantum mechanics and statistical physics to particular problems which have attracted current interest lately. Nevertheless, one can state that he was not completely satisfied with these lectures, even though he gave them in turn with Professor A.I. Anselm, and so didn't have the repetition of one and the same course every year. (I recall how once, at the beginning of 1949,

J.I. said: «Today I have managed to prove the Liouville theorem in an absolutely new way. It's a pity, but the majority would not be able to appraise it!»)

Frenkel gained much greater satisfaction from his lectures to the graduate students. The structure of these lectures was characteristic of J.I. the lecturer as described earlier.

Frenkel also spared much time for the guidance of graduate student work. Among the postwar students who wrote their theses under J.I., many have already become recognized in their fields (Academician V.E. Golant, Professors A.Z. Dolginov, B.E. Kinber, Yu.V. Novozhilov, L.A. Rivlin, Yu.P. Rayzer, and A.S. Tager). How did the graduate students get their advisor? Usually they came by themselves to Frenkel and expressed their desire to work with him. As mentioned, this desire was usually enough for Frenkel to start working with them.

The first steps in such cases were to choose a topic to work on (sometimes preceded by solving exercise problems) and to study the appropriate literature, if it existed. J.I. took out his notebook and suggested the student choose one of the items he kept in what he called his «topic-book». This «topic-book» was replenished as fast as it was exhausted, or even faster when Frenkel got interested in some new set of questions. J.I.'s notebook from the end of the 1940s contains 75 entries of this kind, reflecting the broad range of his scientific interest so many times spoken about here. The main planned investigations were the kinetic theory of the condensed state and those of its aspects which were related to the experimental studies carried out at Frenkel's laboratory in Moscow. We list here several notable excerpts:

1. Should one take into account the Pauli principle when considering interacting particles?
2. Isotropy of space in quantum mechanics.
3. Generalization of quantum mechanics to dissipative systems.
4. Quantum statistics of crystals.
5. Theory of electric wind (related to the operation of the lightning-conductor).
6. Width of the lightning channel.
7. (Photochemical) Origin of condensation centres in the atmosphere and their role in the formation of ball lightning.
10. Temperature of the sun's corona.
11. Optimal density of a fish shoal or bird flock associated with the behaviour of high concentration solutions.

Many of these later became the subject of Frenkel's own papers; the others were addressed by his graduate students. The sense of most of these topics can be easily grasped, except perhaps the problem of the fish shoal density. Not only graduate, but postgraduate students have found their topics in this «topic-book».

In postwar years Frenkel headed, for a period of time, the Council of the Leningrad House of Scientists in Lesnoe. He was one of the initiators of the movement for strengthening the connections between science and industry. In these years consultations were held regularly at the House of Scientists for people working in the city's industry (mostly given by leading scientists of the Polytechnic and PTI).

J.I. himself provided his expert opinion to engineers who built electrical machines and turbines (he advised on the problems of cavitation)[20], and on metallurgy and technical acoustics. He was approached by people from a paper-making plant, and Frenkel carried out a detailed analysis of the operation regime of a paper-making machine to try to increase its productivity. People came to him from the Institute of Artificial Limb Design; J.I. became interested in the mechanics of human motion and suggested some improvements in the structure of an artificial limb. He also invented his own design for an artificial limb; the idea was very simple and, as far as I remember, ran as follows. The suggested device (for cases when a foot had been amputated to above the knee) had a simple electric circuit which switched on when the unit touched the ground. Thus a fairly simple mechanism was activated, and the artificial limb bent, helping in this way to move the foot. The same operation principle was also supposed to be implemented in the case when both feet were replaced by artificial units.

Frenkel, while urging for a relationship between scientists and industry, was simultaneously against too simple an understanding of this relationship. People say that once (back in 1939) somebody criticized sharply, in J.I.'s presence, one of the experimental works on solid state physics on the grounds that this work, submitted as a dissertation, had not yet been implemented in industry. In Frenkel's opinion, the study was at such a stage where it was too early to speak about its industrial implementation, and he felt the criticism was absolutely inappropriate. Otherwise, Frenkel remarked, we would have to reproach, say, Maxwell, for he had not invented the telegraph!

Here is one more (postwar) episode of J.I.'s consulting activity. The Leningrad City Soviet had approached him and some other physicists concerning the following question. During the war, the bells of the St. Peter and Paul Cathedral clock which played the National Anthem at midnight got out of tune. J.I. explained the physics of this phenomenon and gave recommendations for the repair of the mechanism.

In 1946 a group of professors from the Department of Metallurgy of the Leningrad Polytechnic asked J.I. to give a lecture course on metal physics for them. Frenkel responded eagerly, and the course was given and, based on it, published the book «Introduction to the Theory of Metals». Its first edition (1948) instantly sold out, and only a half year later J.I. prepared a new edition, substantially reworked and extended, which came out in 1950. This book was republished in multiple editions in Eastern European countries, its third Russian edition was issued in 1958, the fourth one in 1972, and in 1961 it was translated into Japanese.

---

20) In 1969 I published the corresponding notes found in my father's archives: «Cavitation on the surface of a propeller», Inzhenerno-fizicheskiy zhurnal, 1969, v. 16, # 2, pp. 342–55 (in Russian).

In 1948 Frenkel's fundamental treatise «Statistical Physics» was published. «In the completeness and clarity of the material exposition,» Ya.G. Dorfman wrote, «J.I. Frenkel's book on statistical physics has to be regarded as one of the best in the world.»[21] (In 1957 the book was published in the German Democratic Republic.)

Starting in the fall of 1945, Frenkel began an elective lecture course on the physics of atomic nuclei. In 1946 a short book «Atomic Energy Liberation» based on the lecture notes was published; it seems to be the first piece in our country of the enormous now popular literature on this subject. In the preface to it J.I. wrote: «Atomic explosions in the skies of Hiroshima and Nagasaki caused a natural burst of public interest, especially in the engineering and technical circles, over the question of intratomic energy. Having dozed before in the depths of heavy atoms and leaking from the heaviest of them in the form of radioactivity, this energy was at last released by man, like a genie from the bottle in the Arabian fairytale. The problem now is to control it again and to let it free in a controlled way, making it work for the benefit of mankind.»[22]

In 1950 Frenkel reworked and extended this book considerably and published it as «Principles of the Theory of Atomic Nuclei» (the posthumous edition was published in 1955, and the German translation in 1957).

Frenkel was full of new plans and ideas. He started work on a new edition of «Electrodynamics». At the end of 1949, in his proposal to the Technical-Theoretical Publishing House (GITTL), he wrote: «The basic plan of composition for my book with its subdivision into a part on microscopic (general) and a part on macroscopic electrodynamics of material bodies will be preserved, and the indicated changes will only make it more orderly and accessible.

«In conclusion I would like to note that I regard the construction of electrodynamics on the grounds of the theory of relativity as a poor teaching method. In my book on the latter issued in 1922 I did formulate the theory of relativity with the help of electrodynamics, only later having used a new explanation ... The method of presenting the concrete foundation of electordynamics which I chose for the present book, based on a formal analogy between electric dipoles and elementary currents, seems to provide a much more visual image of the principles and relations of electrodynamics than does its derivation from the abstract and also unspecific (for electrodynamics) principles of the theory of relativity.»

At approximately the same time, the beginning of 1950, J.I. started to work on a new edition of his «Wave Mechanics». He found time to correct and supplement about one half of the first volume of his textbook.

Soon after the end of the war Frenkel was called to the Military Commissariat. He returned home surprised: «You know, my blood pressure proved high; they even

---

21) Ya.G. Dorfman. J.I. Frenkel. In: «J.I. Frenkel. Collected Papers», v. 2. AN SSSR, Moscow-Leningrad, 1958, p. 15 (in Russian).

22) J.I. Frenkel. «Atomic Energy Liberation». AN SSSR, Moscow-Leningrad, 1946, p. 3 (in Russian).

Ya. Frenkel in the summer of 1951.

dismissed me completely from their files.» But he immediately calmed down his relatives and himself: «It is better to have high pressure and feel good than to feel bad and have normal pressure.»

He did not change the pace of his life and work. But in his summer letters of 1947, and later, some disconcerting, and then alarming, tunes began to sound: he had not felt relaxed by the end of his vacation. Unlike before, he now failed to quickly recover his working ability, which was always diminished by the beginning of summer. And this ability to recover rapidly is one of the most significant indications of health. Despite these signals, J.I. continued as before to work in the summer, too; it was on vacations that he worked on his postwar books, but only for two or three hours a day, having no desire to write, in his own words, with a «half-hearted pen».

J.I.'s health had grown noticeably worse by the end of the summer vacation of 1951, which he spent in the Estonian town of Pyarnu, then comfortable and quiet. Having returned to Leningrad and worked about for a month, he decided to take an additional month-and-a-half leave, effective of October. Together with his wife he went to a sanatorium in Repino, a suburb of Leningrad.

Autumn was beautiful that year, and J.I. enjoyed the northern nature, dear to his heart, during long walks. However, the radius of his walks was reduced; back in the spring he had hurt his leg and walked with a small limp, supported by a cane. Once, not finding him at home, I went along one of his favourite routes and soon spotted him. He was sitting on a bench near the highway, next to a small fountain, watching how the jet was playing, going higher and then lower, as if it were breathing. It was still and lonely with only the water whispering.

J.I. stood up to greet me said that he was feeling better. He started to speak about the theory of jets, about the work of Rayleigh. Then he turned to the topic often discussed in our family lately, namely, about coming to live in Moscow.

He spoke about how he was tired of teaching the general course of theoretical physics, that these lectures did not bring him satisfaction, that he would like to confine himself to special courses of the elective type. And that generally, after so many unpleasantries on the part of the Polytechnic Institute's authorities, he was thinking about leaving it without great distress[23]. He said that when A.F. Ioffe was removed from his post of Director of the Physical-Technical Institute, no moral obligations made him stay there. That almost all his closest friends were living in Moscow and that he was going to start negotiations during his upcoming visit there in December. The negotiations were to be with N.N. Semenov, who had expressed his readiness to accept Frenkel at his Institute of Chemical Physics, and with I.G. Petrovsky, Rector of Moscow State University, who thought he would be able to arrange a professorship for Frenkel.

After his leave, Frenkel resumed his work in December of 1951. He was preparing his speech for the All-Union meeting on the history of natural sciences and by 20 December sent to Moscow a summary of his report under the title «Development of electromagnetic views in the XIX century» (the opening of the meeting was planned on January 25, 1952).

The journal «Priroda» had to be published, starting in 1952, under the new Chief Editor, Academician O.Yu. Shmidt. He latter asked J.I., in a letter written on 13 December 1951, to write an article for the journal about the contemporary state of the theory of liquids. Frenkel worked on this paper during the last days of his life. He resumed his advisory activity as well.

In the last period of his life there were some days when J.I. felt very good. Academician B.P. Konstantinov recalled this: «Shortly before the untimely death of J.I., I met him once, around noon, in front of our institute's building and, greeting him, expressed my admiration for his gay and cheerful state. J.I. answered that

---

23) To help account for J.I.'s attitude, I quote here an official document characterizing him which had been written and signed by his superiors in April 1951 (this is just the end of the text, absolutely conforming, however, to the style of the whole document): «Does not take part in the social life of the department. Does not participate in the training of new theoretical physicists. His guidance of his Chair is insufficient. Was awarded the title of the laureate of the State Prize for his work «Kinetic Theory of Liquids».» (Archives of the Leningrad Polytechnic Institute, personal file 3 548, p. 18).

there were some reasons for it, that he had written two scientific papers that morning»[24].

In the middle of January of 1952, Professor Ye.F. Gross told J.I. that while investigating optical absorption in Cu_2O crystals he had found a series of spectral lines which he identified as the absorption spectrum of the Frenkel exciton. Gross explained how interested J.I. had become and how they had agreed that Frenkel would come to Gross's laboratory in a few days ... But these days were not granted to J.I. He died in the early hours of 23 January 1952.

Sarra Isaakovna, his wife, wrote in February of 1952: «J.I. was so unlike other people, and everything was with him unlike with the others. He thought about himself so little, was so afraid of causing any inconvenience for anybody, that he died in five minutes without letting anybody do anything for him ... He always was very quick and determined in all his actions, both in serious and insignificant ones: he bought everything quickly, solved scientific problems rapidly, married without second thoughts and died so decidedly, without betraying his usual manner. All his life of great labour he was running out of time, he was always short of it ...»

Frenkel was a happy man. He loved his motherland and his science. Nature blessed him lavishly with a talent that was recognized by everybody throughout the world during his lifetime. He had yet another fortunate gift in that he was an exceptionally kind and responsive man. This kind of talent, in the language of modern molecular biology, is one of the least preprogrammed in the hereditary code but, similar to other human faculties, is developed and strengthened by systematic training. J.I. was kind to and did good for all those who came to him for help. «Unusual kindness, love and respect, for anyone J.I. could have imagined to bear goodwill, were so organically part of J.I. that he often seemed naive and trusting even of unworthy people. He was alien to the very idea of deception, so far was his spiritual world from any dishonest thought,» wrote A.F. Ioffe[25].

J.I. left behind him a gigantic scientific legacy and a bright memory. Many of his official and unofficial students are working in physics even now. His ideas and works influenced fruitfully the development of physics and, having taken a worthy place in the history of our century, did not grow old; they live on today.

One can envy, in a positive sense, this kind of fate.

---

24) B.P. Konstantinov in: «J.I. Frenkel. Memoirs. Letters. Documents». Leningrad, Nauka, 1986, p. 124 (in Russian).

25) Quoted by: I.E. Tamm. UFN, v. 76, # 3, p. 398, 1962 (in Russian).

Chapter 9
A Mosaic of Reminiscences[1)]

In the previous — especially in the last three — chapters one could find from time to time my personal reminiscences about some events of J.I.'s life. I think and hope, however, that my readers («If», as the great Russian poet Pushkin wrote, «God should give me readers!») will agree with me that I myself, as a rule, «stayed off stage.» I was 21 years old when my father died, so I knew him well and I remember a lot. I would like in conclusion to share some of my reminiscences which are presented as separate (quantum) episodes not necessarily tied together.

One of Father's favourite writers was Anatole France. As I said earlier, Frenkel regarded his «Les dieux ont soif» as one of the best books on the French Revolution; he read and re-read «Les Opinions de M. Jérome Coignard». My mother assembled a very good home library, including many books of memoirs; almost everything written about A. France (in Russian translation) was present there. I remember Father reading these books in the second half of the forties. This collection included two volumes of reminiscences by France's secretary, J.J. Brusson. «Little Jean-Jaques» was hurt by his Maître, maybe not without reason, and decided to write about him with a «bitter pen». «But here is,» as my father noted, «a charm in genius: despite Brusson's aim, France arose from his pages as a fascinating man, though his words and deeds were often somewhat cynical.» Brusson tried to show France not in public but in purely domestic situations. This programme was reflected in his first book's title: «Anatole France in Slippers and Dressing Gown». This title became common for writing about the human side of some distinguished person by putting him in a circle of his family and friends.

As far as I remember myself I did not see Father either in dressing gown or in slippers. He used to wake up rather early and at breakfast was dressed in the same way as at his office — a suit, a shirt with a compulsory stiff collar and a carefully selected necktie.

During breakfast, lunch or supper, always held at a precise time, all our family came together, and it seems to me now that during these very leisurely meals my elder brother Sergei and I used to receive some kind of natural good training without special moral admonitions; we listened to the conversations of our parents and their guests and grew up taking part in them.

These conversations as a rule were carried by Sarra Frenkel, our mother. With her lively sense of humour she could «spur» Father into activity, distracting him from the troubles of a straining working day in the evening or making him smile.

---

1) I translated this chapter into English myself. I would like to thank my brother, Sergei Frenkel, for his kindness in amending and improving my English. Three or four pages of the chapter I took from a paper I published in the Russian journal «The Star» (Zvezda, Nos. 9–10, 1991) which was translated by Ms. Darlene Reddaway and which, I hope, will be published in the United States.

Ya. Frenkel's family. From left to right: The younger son, Victor (b. 1930); J.I. Frenkel, his wife Sarra Frenkel; the older son, Sergei (b. 1923). Leningrad, 1938.

I vividly remember the so-called «rozigrishi» in particular, a special kind of tomfool joke; Mother was a real master in this field! However, more than once she told Father: «It is too easy to play a practical joke on you — you believe everything!» We always sat at the table in the same places. Father sat at the head of the table, and Mother sat on his left in front of two large, bright windows which looked out at the park of the Polytechnic Institute. And often when she had no fresh topics for her jokes she, catching Father's glance, would give a friendly nod as if somebody were standing outside the window. Father invariably took this simple bait, turning to the window, and it made us laugh because we were living, in the post-war years, on the fourth floor. Father's puzzled face gave rise to our hilarity, and he himself laughed most infectiously.

He was a great tea drinker; preparing and drinking tea would turn into an entire ceremony, especially in the evenings, with a samovar. He liked to drink very strong tea with milk, in contrast to the majority of this beverage's disciples. He put some jam on a saucer, and then an asynchronous process began. Either he finished the jam before the glass of tea and so needed an additional portion, or the glass was refilled in order to finish with the remaining jam: it was very difficult to achieve a «synphase sequence».

Father was extremely sensitive to various odours. Sometimes it suddenly seemed to him that a spoon retained the smell of the water in which some of the dinner-plates and dishes had been rinsed. In this case he would explode and even swear. At this point Mother usually intervened: «I am sure that it is your imagination again!» «You tell me! Try it yourself!» He passed to Mother the unfortunate spoon, and then a procedure started which we knew so well but which never lost its charm for all of us. Mother sniffed at the spoon, and in the cases when Father's claims had some justification, stealthily changed his spoon for her own. After that my brother and I were called for help, and we relatively fairly said that there were no odours at all. As for Father, he became more and more angry and urged that the spoon be examined again. He held it to his nose, and here the culmination took place. We saw how his face reflected the whole scale of feelings. At first it was angry: «Why are you trying to pull my leg? It's really disgraceful!» Then a puzzled expression followed: «Deuce take it! What's the matter?» which was followed by a guilty though guiltless, excusing one. He could fire up but he always recovered very fast, even in cases when he was absolutely right. «Like a match,» Mother said of him.

Most often during our meals we discussed the books we had read. Sometimes our parents, in our presence, discussed some of their acquaintances' actions. Father told us about the Council's sessions in the Physics-Technical or in the Polytechnic Institutes. My brother, who began working in the Physics-Technical Institute in 1948, often illustrated Father's stories by skilfully imitating the voice of A. Ioffe, Father and some others.

It sometimes happened that our parents discussed financial matters in our presence. Usually the discussion consisted of making a list of persons in need of immediate help (Father always supported a lot of his relatives, Mother's relatives, and older close acquaintances).

In the summer of 1947, Father was awarded the State Prize of the highest order. It was a very large sum of money: 200,000 roubles. (It should be mentioned for comparison that at that time the newest Soviet car, «Pobeda» [«Victory»], cost 16,000 roubles.) And in October of the same year the government proceeded with a money reform: the old currency was exchanged in the proportion of 10:1. I remember how some of our acquaintances remarked that it would have been better to receive the prize in 1948. J.I. carelessly waved his hand, adding: «It does not matter: we succeeded in spending and giving away nearly all of it.» So it was: due to Father and Mother's efforts the main part of this prize was given away. I remember only two purchases for our home: an excellent grand piano and ... a very nice Parker pen. Of the world of belongings Father perhaps liked only fountain pens. Throughout his life he had to write quite a lot and valued a good pen. And since he used to lose them very frequently, he gave his intimates and close friends a splendid opportunity to give him really useful and pleasant presents. By the way, he managed to lose this particular Parker almost immediately.

I have another example of Father's generosity, also from the post-war times. Once Mother was called over the phone by the cashier from the institute, who knew her quite well. The cashier said that an order had been signed providing Father with a large prize, and she asked Mother to come to get it for the following reason: «You understand that J.I. will not succeed in bringing it home, because he would give it all away along the way.»

It would take me quite a long time to tell how readily and openheartedly my parents helped the poor.

It is difficult to explain, but in my childish consciousness, already in the thirties, I formed an image of a scientist, and especially of a theoretical physicist, as somebody ridiculously clothed, a funny, incapable crank. At the same time I understood that Father was quite different. To begin with, he was always distinguishedly clothed. Further, there was his handwriting. Again, for no reason, I thought that slipshod handwriting was a synonym for talent and was, therefore, necessary for a scientist. But his handwriting was exemplary. Once he related to me its origin. My paternal grandfather was a clerk in the early 1900s. The business papers filled with his accurate entries had a magic influence on Father and his younger brother Vladimir (who was one and a half years younger). When they were seven- or eight-year-old boys they would sit at the table for long hours and copy those papers without understanding the meaning of «debits» and «credits». As a result both brothers had magnificent script.

When in the second grade I brought home a poor grade for calligraphy, I was not seriously scolded for that (nor in later times either). But Father did mention that he had always obtained the highest mark in this discipline, not for good calligraphy per se, but because he had always overwhelmed his teacher by writing so much. This was achieved by a little trick: he inserted two pens into one penholder at the right distance so that two lines in the copybook were filled simultaneously. His productivity increased twofold!

At the end of 1929 my parents moved from a wooden cottage to the First Professor's House of the Polytechnic Institute. Our prewar flat was situated five minutes from the elegant building of the Physics-Technical Institute, and passing out by the main entrance and turning behind the corner of the house, one could also see the austere colonnade of the Polytechnic Institute. For as long as I remember, I always associated Father with them. As for his being a scientist, I knew that from the name of our «Professor's» house and the Club (later the «House») of Scientists in Lesnoe; in those years it was a suburb rather than a district of Leningrad. Since the grown-ups frequently used the word «physics» I also knew that Father was a physicist. One of my earliest coherent memories is as follows. When I was five or six years old I expressed my sorrow that Father was neither a doctor nor a writer: it seemed to me that these two professions were the most honourable ones. Father said: «And you know, I am a doctor and a writer!» This news was too good to believe, and I expressed my doubt.

This discussion occurred one morning in Father's study. I still can see its furniture: in the right corner, near the window, there was a huge desk and an armchair; on the left side were three Swedish bookshelves and a small couch. Closer to the door, near the stove (before the war we heated with a wood-burning stove) stood a music stand made of exquisitely encrustated Karelian birchwood.

Father came up to the desk, opened a case and got out a leather booklet, in which I read that Yakov Ilich Frenkel was a doctor — though a doctor (and I was somewhat confused) of certain «physics-mathematical sciences». Then he led me to the bookshelf, took down the first book, then one more — and on the covers of both I saw the last name and initials I knew so well from the brass plate on our entrance door!

While before that incident I was proud of my Father because he was very kind (he allowed me to do whatever I wanted), and this opinion was supported and shared by friends living in our apartment house, which was called «Professor's building», now my pride substantially increased.

Much later Father told me that with the honorarium he received for his first book («Theory of Relativity») he and Mother had decided to buy furniture for the study. It was then that they bought the huge desk I mentioned. My maternal grandfather came up to have a look at the purchase. According to the stories told by our relatives and acquaintances, he was a wise old man with a phenomenal memory (he knew the whole Talmud by heart). However Grandfather held the opinion that the profession of a physicist was not prestigious enough, and during the first years after his daughter's marriage reckoned that she could have made, as it was said in the old times, «a better match». The purchase of the furniture for the study was the first event to make him doubt this conception. Unwilling to approve of his son-in-law, on being asked a direct question about whether he approved of the furniture, Grandfather described it with a word that in Russian had at least three different meanings (one should keep in mind that Grandfather spoke Russian fairly correctly but with a strong accent). Unfortunately the resulting «jeu de mots» cannot be translated into English, but the essence of Grandfather's epithet was that he thought the desk was «Swedish», «Soviet», or «fashionable». When Father came to this point in his narrative, he started laughing (he always laughed, slightly flinging his head back) and added that his father-in-law's answer was one of the rare triple-equivocalities that he had ever heard.

Earlier I mentioned that our parents, especially Father, did not try to teach their children good behaviour by preaching. As far as I remember, only twice did Father criticize my actions, and this is why I remember the circumstances quite well, although more than 50 years have passed passed since then. Once, when I was a boy of 13, I did not consider it necessary to hide my dislike of a man; to put it bluntly, in Father's presence I quite crudely said something rude to the man's face. It happened in the street; I was meeting Father on his way from the office. On our way home he sharply rebuked me, and this time, in contrast to his usual manner, he did not cool down immediately, as he did when the problem was a trifle, but continued the talk. He told me: «Real culture lies in discretion. We live in a human community, and so we must demonstrate mutual politeness. You are mistaken if you think that I like all the people that I meet or that I approve of all actions, even those of the people who are my close friends or relatives (including you); it is just that I seldom express my opinion about them. If you do not like something, it does not mean that this something is bad: you could be mistaken.

Only betrayal and cruelty are repulsive, and so only in these cases should one be intolerant.»

The second incident took place in the spring of 1942, also during wartime. Every day when Father was leaving for his office at the institute, Mother gave him, as the family member on whose health all others' depended, an extra portion of food. His working day began early in the morning and he returned home more or less late, around 8 p.m. He generally brought home his dinner portion (uncooked), which he received as a member of the Academy of Sciences. On his way home Father usually visited his mother. I often waited for him there. That evening, however, forgetting about my presence, he opened his briefcase, extracted a package with the food and passed it to Grandmother, forgetting that this act was a «secret» between the two of them. When we went out I said, trying not to cry, what I thought about this action and started to accuse Grandmother.[2] This time Father was not angry. He only looked at me sadly and asked: «And you, wouldn't you do the same for your Mother?» And, on having observed my reaction, added: «Well, now you see . . .»

It was Father who taught me to read and to count. I even remember the word that helped me comprehend how letters added up to form syllables and words. He tried to interest me and my brother in physics. As a rule, this kind of talk took place during our walks. In the prewar years he explained to me that a fur coat did not heat a man, but that, on the contrary, the man heated the coat, a paradox which helped me understand the ideas of thermoinsulation and thermoconductivity. During this walk (it was a cold winter evening in the park of the Polytechnic Institute) Father tried to check in the following way whether I had understood what he had explained to me. We came up to a tree, and he showed me a small crater located around the trunk. Why did it appear? On receiving a more or less reasonable (for a nine- or ten-year-old boy) reply he was satisfied. Even then he preferred to answer questions instead of telling some «scientific» story. And he answered in such a masterly way that I usually understood him. However, once he became angry when I said yes instead of rationalizing the essence of his question. It took him little time to choose language appropriate for explaining something to a little boy. Only once did he become a little bit thoughtful. He was telling me something about T.A. Edison's inventions, and I asked: «What is easier: to invent or to make, i.e., to realize the idea that one had in mind?» I do not know why, but he was pleased with this question of mine. One day later, when five or six years had passed, we were walking in the same park. This time I asked him some stupid question, and he was grieved by it. So, not hiding his annoyance, he said: «When you were a little boy you asked me cleverer questions.» Slightly hurt, I asked what he had in his mind and was really struck when he reminded me of my old question. Now I see with a special clarity what a great importance Father attached

---

2) It was not necessary to explain the situation for the Russian readers, but maybe I should explain for foreigners that the food given to Father was taken from a more than modest reserve and that of course Grandmother always had her own share.

to this kind of talk and how close to his heart he took his children's successes and failures.

When, during World War II, in 1942, Father began to work on his «Kinetic Theory of Liquids» and its first pages appeared, the problem of typing them arose. My mother was working at the Academy of Sciences Publishing House, dealing with scientific translation into English and German. The Russian and Latin typewriters which she took when we evacuated Leningrad helped her earn a satisfactory salary. These typewriters for a long time held some kind of attraction for me. I asked her permission to type my father's book. Such permission was given. Father's handwriting was excellent, so I began to work. He allowed me to insert formulae into the text, named for me the Greek letters, and showed me how to write them. I was especially astonished by the Greek sigmas (Σ). Sometimes several of them were written up in a line and were, besides, burdened above and below with signs. It was strange for me, too, to see a lot of exclamation marks in a purely scientific text. In fact, in sentences that sounded quite dull and calm, Father sometimes put two exclamation marks! I told him of my confusion, and he explained to me the meaning of the symbol for summation and about factorials. At the same time he explained the meaning of the integral sign. He drew a curve bounded by vertical lines and showed me how the resulting area might be divided by more and more vertical lines into narrower and narrower strips. The summation and integral signs turned out to be so closely connected!

After the end of the war Father renewed his «physics» wardship of me. When I had to do a report on absorbtion at school, he brought me Eucken's book on the subject. Some time later, being very pleased, he presented me with 2 volumes of Lorentz's book «The Elements of Higher Mathematics», printed in the beginning of the century in the well-known Russian library for self-education — those were the books that he had read himself in his youth. During my school years I read some chapters from Father's book «Structure of Matter» and from his very simple booklet «Electricity and Matter». At that time we were studying the history of the USSR and the GOELRO plan[3], and I was especially glad to find in the author's preface to this booklet a quote in which Father said that during the last decades one could observe a tendency towards the «electrification of matter», the explanation of its properties by means of charges and their interactions. We heard and learned a lot about Lenin's plan of «bringing electricity to all the country», which was known as «electrification», too. Among the other books that Father recommended to me, I would like to mention «Soap Bubbles» by Charles Boys and the wonderful books of Father's young collaborator Dr. M. Bronstein. The first one was his «Sun Substance» (about helium) which was very easy to understand, and then the more difficult one «Atoms, Electrons, and Nuclei»[4], and finally the most complicated

3) The state plan to electrify Russia.

4) When in 1980 I undertook to publish this book (under my editorship) as the first issue of «The Library of the Journal «Quant»» which is so popular now, for me it was a tribute not only to Bronstein's memory but to my father's memory as well.

«Structure of Substance». All these books that Bronstein gave to my Father are still kept in my home library.

In the summer, during the holidays, Father usually crafted for me wonderful guns, sabers, and daggers which he made from planks. He made strong and flexible walking sticks from juniper and decorated them with intricate patterns. Once when we were still in the city he saw me with a very bad bow and said: «What a bow you have — it is not a bow at all! One should make it from oak!» And he promised to make me a «real» bow. I took him at his word and for several days reminded him of my request, until he came home with a very dirty lag, probably from a wine or beer barrel. This lag disappointed me deeply! Father told me that he had picked it up from a puddle in the street! On his next day off the lag was washed, shortened and carefully whittled. On its end Father made accurate incisions for the bow-string and a special small trough for an arrow in the middle. The bow used a real arrow with some feathering on one end and a metal arrowhead (made from a small cover that in those years was usually put upon the pencils) on the other. The bow was excellent; it shot very far and — in Father's hands — very accurately.

With the same skill Father made kites that he used to help me and my friends launch in windy weather under his control. Especially fascinating was the procedure of sending so-called letters (or cables), small sheets of paper which we put on the string and which — in full compliance with the laws of aerodynamics — swooped on to the kite soaring in the high sky!

My admiration for Father's occupation with music and painting was even stronger. He did not acknowledge an idle rest, and I remember him during the holidays either walking, or behind the music stand, or with brushes and an easel. In the countryside during our mushroom-hunting rambles Father, who often accompanied us, did not take part in these «peaceful hunts» but simply wandered in the woods, whistling and condescendingly enjoying the enthusiasm with which all of us were filling up our baskets. As far as I could judge, he was not reckless; he never took part in the card games, simple ones or refined bridge, which were a beloved pastime for his friends and family during the dark summer evenings of the holidays. At these times he sat aside with his sketchbook opened and impressed the gamblers by producing pencil drawings in 20 or 30 minutes each.

The work took longer in the case of an oil portrait. It was «baked» (using Father's own expression) during three or four sittings of approximately two hours each, the longest period endured by his model, whom he tried to entertain by telling various stories. Very often we, his family, were his picture's subjects, and with some of these pictures a certain episode is connected.

He did portraits of my brother and me with a crayon at approximately the same time in, I think, 1936. On the eve of some holiday our apartment had to be put in order. Mother asked our maid Dasha to dust all the pictures (most of them were by Father) which were hanging on the walls. After the work was finished Dasha

Four peasants. Oil portraits by Ya. Frenkel, 1936–1939.

complained: «Sarra Isaakovna, how difficult it was to clean the boys' pictures — I wasted a lot of time with them and still the dust was not removed!»

In 1936 our family was on a vacation in the Ukraine, near its capital, Kiev. The most picturesque among the workers at the rest home was «Grandfather Danilo», who guarded a melon field. His constant three-day-old bristle, tousled hair, sleeveless vest that needed washing, and, above all, his good-natured character, were reflected in the expression of his face, — all this attracted Father's attention, and so Father asked Danilo to pose for him. Well, on the day and hour set, Danilo's entrance into my parents' room was preceded by a strong smell of eau de cologne. He was carefully shaved, had combed his hair, and was dressed in a new sleeveless suit. So, he had lost his former look but had received a new one, and Father painted him as he now looked against a background of wild greenery which surrounded our terrace.

The series of portraits done in the summer of 1939, when we were living in a near-wilderness in a village called Oleshno[5] seems to me very successful. In Oleshno Father went through an energetic artistic period and painted a lot of landscapes and portraits. Three venerable old village men were especially well done — their portraits later decorated the walls of our apartment in the high company of portraits of Dirac, the well-known Russian poet Marshak, biologist Professor Gurvich, and other celebrities. («I remember how your Father worked on my portrait,» Professor Dirac told me in July, 1973. «I was awfully eager to sleep, and he made heroic efforts to wake me up by telling me various funny stories.»)

In the spring of 1949 Father asked me to find him some colourful model. I chose the billiard marker from the pool room in our House of Scientists in Lesnoe, Ivan Ivanovich. One evening the model was «looked over»: Father visited the pool room (the House of Scientists was situated just next to our apartment on the 2nd block of on-campus housing for professors at the Polytechnic Institute), got acquainted with Ivan Ivanovich, and, as if accidentally, asked him to pose. During the following month, on Sundays Ivan Ivanovich came to our home, took his place in a deep armchair in one sitting room, and thus the sitting began. It usually lasted for two or three hours. As a result, the portrait was a very good one. The face of a specifically highly experienced man looks at you from the canvas. And there is no need to explain that Ivan Ivanovich's profession made him a connoisseur of human characters, which were so brilliantly represented around the green table! Father was very pleased with Ivan Ivanovich's stories, and the latter had the deepest respect for him; a gleam of this respect later fell upon me.

Well, here is the last «portrait» story. Once Father did a very good portrait of one of his colleagues. Usually he did not present his portraits to the models, but this time he yielded to the request. He asked, however, what the reason was

5) This village was situated 100 miles from Leningrad and was the homeland of Maria (Masha) and Daria (Dasha) Nikolaevs, two sisters who connected their lives with our family and lived with us for more than 3 decades, until they died.

for such insistence. «In your portrait I look very similar to a lackey, Yakov Ilich. I do not want to be displayed for people to see me looking this way!»

I would like to mention one more long-standing acquaintance with whom Father had a peculiar friendship: Mr. Ivan Mironov, an exceptionally kind and clever man, a hairdresser who worked in Lesnoe. In the prewar years, once every three or four weeks, on his days of rest, he would visit our apartment with his unchanging black case, which was filled with attractive (to me) paraphernalia for grooming all the members of our family. For Father, in addition, he used to sharpen the razor. For an hour the scissors would snip in the room. I remember one interesting talk between Father and Mr. Mironov. The «service procedure» for Father was the most lengthy, and both Mr. Mironov and Father knew the reason. However, they did not confess it; it was a pleasure for them to chat. Once Father asked Mr. Mironov a sly question: «Why, dear Ivan Petrovich, does cutting my hair take you such a long time? Take a look, I do not even have enough hair for combing!» «Right, Yakov Ilich, but you see, as a matter of fact I am cutting your hair very fast; but I spend most of the time just finding what I have to cut!»

When Father was working, he did not like anybody to turn up in his room, and we were trained to this custom from our childhood; despite the activeness characteristic of such unrestful youths, we never disturbed him. The telephone was taken away from his room, and the people who called him were asked to call back later. Everything was done to create silence. Otherwise, even rather late in the evening or on a rest day, Father had to go to his office in the Physico-Technical Institute to avoid disturbances.

But if he had decided to play the violin, he had nothing against my presence, and when I became older he even invited me to his room. He would select a big pile of music, place the first music book on a sliding door of the Swedish bookshelf, and start playing. I mentioned earlier the composers and pieces he loved. Among them there were violin concertos, sonatas for violin and piano, a lot of violin solos, and arrangements of various operas and even Chopin's compositions for violin. Father especially liked such arrangements of Rimskii-Korsakov's operas, above all the «Golden Cockerel». The Frenkel brothers (Yakov, Vladimir and Sergei) used one of the tunes from this opera as a prearranged signal. By whistling it, they used to give notice of their arrival home or to call each other.

Father used to judge people according to his most recent impression, as if forgetting (or sometimes really forgetting) his own previous experience. When this was connected with his friends, everything was all right, but in the case of people whom he knew only superficially this led to rather frequent changes of opinion, especially in the cases when those people's behaviour was not blameless. Father's conversation with Mother in these cases was of the following kind.

Father: «Well, all of you tell me that NN is a complete scoundrel. And just yesterday during the institute meeting he energetically spoke in defense of an unfairly criticized man. Really, he is not so bad. I met him today and directly expressed this positive opinion.»

Mother: «Which NN do you have in mind? Isn't he the one who, as you told me, defamed you as a cosmopolitan and groveler? You know, your friends did not shake hands with him after that, and you — just think — barely show him your love!»

In these opinions of Father's was reflected not only his naiveté, but, more probably, his desire to see something good in every person, his hope that the good side would now overcome and would overcome once and for all.

Unfortunately, Father character judgements were often wrong, and this resulted in many disappointments. Mother tried hard to convince him not to be so hasty in his conclusions and, especially, his actions (speeches) when he received some new information about some person from an untrustworthy source. Father in reply would assure her that such a thing would never happen with him, that he was standing his own ground and judging people exclusively on the basis of his own observations.

I remember quite well how Mother once quite imperceptibly forced Father to change his opinion about one of his colleagues twice in half an hour! He did not even notice it and became very anxious (and sad) after being convinced. Such naiveté, which often had bad consequences, coexisted, however, with perspicacity. I would like to demonstrate this trait of his with two examples which Mother related to me after his death.

In 1950, talking about his friends, he suddenly said to Mother: «When I die, all of them will help you. X will come along tearful at my funeral, Y will compose a moving obituary, but your strongest support will come from Igor.» Father meant Igor Tamm. And he was a prophet in this respect! I remember how Mother in the middle of the 1950s recounted this conversation to X, who was visiting her. He became thoughtful and replied: «Yasha, of course, understood perfectly the real characters of the people with whom he dealt. However, he spoke of his feelings about them very seldom.» And that evening they recollected such a case. Father was nominated the so-called official challenger of a doctoral thesis presented by a young and talented physicist. The procedure was brilliant. However, answering the question about his impression of the new Ph.D., Father replied with Pushkin's words: «Sly courtier!» This assessment turned out to be extremely apt and the characterization very neat, though its correctness was shown only many years later.

I have mentioned here very extreme cases; maybe their extremeness is the reason why I remember them. It would be wrong to conclude from this that Father only seldomly understood human nature and that he, as a rule, was shortsighted. His mistakes were connected with questionable people in whom a lot of «polar» qualities were mixed together — good and evil, audacity and mean-spiritedness, generosity and miserliness, truthfulness and an inclination to deviate from the truth. In various cases one or another of these human qualities prevailed. Yakov Ilich never forgot when the good side manifested itself. «He is not rancorous at all; he does not remember the evil which he did,» Yakov Ilich said once. As for himself, he was peaceable in the traditional sense of the word. I would like to give now one example which especially struck me.

Father had to present a character statement to the District Military Committee in Leningrad. He asked the administration of the Polytechnic Institute to draw up this document and received it in a sealed envelope. I do not know how it happened, but one of Father's friends learned about this. He told Father that such a presentation was a violation of certain rules, i.e., the character statement was to be given in an unsealed envelope and its contents were not to be a secret for the receiver. Father thought that such a deviation from the rules was impossible, so he and his friend went to the administration of the PTI and confirmed there what the friend had told Frenkel. After all this, Frenkel opened the envelope and read his character statement. Everybody to whom he showed it was deeply offended by its content: Frenkel was presented as practically an «enemy of the people» who was dishonouring Soviet students by spreading the propaganda of idealism, and so on and so forth. The statement was signed by a colleague of Frenkel's, quite a «respectable» man (as Frenkel had thought), which was an especially hard shock for Father. He returned the document and urged its revision. He said that otherwise he would call for a public discussion. In addition to this he said exactly what he thought about the document and its author. This time Father's tolerance was exhausted; he broke off greeting that man, especially since, as he was informed, the man had not been blameless in the past.

Later, one autumn evening after the war had ended, this man, who had spent the whole war in the Army and now was back at his job in the Polytechnic Institute, called Father and asked his permission to visit him at home. Soon he came, and Father invited him to his study, where (as he told us later) a moving conversation took place. The guest bitterly repented and criticized his past action. Relations between him and Father were restored and never broke off again. In connection with this reconciliation, as I remember, Father's close friends scolded him a little and predicted that, when it became profitable, the repented sinner would create some new «material» for a new penitence. But this time Yakov Ilich was right: the man's behaviour remained irreproachable in all situations, sometimes in very complicated ones. This story gave my parents the material for exchanges full of aphorisms like: «A good establishment is one where villains behave as honest people from the considerations of their career» and «If a villain permanently — out of any motive — takes on noble actions, is it right to treat him as a villain?»

I hope that now it is easy to understand why Father liked Rudyard Kipling's «If» so profoundly. The ideas that are beautifully formulated within this deeply moral poem corresponded to the ideas and principles that constituted the basis of Father's outlook on the world:

If you can keep your head when all about you
Are loosing theirs and blaming it on you,
If you can trust yourself when all men doubt you,
But make allowance for their doubting too;

If you can wait and not be tired by waiting,
Or being lied about, don't deal in lies,
Or being hated don't give way to hating
And yet don't look too good, nor talk too wise ...

A talent which was not given to Father was that of the diplomat. In everyday life he was uncommonly ingenuous in his conversations, never trying to avoid the reefs, as if forgetful of their existence. We rarely had an opportunity to talk without keeping in mind that almost each concrete situation was connected with various «prohibition principles». Yakov Ilich was never offended when some of his acquaintances or friends expressed an opinion contradicting his own or when, more generally, somebody criticized him. And he himself, in his turn, always said what he thought to the interlocutors' faces, sometimes even without noticing their reactions. He really was a little bit absent-minded. Sometimes Mother attempted to repair the situation at the very moment when the corresponding conversation occurred (especially when her husband was discussing some political situation). If he missed her fixed and expressive looks, she would step on his toes if the talk was taking place at the table. Sometimes this manoeuvre was successful, but at home we often witnessed scenes where Father started to look under the table to see if our pet spaniel was there and was the source of disturbance. Or — much worse — Frenkel would turn to Mother and say: «Please don't bother and do not step on my toes. I know whom I am talking with, and there are no reasons for misgivings.»

Let us return to the prewar years. In 1937, I was 7 years old, but somehow some memories of this drastic time remain with me. Maybe they will be of some interest. The arrests which occurred in the building where we lived became known to me almost immediately, both from the conversations of the adults and from «outward» signs. The last name of the arrested was immediately stricken from the list of residents which hung at the entryway doors. The hunted looks on the faces of the small children come to my mind. The children of the arrested, in the morning after the nighttime arrest, would go out to the park and wait with fear to see if yesterday's friends would play with them.

I was especially shaken by the arrest of a close friend of my parents', a remarkable Soviet physicist and person — Vsevolod Konstantinovich Frederiks — who was also living in our building. There was no way that this kind man, whose face exuded nobility — I saw this and felt it — could be an «enemy of the people»! In answer to my questions, my parents explained that this was a misunderstanding which would soon be dispelled. I remember Father's conversation with Vsevolod Konstantinovich's mother-in-law, Sofia Vasilievna Shostakovich (the composer's mother). We met her at the building's exit — Father had taken me for a walk. He assured her that he had secured consent for the signing of a letter in defense of Frederiks from a number of physicists and said that it would be desirable to add to these signatures that of Dmitrii Dmitrievich Shostakovich.

Ya. Frenkel with his spaniel. Leningrad, towards the end of the 1940s.

The arrest of Frederiks, in whose innocence I along with my parents firmly believed, led me to the simple conclusion that the well-being of every person was threatened, and that meant my Father's, too. This fear, and fear for the life of my mother who was often ill, darkened my otherwise happy and cloudless childhood in a safe and completely solid professor's family.

A certain scene comes to mind. One winter evening, the whole family gathered for dinner. Suddenly, our maid ran to the dining room with a cry and quickly began to say: «Yakov Ilich, they have come for you! A police officer and the maintenance man are calling for you. Dress, run [it must be said that there were two entrances to our apartment — one from the kitchen and the front door from the entryway], run, you are our dear, we will not give you up.»

I will not attempt to recreate what we all felt. Father got up from the table and set off for the back door. After a minute he returned, smiling. It turned out that they were conducting a peculiar kind of dog registration, so the one guilty of causing the alarm was our glorious spaniel. I have stored in my memory a colossal feeling of relief and of the somewhat heightened merriment which reigned at our table after this incident. Now, judging rationally, it is clear that there was no cause for alarm in that situation. However, in those few moments in 1937 there certainly was not enough time for analysis!

I will confess to yet one more feeling I had during that period. The fact that neither Father, nor his brother, nor any of our closest friends (with the exception of Frederiks) were arrested, suggested to my child's mind a comforting thought: well, after all, the truly good people are not touched ...

Much later, already an adult, I asked the question: why did those physicists who bravely stepped out against the representatives of the official philosophy, the party functionaries, not suffer in the 1930s? What were the laws governing the sinister choice of victims? Did their worldwide reputation save them? Hardly, for there are examples when that was not taken into consideration (N.I. Vavilov, L.D. Landau, Yu.A. Krutkov, and many others). I remember how long ago one of our acquaintances put forth the following naive argument: Sergei Vavilov, Ioffe, Tamm, and Frenkel openly stated their refusal of the official philosophy, «before the Party's very eyes», and this openness saved them.

It seems to me that for Father and for the people of his circle (and way of thinking), all the things that happened in 1937–38, the scale of which remained unknown to them, were a gigantic fluctuation which would quickly resolve itself and be replaced by a forward and progressive trend. The arrest of N.I. Ezhov and his myrmidons, along with the halt put to the arrest of ordinary citizens, seemed to confirm this point of view. After all, no one guessed what the true outcome of changing the guard from Ezhov to Berio would be, nor did anyone suspect Stalin's sinister role. In my opinion, it wasn't clear then. As I recall, softened judgements made the rounds: Stalin was good, but those around him (from the NKVD, the People's Comissariat for Internal Affairs) were criminal. In other words, one heard the same motif common in the conversations in the pre-revolutionary years: «The Tzar is good, but the ministers are scoundrels». However, I cannot guarantee the veracity of these impressions, and I even doubt that I heard these things at home. Conversations on domestic politics were not conducted in our home in front of the children (which is also symbolic!) — in contrast to foreign politics. I especially recall a heated wave of solidarity with Republican Spain. For about a month two boys lived with us who had been evacuated from Spain before the «Home for Spanish Children» had been organized.

In these long-past years, under the influence of didactic conversations in kindergarten about Lenin and Stalin, I arrived at a problem. Which of them did I love most? Prominent actor Shchukin's brilliant acting in the famous pair of films produced by Romm tilted the scales in favour of Grandfather Lenin. (In general, I think that these films played a colossal role in building an attractive image of the leader in not only the child's, but also in the adult's consciousness.) Then, no matter if it were Gelovani or Dikii who played Stalin, and no matter how much they desired to do a good job, it, in my opinion, just didn't work out. And so I asked Father: «And can I tell the teacher that I like Lenin more than Stalin?» «You can,» Father answered, «but it's better not to.»

About 30 years after that time I had a notable conversation with my five-year-old son. I went in to him, to say good night before he went to sleep, and by his slightly guilty expression I understood that he had done something wrong and doubted whether he should tell me about it. «Do you want to tell me something?» I asked. Having thought a little, and apparently having wavered, he said to me: «Yes, papa. I wanted to tell you that I love you very much, but that I love Lenin more.» So it is evident that the system of kindergarten education had not changed much.

I present these examples from my childhood in my story about Father to show in a few episodes the world of a family of the intelligentsia in the middle and second half of the 1930s as seen by the eyes of an eight- or ten-year-old boy.

After the end of the war, we learned about the monstrous scale of heinous crimes committed by Hitler's followers in Germany from numerous publications, materials from the Nuremberg trials, newsreels, and stories of people who had been in Germany. We learned how the German people regarded them; after all, weren't all the Germans unbridled fascists? During the year that he stayed in Germany (in 1925–26), Father had become acquainted with many physicists. They were normal people, with normal reactions; the cruelty could not help but call forth in them repulsion and condemnation. And later, when we read that for many Germans it was only after the end of the war that they came to know of the nightmarish pictures of terror and villainy that were committed right in their own back yard, we didn't believe it. And still later, when the picture of the oppression and the destruction of people in our own country reached a mind-boggling clarity, analogous questions had to be put to our countrymen.

Did they know, or did they not? I didn't have a chance to ask Father frankly about this, but I can here formulate a proposition. My point of departure is that he, like many millions of people in his generation, hated the tzarist regime and its decadent leadership. From his letters cited earlier, and above all, those of the spring and autumn of 1917, and those that followed, it can be concluded that he almost to his last years remained an adherent to the newly-formed order, seeing its numerous faults, which he supposed to be «transaction costs», but not forgetting about that system which it replaced. He by no means closed his eyes, but, on the contrary, with all of his heart rejoiced at the apparent achievements: the unsurpassed flourishing of the sciences (above all, of course, the physics which was so dear to him, but also mathematics and chemistry), the gigantic growth of literacy, and industrial development. I am sure that comparing the good and bad which were known to him (and not knowing about the scale of the bad, believing it to be local rather than general), he considered the balance to be positive. I think he found confirmation of this later in the popular character of the war against fascism. Father had doubts at the end of his life, in the late 1940s and early 1950s, a period on which I will briefly touch later. But perhaps I err in these judgements, and Father, simply not wishing that his children lead a double life, didn't say anything to us about those horrible events that we now know in all their details, and about which he could then guess (or knew?) by virtue of being a part of the order of things.

We, Yakov Ilich's closest relatives, often noticed surprising contradictions. For instance, he had a brilliant «professional» memory; his colleagues told me many times that if he read a paper once, he, as a rule, never forgot it. However, they also gave some quite different examples. Once he took part in a students' seminar. The participants presented brief reviews of papers based on publications from various physics journals, and an associate professor (assistant) commented on their reports. Yakov Ilich was a little late to one of these reviews, and after it was finished, he said: «I do not know who the author of this work is, but I do have some objections.» He set them forth, after which, to his own astonishment and to the great pleasure of the other participants, he was informed that the paper was his own. And it was not at all the case that the reviewer had mispresented the ideas; on the contrary, the paper was described quite correctly.

Frenkel often forgot the names of people who were not too close to him. Once when our parents had been invited by some of their rather new acquaintances, Mother specially asked him to put down their names into his note-book and to try to remember them. On having returned home she related us with a laugh that in the cases when Father could not avoid giving a personal salutation, he would take out this note-book, glance in it, and only after that begin to speak, a naive ruse which had been, of course, immediately noticed.

One more contradiction which astonished all of us was his complete lack of financial ability (it was of great value during the war to have such abilities), despite his exceptional energy and initiative. Some of Father's friends called him a «naive struggling through any obstacles».

He was good at finding a way out of very critical situations. The winter of 1941–42 in Kazan was extremely severe, with sharp frost and temperatures sometimes reaching minus fifty-two degrees Celsius. We had no firewood. Firewood was worth its weight in gold. However, even with money one practically could not buy it. It was very cold in our two little rooms, about six or eight degrees Celsius. On one of those days Father went early in the morning to the river (the Volga) harbour in search of firewood, having taken a special order for buying it[6]. Several hours later he returned home quite frozen (he had no warm winter coat and was wearing a light one, completed by a monstrously ugly winter hat (a so-called «ushanka») on his head, and a scarf which was put on over the coat and covered both the chin and mouth), but very satisfied. «Everything is quite O.K.,» he informed us. «Today in the evening we shall have a full truck of birch firewood.» And to justify his confidence, he added: «I have already paid money for it.»

When Mother realized that the order and the money had been given to an old man («an extremely nice one», as Father said) who had promised for some additional money to take all the necessary actions, she didn't share Father's enthusiasm. I remember quite well that evening when all of us awaited — rather hopefully, but fruitlessly — for the truck with the firewood. In answer to Mother's accusations that Father was credulous and careless, he, looking puzzled, shrugged his shoulders and appealed to the impression made upon him by his new acquaintance. He attached the greatest importance to the old man's sad story of his son, who was missing at the front, and of his family that remained in territory, occupied

6) This order gave a nominal right to buy some firewood.

by the Nazis. Father was really greatly depressed, of course, about both his having misjudged someone and about the new difficulties created by the loss of the order and the money. But one can imagine Father's happiness, and even triumph, when the next day, early in the morning, we saw a truck driven up to our house and yesterday's acquaintance alight from it. By our common efforts the firewood was unloaded, and then Father nearly forced our benefactor to have breakfast with us.

A lot of things which seemed to me to be examples of Father's «enterprise» were in fact a simple consequence of his wide fame, which helped him to attend the first show at every theater or arrive at a concert a quarter of an hour before its beginning; it was also easy for him to get railway tickets for Moscow (just after the war when one had a lot of difficulties obtaining them). However when I try to remember some concrete episodes, I realize that my memory for some reason kept the cases when Father's omnipotence misfired. Here is one such example. I was 11 years old when the war began. In these early years children, as a rule, read more than usual, but it was practically impossible to find books after we were evacuated; neither we nor our acquaintances had any. Once during the war, in Kazan, I arrived at Father's office at the PTI. He was getting ready to go home, but before that, he had to drop by the university's main library to turn in his order for the next day. The general catalogue was housed in the same hall where the books were checked out. Before checking out the needed materials, Father asked if I wouldn't want to check out something. Naturally, I wanted to do this very much. Then he showed me how the requests were made out and how to use the catalogue. I started to study the drawer labeled with a *D*. Dumas! I was simply amazed at the abundance of his novels which they had in the library. I filled out the card for two or three and gave the requests to Father. He and I together set off to see the librarian. She was an elderly, gray-haired woman with a pince-nez that had a hoop and a black cord which was inset into the right edge of one of the lenses. She quickly examined Father's requests, came to mine, raised her head, and informed us that these books did not circulate: that was the rule. «But after all, you and I know,» Father said, «that rules exist only to be broken!» «Not in the Soviet Union,» the librarian cuttingly replied. Father could not come up with any answer, and so it was fated that I had not read Dumas. When Father saw how upset I was he said: «It doesn't matter! I'll get you some books on Russian history. You know, it is even more interesting than French history.» Knowing my specific historical interests he added: «We had many more wars, plots and coups!»

When I was older, he told me about Ivan Vasilievich Grozny's times. I remember especially well one such detail. In one of the German encyclopedias Father found a brief paper about this Tzar. It included the following sentence: «Tzar Iwan der Schreckliche wegen seiner Grausamkeit Wasiliewitsch genannt», which is: «Tzar Ivan the Terrible, named Vasilievich because of his bloodthirstiness» (Vasily was his father's first and his own second name). Father knew many such historical anecdotes and liked to recount them to his relatives and guests. He was well acquainted with all the prominent books on Russian history, beginning with Kluchevskii and Soloviev and ending with Valishevskii.

As I have mentioned, I remember perfectly the years spent in evacuation (in Kazan). The letters that Father wrote to my Mother from his trips to Sverdlovsk and Moscow, and, beginning in the spring of 1944, from Leningrad, too, make these reminiscences of mine even clearer. It is the same with his letters to his friends and other relatives. But in almost every letter, amidst fear and joy over the situation at the fronts, beside the details of hard life in the rear, once in a while a smile sparkled. Or one might find some sad lines. On having received in 1942 the news about his cousin's death in blockaded Leningrad, he wrote: «Isn't it strange that in these hard times the people who are strong and adapted to the struggle for survival perished more often than the weak and unadapted ones. I know a lot of examples of this kind. Maybe it is explained by the fact that the strong ones have to care for both themselves and the weak ones.» Not long before the end of the war he wrote: «Looking through Pascal's «Pensées», I think constantly about the vanity of our earthly bustle, and have returned to the comparison of people with soap bubbles. It is interesting to note that these thoughts are with me all the time, even in the minutes that one would consider the most funny.»

I have mentioned that in the 1920s and 1930s Father gave his lectures at the Polytechnic Institute to rather small groups of students. But at the end of the second half of the 1940s the situation changed remarkably. The students of the first class admitted after the war (1946) entered the third year at the Physics-Mechanical Department. More than 150 students attended the lectures. Direct contact between the lecturer and his students became practically impossible, and Father could acquire some impressions of them only during the examination sessions.

I entered the Department in 1947, and my turn to listen to Father's lectures came in the first half of the third year of training, in the fall of 1949. The course of theoretical physics which Father was teaching for the students specializing in physics covered electrodynamics, statistical physics, and quantum mechanics; each subject was taught for two terms, one or two times a week.

For most of the students Frenkel's lectures were very difficult, especially in the beginning, and especially those on electrodynamics. They were the first non-general, or specific, subjects taught after the more or less simple ones such as general physics, chemistry, mathematics, and so on. Father started the course on electrodynamics by briefly explaining the foundations of vector and tensor analysis (these questions were a kind of introduction to the 1st volume of his «Electrodynamics»). All this was quite new to us, for at that point we knew only vector algebra; we knew nothing about vector differential operations and so on (the problem had been no more than briefly mentioned in our mathematics' lectures). Yakov Ilich, after having received from us uncertain answers to his question about whether we knew anything about such operations, told us about them in a few words and demonstrated the symbolic method of executing double differential operations (second derivatives). He immediately began to use all these things, essentially new to us, in his lectures. Frenkel presented electrodynamics in an unorthodox way. He started not with the notion of isolated (positive or negative) charges, but with that of dipoles (in connection with this he postulated the law of total charge conservation). Then, he did not introduce the notion of «magnetic» charges, but in a consecutive order carried out the analogy between the dipole field and the field of an elementary current curcuit. He called this analogy an «equivalence principle». His didactic approach was connected with some difficulties in comprehension. One has to add that for the first time Father did not take into account the fact that, as compared to the prewar times, the average level of the students' knowledge was lower. There were some objective reasons for such a situation: the training at schools was worse because of the hard war conditions, and the students who had recently been soldiers had had a long interruption in their studies.

Father started to prepare his lectures carefully to avoid some of the «idiosyncracies» which his prewar students still recollect with such a warm affection nonetheless. He thought that it was impossible to «experiment» with such a large audience. There are some copybooks and separate sheets in his files in which he sketched the plan and the main points of each lecture. I would like to repeat that he tried to make the essence of his lectures as clear and accessible as possible.

But some of the students still did not understand them and complained of difficulties. It is not possible to accuse them; it was not their fault, but rather their misfortune. Father, with great pain, accepted the fact of such lack of understanding (but not the complaints: he believed that a student must simply work harder). He tried, however without success, to organize additional lessons (we had such lessons in the course on general physics and mathematics). It was very difficult to set up a similar «support» because of the absence of qualified assistant lecturers at that time. The difficulties were even greater because Father's two-volume book on electrodynamics had been published about 10 years earlier and had become hard to find; the institute's library could not provide books to all the students.

I remember how some of my course-mates, especially after very difficult lectures, asked me if I had really understood those lectures. But I was, naturally, in a particular position. First of all, I had my own «Electrodynamics»; but what was more important, I could always ask Father about the difficult parts in his last lecture. When I told him about such difficulties, he, as a rule, was astonished: «Why, it is so easy! Let's go, I shall explain everything to you!» We would go into his room, I would place myself comfortably beside him at the precious desk, and he would sit on a swivel armchair with an adjustable back. He would take a sheet of paper and explain to me the parts I had not understood. Einstein and Langevin were silent witnesses of such explanations; their photos (which they had presented to Father) hung on the wall in front of the desk.

Here I have to mention that during these talks Father did not demonstrate his usual patience, and if I was slow to grasp his explanations, he would start to get irritated rather quickly. This was the reason why I, in my turn, would understand even less; the system was going, so to say, «to explode».

Father always attracted my attention to the main principles and explained them, and did not want to waste time making deductions or a formal transition from one expression to another: «I am not interested in algebra, and you, too, should not waste your time on it.»

Once, when I was in the 1st year, he saw me solving some problem on the dynamics of a material point. He got interested in the problem. I tried to explain how I had solved it, but my way of solving it did not satisfy him. «Let me show you how it should be done.» Unfortunately I do not remember either the problem, or the method that I had chosen for its solution. But I do remember that Father suggested that I use the law of conservation of momentum. In this way, he told me I should obtain the result immediately. I replied that since we still did not know about this law, the problem was to be solved in some other way. «Fine,» said Father. «But please do it for me and solve the problem with the help of that law.» «But we still have not studied it!» I repeated in the manner of a lazy schoolboy. «In this case deduce this formula! Why should you sit in inactivity? Go ahead and deduce it!»

After a short silence, absolutely depressed, I said: «I can't.» «Why can't you? Newton could!»

This «Newton could!» later on became a source for a lot of jokes for us and figured in the home charade performances we put on.

However, I never knew if Father had become very angry with me, for he cooled down soon. Half of an hour later, during supper, when I answered Mother's question about the reason for the war-like shouts that had come from Father's room, he laughed good-naturedly together with all of us.

Not long before the winter session began, I started to prepare an abstract of the electrodynamics course — without such an abstract Father would never examine his students. He asked me about my successes, and again in various different places, I asked him to shed light. I did not attend the special consultations which he gave at the institute before the exams. He asked me how preparation for the exams was going among my friends. Once I told him an anecdote popular with them connected with these preparations. A friend asked one of our students what part of the course he had finished studying. «But look through my book — I have put a match between the last pages.» The friend took the «Electrodynamics», and the match fell out. «Damn it!» screamed the student, «now I have to learn it from the very beginning!» However, Father did not enjoy this joke at all, but became sad.

I was anxious about the outward appearance of the exam procedure. I thought it might be better to be examined by one of Father's assistants at the Theoretical Faculty, whom he could mobilize to help him. He rejected the idea. He told me that I should prepare for the exam thoroughly, and that he was sure that I would be successful with this task. A high evaluation of my answer by one of his aid might only redouble my apprehensions that I would be accused of having received an unfairly high grade. «No,» finished Father, «I shall examine you myself». «And where?» It was this second question that worried me. «What do you mean — where? At home, of course, not at the institute. Why play the fool?»

Finally the day before the exam came. Before supper I dropped in Father's study with my abstract. He began to turn over the pages, and scolded me a little for a detailed reproduction of some computations. «Now, what will be the best question to ask you? What about this formula? Please tell me, what is its meaning and where it is obtained from?» I answered the question, trying to do my best. Then he asked me one more, rather easy, question. «Well, I think that will do. Give me your student's book.» «Could you have asked me something more difficult?» I asked. «And what seems to you the most difficult question?»

I do not remember what it was, but I mentioned a question which I had understood after rather a lot of thought. «But we especially discussed this problem the day before yesterday. You did understand all my explanations, didn't you?» «Yes, I did.» «So what reason is there to ask you about this problem! No, this will be quite enough!»

He gave me my other exams in approximately the same manner. Once Father was in a hurry, and after a cursory survey of my abstract he asked me if I had learned all the material well enough. «It seems to me that yes, well enough,» was my answer. «Well, in that case give me your student's book, please.»

Of course, I knew of Father's trustfulness and saw more proof of it each day, but this manifestation of it not only deeply touched me, but confused me as well. «But I could tell you a lie!» «What nonsense,» he replied, «Do you think I can imagine you telling me falsehood?»

He again demonstrated his trust in me in connection with yet another exam. I was taking the exam for my favourite of Father's courses (statistical physics) on an evening we had arranged earlier. He was in the process of reading through my abstract, slowly and with frequent stops, when some guests arrived quite unexpectedly. Our exam was interrupted. We agreed that Father would finish the exam the next morning. When I woke up, I found a brief note. He had written that he had had to leave early and did not want to wake me up, expressed his confidence that I knew the remaining part of the course just as excellently as the preceding part, and that my grade was a «five» (the highest grade). «So you can rest on your laurels», finished Father, «But shave without fail!» I still have this note.

My course-mates, as a rule, received high grades, too. Every time, on returning home from giving an examination (nuclear physics, thermo-physics, radiophysics, physical electronics, and so on), Father would take out the notebook in which he had put down the names of the students who had answered him the best and comment on their answers. It should be noted that those were the students who had already acquired the reputation of being very talented. From among them he would select his post-graduates, as he had always done.

J.I. had what one calls «iron health». I do not remember him as a bed-patient. Flu, quinsy and other diseases that periodically visited our home did not affect him. Either he did not fall ill with them, or maybe he endured any illness on his feet. But he often worried about his voice: because he was constantly giving loud lectures, his voice sometimes would fall to a whisper. It especially worried him when this happened just before some important public address or before a lecturing trip to some other city. In these cases Father tried to find modern inhalers or to use folk remedies. For instance, he breathed the steam from freshly boiled potatoes.

But in 1945 he began to feel symptoms of high blood pressure. The physicians and his brother Vladimir Ilich suggested that above all he slow the pace of his work. However, in the beginning nothing appeared to change. As usual, Father would leave for his office early and return home about 1 p.m., and then, after lunch and a five-minute nap (here his «biological clock» ruled him: he told us that if he increased, even very insubstantially, the length of such a nap, he would go out of his mind) he would return to the institute. He would come home just before supper. After the supper he would go to his study and occupy himself with reading scientific journals or with some other kind of work. We used to go to bed rather late, after midnight.

Now, however, Father started to use both of his allotted months of vacation, spending them on the Baltic coast, in the country-side around Moscow, or in the Crimea or the Caucasus, choosing a pair and living one month in each of the two chosen places. In winter, when he had a break in his lectures during the students' vacation, he would spend about a week with Mother in the sanatorium «Uzkoe», which was close to Moscow.

One of the summer months our family would spend together, and it was then that I spent the most time with my parents. In the morning hours Father continued to work until approximately lunch time. Each post-war summer he produced a book: 1945 — «Liberating Intra-atomic Energy», 1947 — «Introduction to the Theory of Metals», 1948 — «Theory of Atmospheric Electricity Phenomena». After lunch and a brief nap his time for active rest began. If the weather was good, we usually all went to a beautiful beach. On our way to the shore or at the beach itself Father would be «crowded» by groups of acquaintances. Thus we would enter the shallow waters of Riga Bay as a rather numerous band.

The sea bathings were varied by river bathings in the Lielupa River. In this case they were preceded by a long row in a boat — Father was fond of rowing. And in rainy weather, as I have mentioned, he used to take up his pencil, paint brush, or fiddlestick.

Alas, towards the end of the 1940s, J.I.'s summer rest turned out to be insufficient. In November and December his blood pressure started to increase greatly in spite of his medications (those which were available at that time were not very efficient). Father's health became noticeably worse, though this change for the worse was not catastrophic in the beginning. The real threat arose after the summer of 1951. I have written about this at the end of the previous chapter. My parents left Leningrad for a rest home in October and spent about a month in Repino. Father read a lot there and every day walked in the fresh air, though the range of his strolls was rather limited: in September he took an unfortunate leap off a bus and injured his leg. After this accident he limped a little and used a cane.

Once Mother told me that reading did not always distract Father. In these cases he tried to work, but without success, and as a result he became very unhappy. I found a good moment and said to him: «You have seen such a lot, met such interesting and prominent people. Why don't you try to write about all the corresponding impressions? Here you are living so quietly, nobody disturbs you.» «What are you suggesting,» he answered with unexpected irritation, «that I write memoirs? No, spare me that! It is just like undressing in the presence of strangers». Father's answer astonished me greatly. Many years later in one of his letters I found the line: «I feel shy expressing my feelings on a sheet of paper,» which in some way explained to me his reaction to my normal (as it seemed to me) proposal, though even now I am not able to understand it completely. And even here this remark forces me to select with special care the events, episodes, conversations, remarks I've included.

As much as we all tried to avoid thinking about the unavoidable, it still stole into our minds. Did Father think about this? At any rate, he never mentioned such questions when talking with us, his children. However, once, when I was eight, I confided to him my fear of death, which overpowered me, as it often does with small children. I asked him, how one could live, especially in old age, without thinking of this continuously. I remember quite distinctly his answer: «One gets accustomed to the thought of death over the course of years, and, being reconciled, stops thinking much about these matters.»

My parents returned to Leningrad in December, 1951. Professor N.N. Davidenkov, a colleague of Father's from the PTI, asked his brother, S.N. Davidenkov, a well-known neuropathologist and a member of the All-Union Medical Academy, to examine J.I. Having examined Father, he chose an appropriate moment and said to Mother: «Let Yakov Ilich return immediately to his usual situation and begin to work; do not keep him from that, but, on the contrary, urge him on. Herein lies his only hope.»

Father resumed his work at the PTI. I remember him returning home one December evening. He told us that he had dropped in at a seminar organized by the young collaborators from his department, where a paper on quantum electrodynamics was discussed. He spoke of the review report that he had heard in a rather critical manner and added: «It is finally time to investigate these works myself.» (As I would like to remark, they were quite new in 1951.) The next day he returned home with the latest issue of «Physical Review». All of us literally tiptoed when he went to his study saying that he would like to work. Approximately an hour later Mother asked me to look carefully to see how he was getting on; had he lain down for a short rest? I softly opened the door of his study. Father was sitting very quietly at the desk, leaning his elbows on it. The thumb of his right hand was under the ear lobe, the palm propped up his cheek, and the spread fingers lay on the upper forehead. The face, as it usually was when he worked, was frowning, the lower lip a little pressed up. I carefully closed the door and told Mother: «Everything is O.K., he is working!»

The farewell letter written by J.I.'s father, Ilya Abramovich, and addressed to his relatives — wife, sons, daughters-in-law — is kept in our family. He wrote it in 1931, after a rather slight stroke. In July, 1938, after his death, it was passed to his sons. It was a profoundly humanistic document which reflected the bright qualities of our grandfather's soul. Through all my life I have remembered the last paragraph of this letter: «If it should occur that there is somebody whom we name Lord, I would be your advocate to him. Alas! There is no Lord, so I will not have a chance to make a brilliant display of eloquence!»

It might be that in connection with this letter Mother for a long time had hoped to find a similar letter of Father's with his parting words, but in vain. His not writing such a letter suggests that J.I. was not thinking somber thoughts about his death, but was following the principle according to which wisdom consists in thinking of life.

Once Mother recollected an episode concerning the trip to Paris she and Father took in the spring of 1926. Together with the Mr. and Mrs. Semenov, and P.L. Kapitza and his future wife, A. Krylova, they decided to spend an evening at a fashionable restaurant. They had a table which was beautifully set and lit by candles. The whole company was young: the men were a little older than 30, the ladies were five to seven years younger. A playful idea came to somebody. Each gentleman put a candle in front of his place, and the three of them moved back at some distance and tried to blow out the candles by their common efforts; the one whose candle went out the first would be the first to die. It was an experiment quite forgivable for slightly drunken young men. Father's candle was the first to go out. «Well, Yasha, it is the sign that you will live longer than any of us!» and the whole episode was quickly forgotten. Still, it happened that Father was the first of them to die.

For my brother and myself, J.I. was first our father and only next a physicist. His humanity and dignity and our everyday contact with him in the family circle shadowed us from what was the main content of his life from the point of view of the hundreds of people who came into contact with him exclusively as a scientist who did important research on and taught physics. I have tried here to remember Father's faults for this account, and in this way to convince my readers of my objectiveness in describing his merits, but without success. Father's irascibility seems to me a secondary feature of his character, which was compensated for by the fact that he was so easily appeased. His trustfulness, which often taught him such sad lessons — how much more attractive and noble it is than prudence and suspiciousness! Is the belief in good, in the victory of good in a stumbling human being over evil really so naive? I sincerely believe that all these features are worthy of deep respect, not a condescending smile.

The entrance to the Physico-Technical Institute (St. Petersburg). Main building. (The Memorial plaque is mounted on the left side of the entrance.)

In May, 1973, P. Kapitza visited Leningrad, and, on being acquainted in my apartment with the collection of photostats of Father's paintings, told me: «How many talents Yakov Ilich had! Look, he not only painted brilliantly, but played the violin, too. Well, but his main talent was his unusual goodness!» On another occasion Kapitza remarked: «Never have I met such a kind man!»

If Father's goodness manifested itself through his contacts with people he did not even know well, it is easy to imagine how immense his kindness was towards his family members. That is why in my brother's and my own memory Father is not of an outstanding physicist, but simply of an outstanding human being.

Though, is it so simple?

The memorial plaque established in 1994 on the wall of the main building of the Physico-Technical Institute (in connection with Frenkel's centenary). Inscription: «In this building the outstanding Russian scientist Yakov Ilich Frenkel worked from 1921 to 1952».

Frenkel's books

During his life J.I. Frenkel published more than 200 papers. The first of them was published in 1916, the last ones in 1951; there were a lot of publications that appeared posthumously. I cannot provide here the complete list of these papers[1] and must limit myself to his published books[2].

1. The Structure of Matter. Part 1. Petrograd, 1922, 152 p.
2. Theory of Relativity. Petrograd, 1923. 300 p.
3. The Structure of Matter. Part 2, issues 1 and 2. Petrograd, 1924. 167 and 214 pp.
4. Electrical Theory of Solid Bodies. Moscow, 1924, 285 p.
5. Course of Vector and Tensor Analysis. L.-M.,[3], 1925. 205 p.
6. Electricity and Matter. L., 1925. 176 p.
7. Lehrbuch der Elektrodynamik, Bd. 1. Allgemeine Mechanik der Elektrizität. Berlin: Springer, 1926. 366 S. (Russian editions: L.-M., 1934; L.-M., 1956)
8. Lehrbuch der Elektrodynamik. Bd. 2. Makroskopische Elektrodynamik der materiellen Körper. Berlin: Springer, 1928. 505 S. (Russian Edition: L.-M., 1935)
9. Einführung in die Wellenmechanik. Berlin: Springer, 1929. 317 S.
10. Wave Mechanics, Vol. 1. Elementary Theory. Oxford, 1932. 278 p. (L.-M., 1933; Oxford, 1936; N.Y., 1950)
11. Statistical Physics. Part 1. Classical Theory of Statistical Equilibrium. L., 1933, 264 p.
12. Wave Mechanics, Vol. 2. Advanced General Theory. Oxford, 1934. 524 p. (L.-M., 1934; N.Y., 1950)
13. Theory of Solid and Liquid Bodies. L.-M., 1934, 120 p.
14. Analytical Mechanics. L., 1935. 197 p.
15. Course of Theoretical Mechanics on the Basis of Vector and Tensor Analysis. L.-M., 1940. 436 p.
16. Kinetic Theory of Liquids. M.-L., 1945. 424 p. (Oxford, 1946; Oxford, 1947; N.Y., 1955; Berlin, 1957; L.-M., 1959; N.Y., 1968; L.-M., 1975)
17. Liberating of Intra-atomic Energy. M.-L., 1946. 122 p.
18. Statistical Physics. M.-L., 1948. 756 p.[4] (Berlin, 1957)

---

1) The list is contained in the book (in Russian) «Yakov Ilich Frenkel. Materials for Bibliography of USSR Scientists», M.: Nauka, 1984.

2) All the books printed in Leningrad (Petrograd) are in Russian. Their titles are given here in English. The titles of the other books are presented in the language of the editing country.

3) Hereinafter «L» is for Leningrad, and «M» is for Moscow.

4) This is a virtually new work compared to the 1933 edition of «Statistical Physics».

19. Introduction to the Theory of Metals. L.-M., 1948, 291 p. (2nd ed.: L.-M., 1950. 384 p.; Budapest, 1951; Praha, 1953; Bucuresti, 1953; Warszawa, 1955; Berlin, 1957; L.-M., 1958; Peking, 1958; Tokyo, 1961; L., 1972)
20. Principles of the Theory of Atomic Nuclei. M.-L., 1950. 294 p. (M.-L., 1955; Berlin, 1957)
21. Collection of Scientific Works, Vol. 1 (see: Electrodynamics, 1956)
22. Collection of Scientific Works, Vol. 2. Selected Papers. L.-M., 1958. 600 p.
23. Collection of Scientific Works, Vol. 3 (see: Kinetic Theory of Liquids, 1959)
24. Abram Fedorovich Ioffe. L., 1968. 26 p.
25. On the Eve of the New Physics. Collection of the Scientific-Popular Papers. L., 1970. 384 p. (Tokyo, 1972).

The bibliography of works and papers dedicated to J.I. Frenkel contains more than 100 entries (including reviews of his books printed in Russia and abroad). Here we shall limit ourselves to six items:

Yakov Ilich Frenkel. Reminiscences. L., 1976. 278 p.

Ya.I. Frenkel. Reminiscences, Letters, Documents. L., 1986. 491 p.

V.Ya. Frenkel. Ya.I. Frenkel (1894–1952): Materials for his Scientifical Biography. Archs. Hist. Exact. Sci., 1974, Vol. 13, No 1, pp. 1–26.

R. Peierls. Yakov Il'ich Frenkel. Physics Today, June 1994, pp. 44–49.

The Search for the Challenge of Theoretical Physics (Voprosi Teozeticheskoi Fisiki) (published on the occasion of Ya.I. Frenkel's centenary). SPb. PIYaF Publishers. 1995, 258 p.

Physics-Uspekhi, Special issue dedicated to Ya. Frenkel in connection with the centenary of his birth. Vol. 37, N4, 1994; see also N6 (pp. 609–612).

Index

Science Networks • Historical Studies

The publications in this series are limited to the fields of mathematics, physics, astronomy, physical chemistry, and their applications. The publication languages are English preferentially, German, and in exceptional cases also French. The series is primarily designed to publish monographs. However, special editions featuring collected letters as well as thematic groupings of smaller individual works or proceedings can be taken into consideration. Annotated sources and exceptional biographies might be accepted in rare cases. The series is aimed primarily at historians of science and libraries; it should also appeal to interested specialists, students, and diploma and doctoral candidates. In cooperation with their international editorial board, the editors hope to place a unique publication at the disposal of science historians throughout the world.

SN 10 Benoit, P., Chemla, K., Ritter, J.: Histoire de fractions, fractions d'histoire, 1992 (ISBN 7643-2693-X)

SN 11 Reich, K.: Die Entwicklung des Tensorkalküls. Vom absoluten Differentialkalkül zur Relativitätstheorie, 1992 (ISBN 3-7643-2814-2)

SN 12 Gorelik, G.E /Frenkel, V.Ya.: Matvei P. Bronstein and the Soviet Theoretical Physics in the Thirties, 1994 (ISBN 3-7643-2752-9)

SN 13 Vizgin, V.P.: Unified Field Theories in the first third of the 20th century Translated from the Russian by Julian B. Barbour, 1994 (ISBN 3-7643-2679-4)

SN 14 Klein, U.: Verbindung und Affinität, Die Grundlegung der neuzeitlichen Chemie an der Wende vom 17. zum 18. Jahrhundert, 1994 (ISBN 3-7643-5003-2)

SN 15 Sasaki, Ch. / Sugiura, M. / Dauben, J.W. (Eds): The Intersection of History and Mathematics, 1994 (ISBN 3-7643-5029-6)

SN 16 Yavetz, I.: From Obscurity to Enigma, The Work of Oliver Heaviside, 1872–1891, 1995 (ISBN 3-7643-5180-2)

SN 17 Corry, L.: Modern Algebra and the Rise of Mathematical Structures, 1996 (ISBN 3-7643- 5311-2)

SN 18 Hentschel, K.: Physics and National Socialism. An Anthology of Primary Sources, 1996 (ISBN 3-7643-5312-0)

SN 19 Ullmann, D.: Chladni und die Entwicklung der Akustik, 1750 - 1860, 1996 (ISBN 3-7643-5398-8)

Vita Mathematica

Herausgegeben von
Emil A. Fellmann, Basel, Schweiz

*Die Reihe **Vita Mathematica** bringt unter einheitlichen Gesichtspunkten verfasste Werkbiographien bedeutender Mathematiker von der Antike bis in unsere Zeit – unter Berücksichtigung der wissenschaftshistorischen Forschung der letzten Jahrzehnte. Diese Bücher sind nicht primär für professionelle Mathematikhistoriker geschrieben, sondern wenden sich an Studierende der Mathematik, der Physik und der technischen Wissenschaften in den ersten bis mittleren Semestern, an Mathematik- und Physiklehrer, Mathematiker und Physiker, die ihre Fachdisziplinen in ihrer Einbettung in die Kultur- und Geistesgeschichte kennenlernen und studieren möchten. Die Publikationssprachen sind Deutsch, Englisch oder Französisch.*

Previously puiblished in this Series

VM 1 **Georg Cantor,** von Walter Purkert und Hans Joachim Ilgauds (deutsch)
1987, ISBN 3-7643-1770-1

VM 2 **Blaise Pascal,** von Hans Loeffel (deutsch)
1987, ISBN 3-7643-1840-6

VM 3 **Heinrich Heesch**, von Hans-Günther Bigalke (deutsch)
1988, ISBN 3-7643-1954-2

VM 4 **George Berkeley**, von Wolfgang Breidert (deutsch)
1989, ISBN 3-7643-2236-5

VM 5 **Norbert Wiener**, von Pesi R. Masani (englisch)
1990, ISBN 3-7643-2246-2

VM 6 **André Weil**, Autobiographie (französisch)
1991, ISBN 3-7643-2500-3

VM 7 **Johannes Faulhaber**, von Ivo Schneider (deutsch)
1993, ISBN 3-7643-2919-X

VM 8 **Christian Goldbach**, von A.P. Juskevic und Ju.Kh. Kopelevic (deutsch)
1993, ISBN 3-7643-2678-6

VM 9 **Friedrich Wilhelm Bessel**, von K.K. Lawrinovic (deutsch)
1994, ISBN 3-7643-5113-6

VM 10 **Bernhard Riemann**, von B. Laugwitz (deutsch)
1995, ISBN 3-7643-5189-6

GPSR Compliance

The European Union's (EU) General Product Safety Regulation (GPSR) is a set of rules that requires consumer products to be safe and our obligations to ensure this.

If you have any concerns about our products, you can contact us on ProductSafety@springernature.com

In case Publisher is established outside the EU, the EU authorized representative is:

Springer Nature Customer Service Center GmbH
Europaplatz 3
69115 Heidelberg, Germany

Batch number: 10800372

Printed by Printforce, the Netherlands